Computational Intelligence Methods and Applications

The books in this series explain methods developed in computational intelligence (including evolutionary computing, neural networks, and fuzzy systems), soft computing, statistics, and artificial intelligence, and their applications in domains such as heuristics and optimization; bioinformatics, computational biology, and biomedical engineering; image and signal processing, VLSI, and embedded system design; network design; process engineering; social networking; and data mining.

This series includes monographs, introductory and advanced textbooks, contributed volumes and state-of-the-art collections. Furthermore, it supports Open Access publication mode.

Bassem Jarboui • Said Toumi • Patrick Siarry
Editors

Data-Driven Waste Revolution

Springer

Editors
Bassem Jarboui
Hamdan Bin Mohammed Smart University
Dubai, United Arab Emirates

Said Toumi
College of Business Administration
Majmaah University
Al-Majmaah, Kingdom of Saudi Arabia

Patrick Siarry
LiSSi Lab
Paris-Est Créteil University
Paris, France

ISSN 2510-1765 ISSN 2510-1773 (electronic)
Computational Intelligence Methods and Applications
ISBN 978-981-96-5889-3 ISBN 978-981-96-5890-9 (eBook)
https://doi.org/10.1007/978-981-96-5890-9

This Springer imprint is published by the registered company Springer Nature Singapore Pte Ltd.
The registered company address is: 152 Beach Road, #21-01/04 Gateway East, Singapore 189721, Singapore

Preface

The **Circular Economy** represents a transformative framework addressing environmental challenges while creating economic opportunities. By promoting resource efficiency, cost savings, and innovation in recycling and remanufacturing, it enables societies to transition toward sustainable systems of production and consumption. At its core, the circular economy aims to extend the lifecycle of resources, minimize environmental harm, and align economic growth with ecological stewardship. Within this context, **waste management** plays a pivotal role, marking a paradigm shift from traditional linear models to circular approaches that emphasize sustainability, circularity, and resource efficiency. Leveraging advanced data and technology, waste management aims to optimize resource use, reduce waste generation, and foster a system that aligns with the principles of the circular economy. This evolution addresses critical challenges while fostering innovation and economic opportunities. This book brings together diverse perspectives on waste management within the framework of the circular economy, highlighting the intersection of **operations research** and **data science** as key enablers of this transition. It is organized into two distinct sections that provide complementary insights:

Part I: Optimization Techniques in Waste Management

This section includes six chapters that highlight the application of **operations research methodologies** to optimize decision-making processes and enhance the efficiency of waste management systems.

Chapter 1 "Decision-making support for circular construction and deconstruction: a literature review" (T. Boukherroub et al.) explores decision-making approaches for planning deconstruction operations. It presents an overview of supply chain operations, planning levels, and the objectives considered in optimization models. The chapter also discusses real-world case studies, identifying gaps in current research and suggesting future directions, such as incorporating sustainability metrics and integrating advanced planning tools.

Chapter 2 "A Novel Optimization Model for Sustainable Life Cycle Design: Circular Economy Case Study in the Automotive Industry" (H. R. Yazgan et al.) addresses the challenge of hybrid circular network design in the automotive industry. This chapter reviews reverse supply chain studies and presents a linear programming model aimed at maximizing profits while adhering to sustainability principles. The proposed model is applied to real-world data, showcasing its effectiveness in optimizing resources, reducing waste, and meeting circular economy objectives.

Chapter 3 "Exploring Options for Designing an Optimal Circular Spent Coffee Grounds (SCG) Reverse Logistics Network" (A. Chaabane et al.) develops an optimization-based framework for managing spent coffee grounds. The chapter highlights the role of circular coffee shops equipped with pre-drying technology, reducing contamination risks and improving logistics efficiency. It evaluates the environmental and economic impacts of this innovative approach and provides a blueprint for scalable implementation.

Chapter 4 "Optimizing Circular Economy: Simulation and Optimization in Stochastic Inventory Control of Perishable Dairy Products" (Ç. Sel and T. Bayraktar) introduces a simulation-optimization model for managing perishable goods with fixed shelf lives. The chapter provides a heuristic approach integrated with simulations to assist decision-makers in minimizing waste, improving inventory management, and promoting sustainability in the dairy sector.

Chapter 5 "Sustainability with 4R Principle on Multi-objective Waste Management Under Fermatean Fuzzy Environment" (S. K. Roy et al.) proposes a comprehensive model that incorporates circular economy principles, carbon emissions, and time management. The chapter uses advanced fuzzy programming techniques to solve multi-objective waste management problems, offering decision-makers robust and practical solutions that balance environmental and economic priorities.

Chapter 6 "Optimizing Municipal Waste Management to Support Circular Economy in Banyumas, Indonesia" (E. Rianawati et al.) examines innovative waste management practices in Banyumas Regency. Through a combination of qualitative and quantitative approaches, including linear programming and mobile applications like Salinmas, the chapter showcases effective strategies for optimizing waste collection and transportation, offering valuable insights for policymakers and practitioners.

Part II: Data Science Applications in Waste Management

This section comprises five chapters that emphasize the use of data science and analytics in addressing the challenges of transitioning to a circular economy, highlighting innovative tools and technologies.

Chapter 7 "Transition Pathway from Linear to Circular Economy in Waste Management in Indonesia" (E. Rianawati et al.) provides an in-depth analysis of waste management practices in Semarang and Makassar, Indonesia. Using qualitative methods such as interviews, field surveys, and focus group discussions, the chapter explores stakeholder dynamics, operational challenges, and policy frameworks required to facilitate the transition to a circular economy.

Chapter 8 "The Price and Attitude to Recycling: Manufacturers and Consumers Points of View" (A. Y. Nikitaeva and O. I. Dolgova) examines consumer and manufacturer attitudes toward recycling in Russia through semantic analysis of mass media and social networks. The chapter identifies key drivers and barriers, offering insights into how public perceptions influence recycling behaviors and providing recommendations for improving communication strategies.

Chapter 9 "Multicriteria Sustainability Index (MSI) in the Chico Mendes Extractive Reserve, Brazil" (A. R. Romeiro et al.) evaluates sustainability metrics using the Social Carbon concept and Multi-Criteria Decision Support Analysis. Based on field research, workshops, and interviews with local stakeholders, the chapter proposes actionable strategies to improve the sustainability of the Chico Mendes Reserve, focusing on social, environmental, and economic dimensions.

Chapter 10 "Plastics Recycling Dilemma: Integrating Mechanical and Chemical Recycling to Close the Plastic Loop" (P. Trucco et al.) explores the challenges of plastic recycling in Lombardy, Italy. Surveying 115 stakeholders, the chapter discusses how chemical recycling can complement mechanical recycling to overcome economic, technical, and environmental barriers. It provides actionable recommendations for aligning the plastic value chain with circular economy principles.

Chapter 11 "BCpbP—Blue Circular PostBranding Project: A Portuguese Case Study" (C. A. M. Duarte et al.) highlights a case study on marine waste management in Portugal. The chapter describes how discarded fishing gear and other marine debris are transformed into sustainable products, promoting environmental education and fostering community engagement.

Together, these chapters provide a comprehensive exploration of waste management through the lenses of **optimization techniques** and **data science applications**. By presenting theoretical frameworks, case studies, and practical implementations, this book underscores the transformative potential of operations research and data-driven strategies in advancing sustainability, fostering innovation, and achieving circular economy goals. It serves as a valuable resource for researchers, practitioners, and policymakers seeking to align waste management practices with the principles of the circular economy.

Dubai, United Arab Emirates — Bassem Jarboui
Al-Majmaah, Kingdom of Saudi Arabia — Said Toumi
Paris, France — Patrick Siarry
January 2025

Contents

List of Contributors

Derya Eren Akyol Department of Industrial Engineering, Faculty of Engineering, Dokuz Eylul University, Izmir, Turkiye

Gulseli Altıntaş Department of Industrial Engineering, Faculty of Engineering, Sakarya University, Sakarya, Turkiye

Firdha Laila Amalia Resilience Development Initiative, Bandung, Indonesia

Taha Arbaoui INSA Lyon, Université Lumière Lyon 2, Universite Claude Bernard Lyon 1, Université Jean Monnet Saint-Etienne, DISP, Villeurbanne, France

Muhammad Asa Resilience Development Initiative (RDI), Bandung, Indonesia

Fikri Aswan USAID Clean Cities Blue Ocean, Jakarta, Indonesia

Jean-François Audy Interuniversity Research Centre on Enterprise Networks, Logistics and Transportation (CIRRELT), Université de Montréal, Montréal, QC, Canada

Subkhi Abdul Aziz Research Organization for Governance, Economy, and Community Welfare, National Research and Innovation Agency, Jakarta, Indonesia

Tuğrul Bayraktar Department of Industrial Engineering, Karabuk University, Karabuk, Turkiye

Tasseda Boukherroub Interuniversity Research Centre on Enterprise Networks, Logistics and Transportation (CIRRELT), Université de Montréal, Montréal, QC, Canada

Amin Chaabane Ecole de technologie superieure (ETS), Montréal, QC, Canada

Evita Mahar Dewi Resilience Development Initiative (RDI), Bandung, Indonesia

Olga I. Dolgova Department of Information Economics, Faculty of Economics, Southern Federal University Rostov-on-Don, Russia

Carlos A. M. Duarte Universidade Europeia, UNIDCOM/IADE, Lisbon, Portugal

Isabel Farinha Universidade Europeia, UNIDCOM/IADE, Lisbon, Portugal

Julien Fondrevelle INSA Lyon, Université Lumière Lyon 2, Universite Claude Bernard Lyon 1, Université Jean Monnet Saint-Etienne, DISP, Villeurbanne, France

Shyamali Ghosh Department of Applied Mathematics, West Bengal, India

Anthony Halog School of Earth and Environmental Sciences, University of Queensland, St Lucia, QLD, Australia

Febrianty Hasanah Resilience Development Initiative, Bandung, Indonesia

Almira Rahma Hinanto Resilience Development Initiative, Bandung, Indonesia

Akhirul Insan Resilience Development Initiative, Bandung, Indonesia

Armin Jabbarzadeh Ecole de technologie superieure (ETS), Montréal, QC, Canada

Josef Jablonsky Department of Econometrics, Faculty of Informatics and Statistics, Prague University of Economics and Business, Prague, Czech Republic

Yulia Lapko Department of Management, Economics and Industrial Engineering, Politecnico di Milano, Milan, Italy

Maya Larasati Resilience Development Initiative, Bandung, Indonesia

Nuno Sá Leal Universidade de Tras-os-Montes e Alto Douro, Vila Real, Portugal

Lucas Ferreira Lima Institute of Economics, University of Campinas (UNICAMP), Campinas, Brazil

Mitali Daian Alves Maciel Institute of Economics, University of Campinas (UNICAMP), Campinas, Brazil

Raimundo Cláudio Gomes Maciel Center for Applied Legal and Social Sciences, Federal University of Acre (UFAC), Rio Branco, Brazil

Carol Maione Department of Management, Economics and Industrial Engineering, Politecnico di Milano, Milan, Italy

João Alfredo Carvalho Mangabeira Ministry of Agrarian Development and Family Farming (MDAF), São Paulo, Brazil

Yosef Barita Sar Manik Environmental Engineering Program, President University, Bekasi, Indonesia

Rui Miguel Universidade da Beira Interior, Covilha, Portugal

Baihaqi Muhammad Resilience Development Initiative (RDI), Bandung, Indonesia

Atik Munggiarti Banyumas Regency Regional Development Planning, Research and Development Agency, Banyumas, Indonesia

Anastasia Y. Nikitaeva Department of Information Economics, Faculty of Economics, Southern Federal University, Rostov-on-Don, Russia

Oleides Francisca Oliveira Center for Applied Legal and Social Sciences, Federal University of Acre (UFAC), Rio Branco, Brazil

Lauro Charlet Pereira Brazilian Agricultural Research Corporation (EMBRAPA), Jaguariúna, Brazil

Gabriel Pestana Escola Superior de Tecnologia - Instituto Politecnico de Setúbal, Setúbal, Portugal

Didit Okta Pribadi Research Organization for Governance, Economy, and Community Welfare, National Research and Innovation Agency, Jakarta, Indonesia

Pradnya Rahmani USAID Clean Cities Blue Ocean, Jakarta, Indonesia

Elisabeth Rianawati Resilience Development Initiative, Bandung, Indonesia

José Manuel Robalo HR Consultors, Guarda, Portugal

Ademar Ribeiro Romeiro Institute of Economics, University of Campinas (UNICAMP), Campinas, Brazil

Sankar Kumar Roy Department of Applied Mathematics, West Bengal, India

Ali Sabaghpourfard Air Canada Vacation, Montréal, QC, Canada

Nabila Safitri Resilience Development Initiative, Bandung, Indonesia

Oscar Sarcinelli Institute of Economics, University of Campinas (UNICAMP), Campinas, Brazil

Sónia Seixas Universidade Aberta, Lisbon, Portugal

Çağrı Sel Department of Industrial Engineering, Karabuk University, Karabuk, Turkiye

Sérgio Gomes Tôsto Ministry of Agrarian Development and Family Farming (MDAF), São Paulo, Brazil

Paolo Trucco Department of Management, Economics and Industrial Engineering, Politecnico di Milano, Milan, Italy

Félix Veillette Interuniversity Research Centre on Enterprise Networks, Logistics and Transportation (CIRRELT), Université de Montréal, Montréal, QC, Canada

Yudhistira Jalu Waskitha Resilience Development Initiative, Bandung, Indonesia

Harun Resit Yazgan Department of Industrial Engineering, Faculty of Engineering, Sakarya University, Sakarya, Turkiye

Hanieh Zohourfazeli Ecole de technologie superieure (ETS), Montréal, QC, Canada

Acronyms

3R	Reduce, Reuse, Recycle
ABS	Acrylonitrile Butadiene Styrene
APP	Permanent Preservation Areas
BCpbP	Blue Circular PostBranding Project
BIM	Building Information Model
BOU	Budgetary Educational Institution
CBM	Circular Business Models
CC	Circular Coffee shops
CE	Circular Economy
CLD	Causal Loop Diagram
CLSC	Closed-Loop Supply Chain
CS	Coffee Shops
CU	Conservation Units
DGVRP	Dynamic Green Vehicle Routing Problem
DLH	Dinas Lingkungan Hidup
DSS	Decision Support System
DTS	Dynamic Tabu Search
EoL	End-of-Life
EPS	Expanded Polystyrene
FFN	Fermatean Fuzzy Number
FFP	Fermatean Fuzzy Programming
FFS	Fermatean Fuzzy Set
FP	Fuzzy Programming
GCM	Global Criterion Method
GHG	Green House Gas
GP	Goal Programming
GPS	Global Positioning System
GVNS	Guided Variable Neighborhood Search
IAA	Aggregate Environmental Index
IAAG	Aggregate Agronomic Index
IAE	Aggregate Economic Index

IAG	Aggregate Governance Index
IAS	Aggregate Social Index
IFS	Intuitionistic Fuzzy Set
IoT	Internet of Things
LA	Location Allocation
LCA	Life Cycle Assessment
LIR	Location-Inventory-Routing
LLC	Limited Liability Company
LP	Linear Programming
MCDA	Multicriteria Decision Analysis
MCDM	Multicriteria Decision Making
MCGP	Multi-choice Goal Programming
MLOR	Minimum Life of Receipt
MILP	Mixed Integer Linear Programming
MINLP	Mixed-Integer Non-Linear Programming
MOLP	Multi-Objective Linear Programming
MOMILP	Multi-Objective Mixed-Integer Linear Programming
MOMINLP	Multi-Objective Mixed-Integer Non-Linear Programming
MOWM	Multi-Objective Waste Management
MPP	Marine Plastic Pollution
MSI	Multicriteria Sustainability Index
MSW	Municipal Solid Waste
NIS	Negative Ideal Solution
NLP	Non-Linear Programming
PA	Polyamide
PC	Polycarbonate
PCTW	Penalty Charge of TW
PDU	Pusat Daur Ulang / Recycling Center
PED	Primary Energy Demand
PET	Polyethylene Terephthalate
PFS	Pythagorean Fuzzy Set
PIS	Positive Ideal Solution
PMMA	Polymethyl Methacrylate
PP	Polypropylene
PS	Polystyrene
PUR	Polyurethane
PVC	Polyvinyl Chloride
QGIS	Quantum Geographic Information System
RDF	Refuse-Derived Fuel
REO	Regional Environmental Operator
RESEX	Chico Mendes Extractive Reserve
FID	Radio Frequency Identification
RL	Reverse Logistic
RMC	Remote Monitoring and Control
RWM	Revenue of Waste Management

SAF	Agroforestry Systems
SCG	Spent Coffee Grounds
SCLSC	Sustainable Closed-Loop Supply Chains
SDG	Sustainable Development Goal
SDP	Stochastic Dynamic Program
SGAC	Smart Green ASEAN Cities
SRA	Secondary Reducing Agent
T4GC	Tech for Green Cities
TFFN	Triangular Fermatean Fuzzy Number
TOPSIS	Technique for Order Preference by Similarity to the Ideal Solution
TPS	Tempat Penampungan Sementara / Temporary Garbage Disposal Sites
TPS3R	Tempat Pengolahan Sampah Reduce, Reuse, dan Recycle / Reduce, Reuse, and Recycle-Waste Management unit
TPST	Tempat Pengolahan Sampah Terpadu / Integrated Waste Management Facilities
TW	Time Window
UHMW	Ultra-High Molecular Weight
UHT	Ultra-High Temperature
UMKM	Usaha Mikro, Kecil, dan Menengah / Micro, Small, and Medium Enterprises
USAID	United States Agency for International Development
VRP	Vehicle Routing Problem
VRPTW	Vehicle Routing Problem with Time Windows
WM	Waste Management
WMC	Waste Management Charge

Part I
Optimization Techniques in Waste Management

Decision-Making Approaches for Deconstruction Operations Optimization: A Literature Review

Tasseda Boukherroub, Félix Veillette, Jean-François Audy, Julien Fondrevelle, and Taha Arbaoui

Abstract The construction industry needs to rethink its current waste management practices. One solution path is applying circular economy (CE) principles to building demolition such as deconstruction practice. Considered as a more-resource-friendly alternative compared to standard demolition, deconstruction involves dismantling a building with the aim of maintaining the highest possible value for its materials and maximizes their recovery potential. This chapter reviews the literature to identify and analyze recent research studies reporting on decision-making approaches developed for deconstruction operations planning (e.g., project scheduling, material transportation planning, and logistics network design). Specifically, the decision-making approaches are presented through the description of the supply chain operations, the planning level, the decisions to be made, as well as the objectives considered in the optimization models proposed. The application of some of these approaches on case studies is also discussed. Finally, this chapter identifies the limitations of current research in the field and suggests future research perspectives.

Keywords Construction · Renovation and demolition (CRD) · Circular economy · Deconstruction · Logistics · Planning · Decision-making · Optimization

T. Boukherroub (✉) · F. Veillette · J.-F. Audy
Interuniversity Research Centre on Enterprise Networks, Logistics and Transportation (CIRRELT), Université de Montréal, Montréal, QC, Canada
e-mail: tasseda.boukherroub@etsmtl.ca; felix.veillette.1@ens.etsmtl.ca; jean-francois.audy@uqtr.ca

J. Fondrevelle · T. Arbaoui
INSA Lyon, Université Lumière Lyon 2, Universite Claude Bernard Lyon 1, Université Jean Monnet Saint-Etienne, DISP, Villeurbanne, France
e-mail: julien.fondrevelle@insa-lyon.fr; taha.arbaoui@insa-lyon.fr

B. Jarboui et al. (eds.), *Data-Driven Waste Revolution*, Computational Intelligence Methods and Applications, https://doi.org/10.1007/978-981-96-5890-9_1

1 Introduction

The construction industry is not only a major consumer of virgin materials [4] but also a major contributor to waste generation [34]. As an example, in the European Union (EU), it constitutes more than 30% of all waste generated (EU, 2023). It is also one of the biggest contributors to greenhouse gas (GHG) emissions in the last years, being responsible for about 40% of carbon emissions yearly [1]. To reduce this significant amount of waste and GHG emissions, the construction, renovation, and demolition (CRD) sector is now incorporating circular economy (CE) principles in building demolition projects leading to what is known as deconstruction, disassembly, or selective demolition [5]. Deconstruction is the practice of dismantling a building with the aim of maintaining the highest possible value for its materials and maximizing their recovery potential [9, 16]. Shifting from demolition to deconstruction is an enabler for the CE [6]. Allam and Nik-Bakht [5] distinguished three main periods in the evolution of deconstruction practice in the literature: the demolition age (1974–1999) characterized by building demolition practice and the management of the resulting waste; the nascence of deconstruction (2000–2014), during which the term "deconstruction" was introduced and the feasibility of CRD waste management and recycling was investigated; and finally, the circular construction period (2015–2021) as a result of the adoption of CE in the construction sector and various international initiatives.

Reducing waste generation through prevention, reduction, reuse, recycling and using more efficiently natural resources are among the key targets of the United Nations' sustainable development goal (SDG) 12: "Ensure sustainable consumption and production patterns" [22]. The practice of deconstruction can contribute to achieve this goal while providing socioeconomic benefits [20]. However, shifting from standard demolition to deconstruction requires thorough changes [14]. For example, time and labor have a major impact on the feasibility of a deconstruction project [15]. Material dismantling and collection through deconstruction can take weeks, while in standard demolition, it is carried out in much less time [16]. The logistics related to selling the salvaged materials (time, quantity, and location of the materials to collect and to send to the clients) is a major source of uncertainty [16, 32]. Items with undefined destination would require an on- or off-site storage location, leading to more costs [16]. On-site storage could even affect the project schedule [12]. These issues can be addressed through better deconstruction operations and transportation planning. As recommended by Marzouk and Elmaraghy [16], preparing the building deconstruction plan with sufficient time before operations start is crucial. EU [11] suggested organizing transportation and creating a materials' management plan very early on in the planning phase. However, as stated by Quéheille et al. [23], existing waste management software are limited to providing support to decision-makers when the waste has already been generated, not prior to waste generation.

This chapter presents a literature review on recent research studies reporting on decision-making approaches developed for deconstruction operations optimization (e.g., project scheduling, material transportation planning, and logistics network design). It is an extension of the work presented in [33]. Despite recent literature review papers such as [5, 8, 10], there is a lack of focus on deconstruction operations planning using optimization approaches. Allam and Nik-Bakht [5] conducted a bibliometric analysis of the literature reporting on various aspects related to deconstruction such as qualitative studies discussing the application of deconstruction or the contribution of technologies in improving deconstruction practice. Ding et al. [10] analyzed through a systematic literature review studies reporting on forward and reverse logistics in the construction sector. They compared the operations involved in the phases of design, manufacturing, construction, operations (forward logistics) and deconstruction, product reuse, waste distribution, and material reprocessing (reverse logistics). Chen et al. [8] conducted a systematic literature review on construction circularity. They analyzed different circular construction strategies used in the phases of design, manufacturing, construction, operations and maintenance, and end of life as well as circular construction drivers. The authors considered logistics network optimization as an external driver and found that it is not well addressed in the literature. Hübner et al. [13] reviewed the literature on project planning methods for deconstruction projects. This chapter presents some overlap with ours. However, relevant studies cited by the authors have been published before 2015. A significant part of these studies are conference papers (excluded in our review), all of them focus only on project planning (while our study also includes transportation and network design problems), and some of them are qualitative studies or present MCDM (multicriteria decision-making)-based models, which are not the focus in our work.

The remainder of this chapter is as follows: Sect. 2 provides a description of the literature search and selection method. Section 3 presents our analysis of the selected articles. Section 4 provides a discussion, and finally, Sect. 5 presents our conclusions and some future research avenues.

2 Literature Search and Selection Method

The research method used in this chapter is a systematic literature review combined with the snowballing technique. First, we defined the scope of the literature review by using the life cycle phases of a building in the context of a circular construction supply chain (Fig. 1). Based on the study of [10], these phases are separated into two main parts, the forward logistics where materials are consumed for the establishment and the use of a building and the reverse logistics where waste materials are generated in the end of life of the building. As highlighted in Fig. 1, the review focuses on three main phases in the reverse logistics part: deconstruction, on-site material storage and treatment, and off-site material storage and treatment. We investigate strategic and operational planning involving these three phases as well as the storage, treatment, and transportation of the material flows generated during these phases. Therefore, the planning problems dealt with in this chapter include,

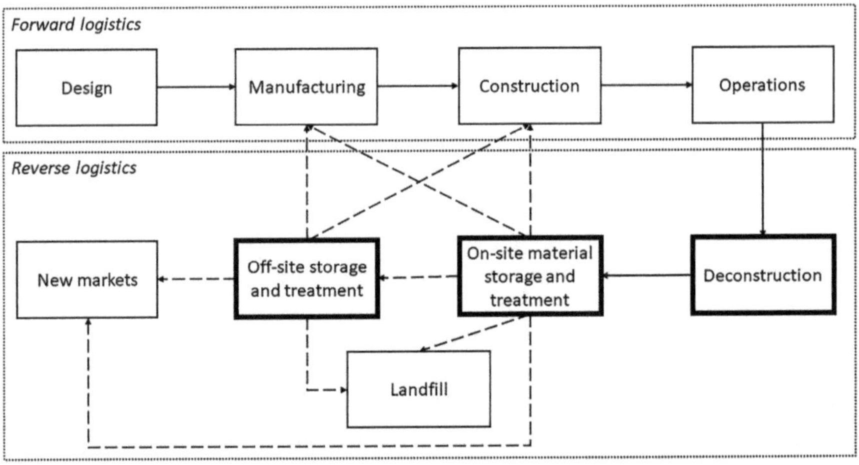

Fig. 1 Life cycle phases of a building in a circular construction supply chain, adapted from [10]

Table 1 List of the concepts and main keywords used for searching the articles

Concepts	Synonyms
Waste	Material—End-of-life product—End-of-life product
Optimi* (Optimization)	Maxim*—Minim*—Management—Planning—Scheduling—Routing—Design
Deconstruction	Selective demolition—Dismantl* building—Disassembl* building—recover* building—reuse building—recycle building—reduce building—circular economy building—circular economy construction—circular economy renovation

among others, deconstruction operations scheduling, reverse logistics network design, post-deconstruction material storage, treatment (e.g., on-site or off-site conditioning or recycling), allocation to a destination and transportation.

Three main keywords, referred to as Concepts in Table 1, were used for the article search process: waste (Concept 1), optimization (Concept 2), and deconstruction (Concept 3). Synonyms were set out for each concept using the Termium Plus, Vitrine linguistique websites (i.e., linguistic databases of the Canadian and Québec governments), and ChatGPT. Specifically for deconstruction concept, the recommendation of [5] to use the terms selective demolition, selective dismantling, and disassembly as synonyms was followed. The inclusion period for this review was set to 2000 up to July 2024. The article search was carried out on Engineering Village (Compendex, Inespec, Knovel databases) and Scopus Websites, between June and July 2024. The search equations were constructed based on all combinations of the three main concepts and their synonyms (see Table 1). Waste AND planning AND "selective demolition" is an example of a search equation used.

As the article search progressed, the search formula was updated to add the not option and remove keywords such as weapon, nuclear, e-waste, food, and energy. Only English-language articles from scientific journals, book chapters, and books were considered, resulting in a total of 1810 studies (from all databases used). PRISMA (Preferred Reporting Items for Systematic Reviews and Meta-Analyses) method [21] was used for selecting the final studies (Fig. 2). First, duplicates were removed, leading to 1175 documents. To this end, Rayyan platform was used. Titles and abstracts were filtered in Rayyan, leading to 88 studies retained for further analysis. For a study to be included, the main focus should be on deconstruction and planning. Therefore, articles using BIM (Building Information Model) or focusing on forward logistics phases only (design, manufacturing, etc.) were excluded. As an example, many studies focus on integrating deconstruction practice in the phase of building design, which is beyond the scope of our study, since we exclude forward logistics. Literature review studies are not included neither. After a more in-depth analysis of the studies, 12 papers were finally selected. For a study to be eligible, an optimization model should be presented. In addition, we have selected five articles

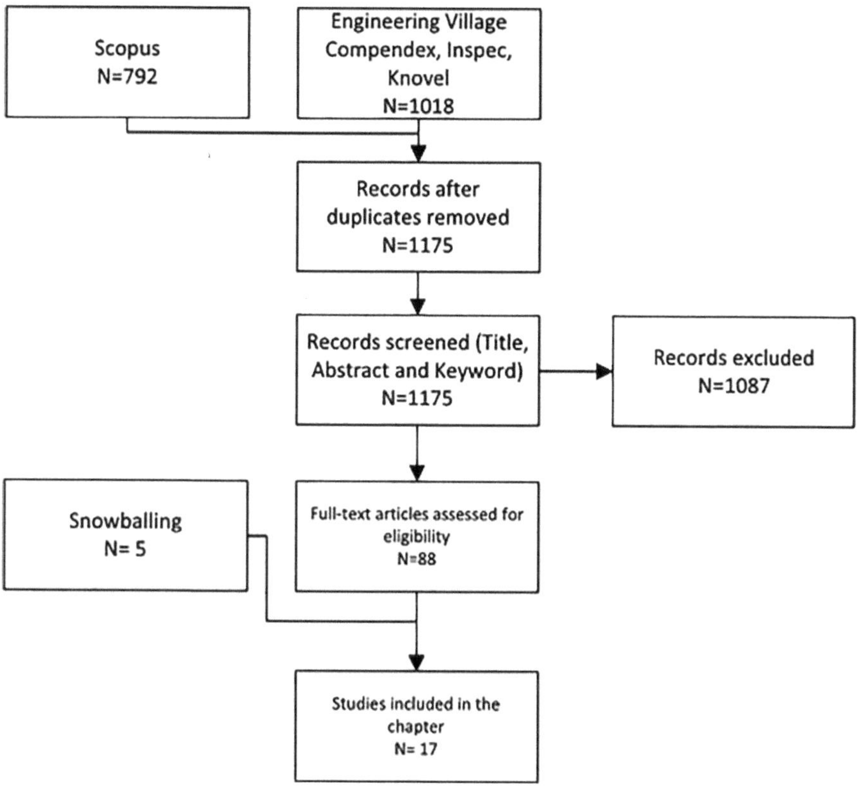

Fig. 2 Selection process performed following PRISMA method

found with the snowballing method in the final 12 selected studies and the literatures review studies by Allam and Nik-Bakht [5], Chen et al. [8], Hübner et al. [13], and Ding et al. [10], leading to a total of 17 final selected studies.

3 Analysis of the Selected Studies

Figure 3 shows that all the selected studies were published between 2001 and 2024. More than half were published in the last five years. Figure 4 specifies the countries or regions where case studies have been conducted. Canada and France are the most frequently mentioned countries (see Table 2 for more details). Chen et al. [19] analyzed building deconstruction in five different countries: Brazil (BR), Canada (CA), China (CN), Nigeria (NG), and Switzerland (CH). Case studies play a crucial role for testing and validating theoretical optimization models. As an example, when [23] presented the model's theoretical results to the project's head engineer (case study in France), this latter stated that the solution was not feasible. Specifically, the recommendation was to install two cranes, but the site could only accommodate one. The model was adjusted and then yielded a feasible solution (i.e., new task scheduling). On the other hand, often in practice, decision-makers are faced with a lack of tools to support deconstruction planning. This is observed in the province of Québec (Canada) as mentioned in [7, 33]. Finally, two articles do not have a country or region specified in the case study.

Table 2 presents the main aspects of the decision-making approaches proposed in the 17 selected studies. These aspects include whether the study concerns on-site or off-site operations, the planning level (strategic or operational), the decisions of the problem considered, the type of the optimization model proposed (mono-, bi-, or multi-objective), the objective function(s) considered, and the region of the case study when it is specified. The references are classified in a chronological order, starting from the most recent one.

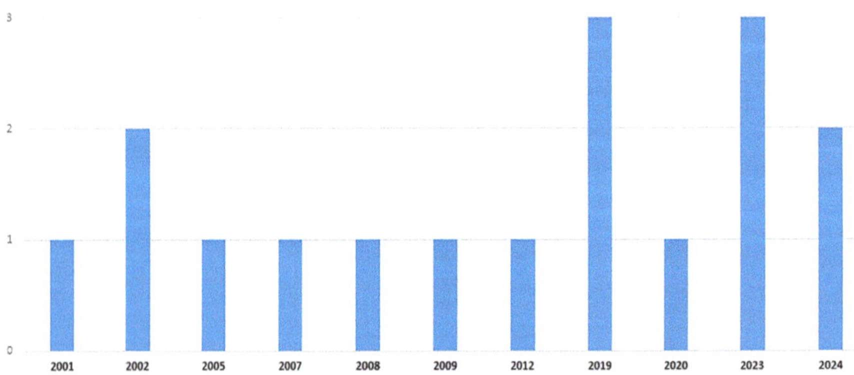

Fig. 3 Publication year of the selected studies

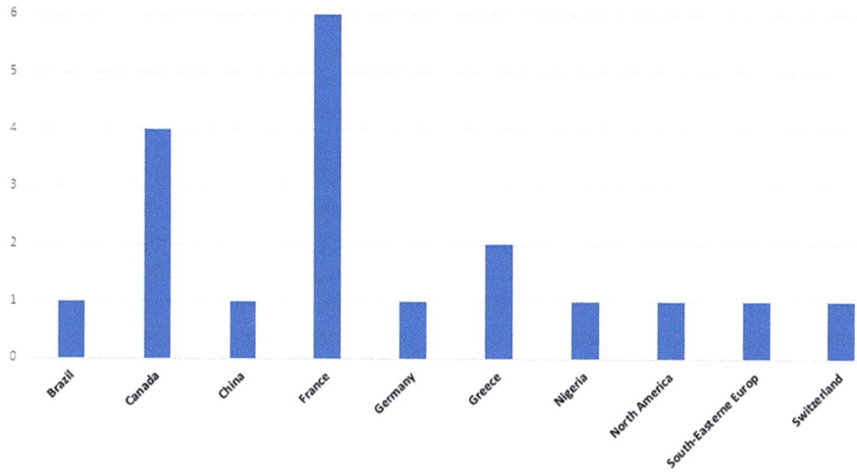

Fig. 4 Countries and regions where case studies were conducted

In Table 2, we can distinguish five groups of studies conducted by the same main researchers and focusing on one typical planning problem with some variations from one study to another [2, 3, 17–19, 23, 24, 26, 27, 29, 30, 32, 35, 36]. The study by Schleier and Walther [28] is the unique exception. This indicates that, up to date, deconstruction operations planning is a research niche in the literature. We can also see from Table 2 that almost all studies reported on operational planning problems (16 over 17 studies). Again, the study by Schleier and Walther [28] is the exception. The authors are the only ones considering strategic decisions in their work where they addressed the issue of EPS-ETICS (Expanded PolyStyrene in External Thermal Insulation Composite Systems) waste management. Since these are composite materials, their efficient recycling is particularly challenging. The authors aim at analyzing reverse logistics involving alternative recycling options like pyrolysis or solvent-based recycling rather than co-incineration in waste incineration plants, to enhance circularity. The possible solutions to manage the end of life of EPS-ETICS are presented, discussed, and evaluated from economic (investment, fixed, and variable operating costs, material-related costs/disposal costs, and achievable revenues from selling recycled products), technological (yield rates of the processes), and environmental (ability to achieve circularity quotas and yield rates) points of view. For each option, successive steps are considered, including preprocessing and recycling/recovery and transportation. An optimization model for strategic network design based on mixed-integer linear programming (MILP) is proposed, with the objective of maximizing the net present value in a multi-period setting; the salvage value of the installed technologies is incorporated using a linear depreciation approach to avoid side effects at the end of the horizon. Decision variables are related to technology selection, capacity, and location for preprocessing and recycling facilities (strategic level) as well as material flow allocation (operational level) in a multi-echelon network and a multi-period setting (with strategic periods longer than operational periods). Economies of scale related to the processing capacity are taken into account (for each technology, multiple capacity classes are considered). But storage of materials across periods is not allowed. Among the specificities of the problem under study, which require developing a new model, the authors identified the spatially widespread distribution of mixed waste materials and the need to accurately model the recycling of contaminated plastics, using chemical processes. The study also assesses the potential impact of regulations/policies and uncertain product prices and transportation costs on the system design and global cost. The proposed approach is illustrated on a case study considering the situation in Germany, from 2020 to 2039, with operational decisions made on a yearly basis, while strategic decisions are made every 5 years, and a focus on EPS-ETICS waste generated during demolition and renovation of the residential building stock.

Table 2 also shows that most of the studies focus on on-site operations (11 studies). This indicates a lack of integrated and coordinated planning between deconstruction activities (performed on-site) and transportation operations or recycling activities (performed off-site). Among the studies which considered both on- and off-site operations planning, we can cite [36] where the developed model aims at determining the deconstruction depth before demolition (when deconstruction and demolition should be applied) and designing the transportation network for moving deconstruction materials and waste from the deconstruction site to potential customers, recycling/waste disposal sites as well as the destinations for the materials such as recycling centers. Quéheille et al. [24] also considered decisions related to both deconstruction/demolition and materials transportation. In particular, their model aimed at identifying which deconstruction or demolition technique to use, the number of employees and quantity of equipment needed for deconstruction/demolition operations as well as the size of the containers required for each material type, their recovery option, and destination. Studies focusing only on on-site operations all deal with operational planning, more precisely, deconstruction project scheduling. Most of them minimize the duration of the project (makespan), which is common in classic project scheduling problems. This is the case of [29, 30, 32] where an approach that considers both material flow management and project scheduling for dismantling domestic buildings is proposed. The material flow management, conducted prior to the project scheduling, consists in identifying the building's elements to dismantle, the presence of pollutants, the possible dismantling activities to apply, the resulting materials as well as the container types to use for each material type. The project scheduling problem is formulated as an optimization problem with the objective of minimizing the total duration of the project. It allows determining the completion time of each activity, which mode to apply (each mode consumes different resource types and results in a specific duration), and the duration of the entire project. Three different scenarios have been considered. In the first one, manual dismantling techniques are used (base-case scenario). In the second scenario, dismantling is performed using partly automated equipment (improved dismantling techniques). The third scenario strictly focuses on separating and recycling as many materials as technically possible (based on the results of the material flow management). The results showed that the

Table 2 Main aspects of the decision-making approaches reviewed

References	On-site vs. off-site operations	Strategic vs. operational planning level	Decisions	Model type	Objective function(s)	Region
Quéheille et al. [24]	On- and off-site	Operational	• Deconstruction or demolition technique to apply • Number of employees required • Quantity of equipment to use • Container types to use and capacities • Recovery option for each material type • Destination of each material type	Multi-objective	• Minimization of the project duration • Minimization of the project cost • Maximization of the recovery rate • Minimization of three indicators measuring environmental impacts	France
Schleier and Walther [28]	On- and off-site	Strategic and Operational	• Technology type to select and capacities • Location of preprocessing and recycling facilities • Material flow allocation	Mono-objective	Maximization of the net present value (NPV)	Germany
Mollaei et al. [19]	On-site	Operational	• Extraction method to use for each component • Starting and finishing times for each activity	Mono-objective	Minimization of the project cost	Canada, Nigeria, Brazil, Switzerland, China

(continued)

Table 2 (continued)

References	On-site vs. off-site operations	Strategic vs. operational planning level	Decisions	Model type	Objective function(s)	Region
Mollaei et al. [18]	On-site	Operational	• Extraction method for each component • Starting and finishing times for each activity	Bi-objective	• Minimization of the project cost • Minimization of the duration of the project	Canada
Mollaei et al. [17]	On-site	Operational	• Extraction method for each component • Starting and finishing times for each activity	Bi-objective	• Minimization of the project cost • Minimization of the duration of the project	North America
Sanchez et al. [27]	On-site	Operational	• Deconstruction method to apply to each component	Bi-objective	• Minimization of the project cost • Minimization of the environmental impact of the project	Canada
Aidonis [2]	On-site	Operational	• Selection of deconstruction or demolition options • Selection of recovery and recycling options	Multi-objective	• Minimization of the duration of the project • Maximization of the profit generated	Greece

Quéheille et al. [23]	On- and off-site	Operational	• Deconstruction or demolition technique to use • Number of employees required • Quantity of equipment to use • Container types and capacities • Recovery option for each material type • Destination of each material type	Multi-objective	• Minimization of the duration of the project • Minimization of the project cost • Maximization of the recovery rate of materials	France
Sanchez et al. [26]	On-site	Operational	• Deconstruction method to apply to each component	Bio-objective	• Minimization of the cost of the project • Minimization of the environmental impact of the project	Canada
Xanthopoulos et al. [36]	On- and off-site	Operational	• Selection of the dismantling method to use on each material type • Selection of the destinations of the matreials (send or not to recyclers)	Mono-objective	Maximization of the profit generated	Greece

(continued)

Table 2 (continued)

References	On-site vs. off-site operations	Strategic vs. operational planning level	Decisions	Model type	Objective function(s)	Region
Xanthopoulos et al. [35]	On- and off-site	Operational	• Selection of deconstruction or demolition options • Quantity of each material type to deconstruct at each stage • Quantity of each material type in a container • Number of containers required for each material type • Destination of the containers	Mono-objective	Maximization of the profit generated	South-Eastern Europe
Aidonis et al. [3]	On- and off-site	Operational	• Selection of deconstruction or demolition options • Quantity of each material type to deconstruct at each stage • Quantity of each material type in a container • Number of containers required for each material type • Destination of the containers	Mono-objective	• Maximization of the profit generated	N/A

Schultmann and Sunke [32]	On-site	Operational	• Dismantling mode to apply to each activity • Activity completion time	Mono-objective	Minimization of the duration of the project	France
Schultmann [31]	On-site	Operational	• Dismantling mode to apply to each activity • Activity completion time	Mono-objective	Minimization of the duration of the project	N/A
Schultmann and Rentz [29]	On-site	Operational	• Dismantling mode to apply to each activity • Activity completion time	Mono-objective	Minimization of the duration of the project	France
Schultmann and Rentz [30]	On-site	Operational	• Dismantling mode to apply to each activity • Activity completion time	Mono-objective	Minimization of the duration of the project	France
Schultmann and Rentz	On-site	Operational	• Dismantling mode to apply to each activity • Activity completion time	Mono-objective	Minimization of the duration of the project	France

project duration has been reduced in all scenarios compared to the current situation. Compared to the first scenario, the second one leads to more time reduction (activities using sophisticated machines could be accelerated). In the third scenario, the makespan is higher compared to the second one because additional activities are needed; however, the total costs are lower due to the higher quality of the materials resulting from dismantling. This scenario leads to a recycling rate (i.e., proportion of materials not sent to landfills) of 98%, whereas the first and second ones lead to a rate of 95%. The optimization models presented in [17, 19] also minimize the total project completion time. The generated schedules determine the best method to extract the materials. Xanthopoulos et al. [36] mention that schedules make it possible to extract components with minimum loss, in addition to facilitating their recovery and reclamation. To avoid improbable cases (e.g., deconstructing the foundation before deconstructing the roof), following an order of priority in the extraction chronology is key [17]. However, [36] indicate that a comprehensive assessment of the building's components is required in the initial study phase to optimize the deconstruction process. The authors place significant emphasis on this initial phase, since a deconstruction project generates a great deal of waste and the sorting and handling of the materials must be organized efficiently. Regarding the types of mathematical models proposed, more than half consider only one objective function (10 studies). Four studies have proposed bi-objective optimization models, and three considered more than two objective functions. Minimizing the total duration of the project is the objective that is considered the most (10 studies) followed by minimizing the total cost of the project (eight studies). Profit (and net present value [28]) maximization is considered in only five studies. Overall, the financial performance (both cost and profit aspects) is considered as objectives in the majority of the models proposed (13 over 17 studies). Aidonis [2], Aidonis et al. [3], and Xanthopoulos et al. [35, 36] all developed models maximizing the profit generated from selling recycled and reused materials (deconstructed components and demolition waste). The scarcity of studies considering profit maximization can be explained by the absence of a clear central market to sell reused or recycled materials and the difficulty to determine applicable resale prices and the demand for the materials as pointed out by Chen et al. [19]. The authors developed a model minimizing the total cost of the project, but also calculated the revenue that could be generated from selling the recovered materials. The model was applied to five different deconstruction projects in Brazil, Canada, China, Nigeria, and Switzerland. The authors mentioned that Switzerland and Canada are the countries with the most support for deconstruction projects owing to the current standards in effect. Switzerland has the highest extraction costs, followed by Canada. In terms of resale profits, Switzerland and Canada were at the top of the list. The authors conclude that it is important to consider the total volume reused and not the quantity, which could be misleading since smaller materials are reused more than larger ones (e.g., reusing 1000 screws is less significant than reusing two doors in terms of weight). One can observe that environmental aspects are formulated as objective functions in only three recent studies [23, 24, 26] that proposed bi- or multi-objective optimization models. Quéheille et al. [24] and Sanchez et al.

[26, 27] estimated the environmental impacts based on the life cycle assessment (LCA) method. As an example, impact categories considered in [27] are global warming potential (GWP) and primary energy demand (PED). In this chapter, the authors developed bi-objective optimization models that minimize the total cost of a deconstruction project and its environmental impact. The generated costs and environmental impacts are both related to the choice of the deconstruction methods to apply to the building's components, each deconstruction method having an associated environmental impact and cost. The optimization model is combined with a BIM, which prior to the use of the optimization model generates disassembly plans for retrieving single or multiple target components from the building based on what is known as the sequential disassembly planning for buildings—SDPB [27]. The weighted sum was used to solve the bi-objective optimization problem. As a case study, the authors used a building of the University of Waterloo campus (office rooms and old labs repurposed in 2017 within a renovation project). Pareto solutions were obtained by varying the weights associated with the two objectives. The authors identified eco-friendly and cost-effective solutions and mentioned that different deconstruction methods per building component influence the environmental and economic aspects. In addition to the LCA-based environmental impacts, [24] also considered maximizing the recovery rate of materials in their model. This objective was also considered in [23] who referred to the European Union's target minimum construction and deconstruction waste recovery rate of 70%. Note that [29, 30] estimated the proportion of materials not sent to landfills (referred to as the recycling rate) obtained after solving the mathematical model (that minimizes the total duration of the project) but did not formulate it as an objective function. Chen et al. [19] also calculated the total mass of recovery but did not formulate a specific objective related to this aspect in their mathematical model. By doing this, the solutions found are not "dictated' or directly impacted by the recovery rate. This does not allow finding trade-off solutions as performed in [27] for instance. In [17], the authors provided a methodology to generate a quantitative estimate of the impact of construction waste diversion and reduce the project's carbon impact. The methodology is based on North American standards and regulations to determine the impact of current policies. To do so, the experts developed a decision-support tool to test six different policy scenarios. For example, one policy imposes the reuse of certain materials or a certain mass of building, and another requires that deconstruction projects recover the materials that are extracted. To that end, the authors analyzed the BIM of five different buildings with unique characteristics including a North American-style single-family home and a low-rise office building. They found that it was possible to drastically reduce waste and the carbon footprint and still achieve a positive financial return. Even so, they concluded that current policies have a different impact on every building type. Moreover, the authors pointed out that there is no evidence public policies (local, regional, national, or otherwise) that have any real impact and that it generally takes several years to assess their effectiveness.

4 Discussion

This literature review allows us to identify some challenges and promising perspectives in the optimization of building deconstruction operations; this field is pivotal in advancing sustainable construction practices. However, several lacks and limitations currently impede the development of robust methodologies. As highlighted by Quéheille et al. [24], one significant challenge is the inability of existing tools to estimate the uncertainties associated with the calculations of some indicators. This issue is compounded by several factors, including uncertainties in LCA data and actual machine/equipment energy consumption. Schleier and Walther [28] further emphasize that reverse logistics networks face primary uncertainties related to unpredictable demand for recovered products and fluctuating supply in terms of both quantity and quality of returned products or waste streams.

Current research predominantly focuses on new building designs, sometimes preconstructed, to facilitate their future deconstruction. While this approach is beneficial for upcoming projects, it overlooks the fact that the majority of existing buildings were constructed with a sole focus on the construction phase, neglecting future deconstruction needs. Given that buildings typically have long lifespans, the scientific community must intensify research efforts on deconstructing the existing building stock. This would require adapting deconstruction strategies to the unique challenges posed by older structures, such as hazardous material management, structural integrity, and varying building standards over time. Also, this would help for better development of BIM for buildings. BIM holds significant potential in improving the efficiency of deconstruction projects, as it provides detailed information about building components, materials, and their conditions, thereby enabling better scheduling and execution of deconstruction activities as shown for instance in [27]. In other words, BIM provides detailed information, which is valuable input for optimization approach, allowing the exploration of a variety of relevant deconstruction scenarios. BIM also facilitates the identification and documentation of reusable materials, supporting the development of CE practices in construction. Despite these advantages, the integration of BIM in deconstruction is still in its nascent stages, and more research is required to explore its full potential, especially in the context of existing buildings where new technologies can help to fill the gap in historical information. Another key challenge in deconstruction is the lack of standardized bill of materials for components and items recovered from deconstructed buildings. Aligning the classification of these materials with those used for new products would enhance their marketability, streamline inventory management, and facilitate their reintegration into production systems. This standardization would also support the development of digital marketplaces for recovered materials, promoting CE by making it easier for buyers to find and procure these materials. Also, this would make it possible to solve the problem raised by Mollaei et al. [18] and other authors, which is the absence of a resale market for materials. Machine learning could be a valuable approach for predicting material inventories in deconstruction projects. By analyzing historical data on material recovery, machine learning models could forecast the types and quantities of materials likely to be available from future deconstruction projects. This predictive capability would enable better planning for material reuse and recycling, anticipating the destination of the materials and their price, reducing waste and promoting sustainable construction practices. The diversity of materials used in construction presents both challenges and opportunities for deconstruction. Each material type (concrete, steel, wood, glass, etc.) requires specific handling, processing, and reuse strategies. Current research often focuses on a limited range of materials, particularly those that are easy to recycle or reuse. Future research should expand to include a broader spectrum of materials, particularly those that are currently difficult to process, to develop comprehensive strategies for their recovery and reuse. A coherent policy framework is crucial for advancing deconstruction practices. The hierarchical 10R framework proposed by Reike et al. [25] offers a structured approach to maximizing material value retention through strategies such as reuse, refurbishment, and recycling. However, applying this framework in practice requires a systemic perspective that considers the interdependencies of deconstruction operations. Misalignment between individual decision-making and systemic optimization can lead to local optima that undermine global sustainability goals. Furthermore, regulatory coherence is essential to support sustainable deconstruction practices. As noted by Quéheille et al. [24], the current European waste policy's recovery rate metric does not differentiate between high- and low-quality recovery. For example, while concrete waste can be processed into high-quality aggregates for concrete production, it can also be used as low-quality backfill material. The former option, although more desirable, involves a more complex and potentially less efficient process. Therefore, policy instruments must be refined to promote high-quality recovery and avoid incentivizing lower-value uses of recovered materials. Deconstruction requires a higher level of skills and training compared to traditional demolition due to the need for careful material recovery and hazardous material management. Exploring the potential of digital technologies, such as augmented reality and advanced robotics, could support workforce training and improve deconstruction practices. These technologies could provide real-time guidance, enhance safety, and increase the efficiency of deconstruction activities. A PESTEL (Political, Economic, Social, Technological, Environmental, and Legal) analysis can provide a comprehensive framework for assessing the external factors influencing deconstruction practices. For instance, political factors such as government incentives for recycling and reuse, economic factors like the market value of recovered materials, and social factors including public perception of sustainability can significantly impact the adoption of deconstruction practices. Technological advancements, environmental considerations, and legal requirements further shape the deconstruction landscape, necessitating a multifaceted approach to policy and research development. In conclusion, addressing the uncertainties and limitations described above requires a multidisciplinary approach that integrates advanced technologies, standardized practices, and coherent policy frameworks. The scientific community must focus on both the future of new building designs and the challenges posed by existing building stock to develop comprehensive deconstruction strategies. By doing so, the field can contribute significantly to the circular economy and sustainable development goals.

5 Conclusion

This review highlights the emerging focus on deconstruction operations planning as a distinct research niche, primarily characterized by a concentration on operational planning issues, with one notable exception addressing strategic decision-making. Most studies emphasize on-site operations, often overlooking the need for integrated planning that coordinates on-site deconstruction with off-site recycling and transportation activities. It is worth noting that only one study analyzed the management of EPS-ETICS waste through alternative recycling methods, emphasizing the importance of developing comprehensive optimization models that incorporate both operational and strategic perspectives. The varied decision-making approaches reviewed indicate a predominant focus on minimizing operational costs, while environmental considerations remain less emphasized. Although some studies integrate environmental impacts and recovery rates into their objectives, there is a clear opportunity for future research to adopt more holistic approaches that balance economical and environmental goals. This could involve enhancing the role of LCA methodologies and developing decision-support tools that better align with evolving regulatory frameworks and market demands. Addressing these gaps can advance the effectiveness and sustainability of deconstruction practices, ultimately supporting the transition toward CE in construction.

Acknowledgments This research is funded by the Research Network on Circular Economy of Québec (RRECQ) and the Circular Economy Acceleration Lab for the construction sector of the Centre for Intersectoral Studies and Research on the Circular Economy (CERIEC) of ÉTS. The lab is supported by Desjardins Group and the Government of Québec. The authors also thank Mr. Mikel González Avilés for his help in the search for relevant research papers.

References

1. Ahmed N, Abdel-Hamid M, Abd El-Razik MM, El-Dash KM (2021) Impact of sustainable design in the construction sector on climate change. Ain Shams Eng J 12(2):1375–1383. https://doi.org/10.1016/j.asej.2020.11.002
2. Aidonis D (2019) Multiobjective mathematical programming model for the optimization of end-of-life buildings' deconstruction and demolition processes. Sustainability 11(5):1426. https://doi.org/10.3390/su11051426
3. Aidonis D, Xanthopoulos A, Vlachos D, Iakovou, E (2008) An analytical methodological framework for managing reverse supply chains in the construction industry. WSEAS Trans Environ Develop 4(11):1036–1046
4. Ajayebi A, Hopkinson P, Zhou K, Lam D, Chen HM, Wang Y (2020) Spatiotemporal model to quantify stocks of building structural products for a prospective circular economy. Resour Conser Recycling 162:105026. https://doi.org/10.1016/j.resconrec.2020.105026

5. Allam AS, Nik-Bakht M (2023) From demolition to deconstruction of the built environment: a synthesis of the literature. J Build Eng 64:105679. https://doi.org/10.1016/j.jobe.2022.105679
6. Bertino G, Kisser J, Zeilinger J, Langergraber G, Fischer T, Osterreicher D (2021) Fundamentals of building deconstruction as a circular economy strategy for the reuse of construction materials. Appl Sci 939:30pp https://doi.org/10.3390/app11030939
7. Boukherroub T, Nganmi Tchakoutio A, Drapeau N (2024) Using lean in deconstruction projects for maximising the reuse of materials: a Canadian case study. Sustainability 16(5):Article 5. https://doi.org/10.3390/su16051816
8. Chen Q, Feng H, Garcia de Soto B (2022) Revamping construction supply chain processes with circular economy strategies: a systematic literature review. J Clean Product 335:130240. https://doi.org/10.1016/j.jclepro.2021.130240
9. Chini AR, Bruening S (2003) Deconstruction and materials reuse in the United States. The future of sustainable construction, 14 October, 1–22
10. Ding L, Wang T, Chan PW (2023) Forward and reverse logistics for circular economy in construction: a systematic literature review. J Clean Product 388:135981. https://doi.org/10.1016/j.jclepro.2023.135981
11. EU (2023) Construction and demolition waste—European Commission. https://environment.ec.europa.eu/topics/waste-and-recycling/construction-and-demolition-wasteen. Guy B, Gibeau E M (n.d.) A guide to deconstruction
12. Guy B, Gibeau EM (2003) A guide to deconstruction. Deconstruction Institute, Florida
13. Hosseini M, Chileshe N, Rameezdeen R, Lehmann S (2013) Sensitizing the Concept of Reverse Logistics (RL) for Construction Context (Doctoral dissertation, Islamic Azad University)
14. Jansen BW, van Stijn A, Gruis V, van Bortel G (2020) A circular economy life cycle costing model (CE-LCC) for building components. Resour Conserv Recycling 161:104857 (11pp). https://doi.org/10.1016/j.resconrec.2020.104857
15. Jansen BW, van Stijn A, Gruis V, van Bortel G (2020) A circular economy life cycle costing model (CE-LCC) for building components. Resour Conserv Recycl 161:104857
16. Lund E, Yost P (1997) Deconstruction–building disassembly and material salvage: the Riverdale case study. NAHB Research Center, Upper Marlboro
17. Mollaei A, Bachmann C, Haas C (2023) Assessing the impact of policy tools on building material recovery. Resour Conser Recycling 198:107188. https://doi.org/10.1016/j.resconrec.2023.107188
18. Mollaei A, Bachmann C, Haas C (2023) Estimating the recoverable value of in-situ building materials. Sustain Cities Soc 91:104455. https://doi.org/10.1016/j.scs.2023.104455
19. Mollaei A, Byers B, Christovan C, Olumo A, De Wolf C, Bachmann C, Haas C (2023) A global perspective on building material recovery incorporating the impact of regional factors. J Clean Product 429:139525. https://doi.org/10.1016/j.jclepro.2023.139525
20. Munaro MR, Tavares SF (2023) A review on barriers, drivers, and stakeholders towards the circular economy: the construction sector perspective. Clean Respons Consum 8:100107. https://doi.org/10.1016/j.clrc.2023.100107
21. Page MJ, McKenzie JE, Bossuyt PM, Boutron I, Hoffmann TC, Mulrow CD, Shamseer L, Tetzlaff JM, Akl EA, Brennan SE, Chou R, Glanville J, Grimshaw JM, Hróbjartsson A, Lalu MM, Li T, Loder EW, Mayo-Wilson E, McDonald S, et al. (2021). The PRISMA 2020 statement: an updated guideline for reporting systematic reviews. Brit Med J 372:n71. https://doi.org/10.1136/bmj.n71
22. Peiris PAN, Herath N, Ngo T, Duffield C, Hui FKP (2023) Lean and green: how the synergy can promote sustainable construction. In: Dissanayake R, Mendis P, Weerasekera K, De Silva S, Fernando S, Konthesingha, C, Gajanayake P (eds), ICSBE 2022. Springer, Berlin, pp 219–228. https://doi.org/10.1007/978-981-99-3471-3-16
23. Quéheille E, Taillandier F, Saiyouri N (2019) Optimization of strategy planning for building deconstruction. Autom Construct 98:236–247. https://doi.org/10.1016/j.autcon.2018.11.007

24. Queheille E, Taillandier F, Saiyouri N (2019) Optimization of strategy planning for building deconstruction. Autom Constr 98:236–247
25. Reike D, Vermeulen WJ, Witjes S (2018) The circular economy: new or refurbished as CE 3.0? - exploring controversies in the conceptualization of the circular economy through a focus on history and resource value retention options. Resour Conserv Recycl 135:246–264
26. Sanchez B, Rausch C, Haas C (2019) Deconstruction programming for adaptive reuse of buildings. Autom Construct 107:102921. https://doi.org/10.1016/j.autcon.2019.102921
27. Sanchez B, Rausch C, Haas C, Saari R (2020) A selective disassembly multi-objective optimization approach for adaptive reuse of building components. Resour Conservation Recycling 154:104605. https://doi.org/10.1016/j.resconrec.2019.104605
28. Schleier J, Walther G (2024) Strategic network design for recycling of EPS insulation material – insights from a German case study. Int J Product Res 62(16):5878–5904. https://doi.org/10.1080/00207543.2024.2302041
29. Schultmann F, Rentz O (2004) Environment-oriented project scheduling for the dismantling of buildings. Produktion und Umwelt: Festschrift für Otto Rentz, 287–314
30. Schultmann F, Rentz O (2002) Resource-constraint project scheduling for deconstruction projects. Proc CIB Task Group 39:1–15
31. Schultmann F (2005) Sustainable deconstruction of buildings. Smart Sustain Built Environ 148–159
32. Schultmann F, Sunke N (2007) Energy-oriented deconstruction and recovery planning. Build Res Inform 35(6):602–615
33. Veillette F, Boukherroub T, Audy JF (2024) An overview of the literature on decision-making support models for deconstruction logistics. In: IISE Annual Conference. Proceedings. Institute of Industrial and Systems Engineers (IISE), pp 1–6
34. Wang J, Yuan H, Kang X, Lu W (2010) Critical success factors for on-site sorting of construction waste: a China study. Resour Conserv Recycling 54(11):931–936. https://doi.org/10.1016/j.resconrec.2010.01.012
35. Xanthopoulos AX, Aidonis DA, Vlachos DV, Iakovou EI (2009) Reverse logistics processes of multi-type end-of-life buildings/construction sites: an integrated optimization framework. WSEAS Trans Environ Develop 5(3):250–259
36. Xanthopoulos A, Aidonis D, Vlachos D, Iakovou E (2012) A planning optimisation framework for construction and demolition waste management. Int J Ind Syst Eng 10(3):257–276. https://doi.org/10.1504/IJISE.2012.045675

A Novel Optimization Model for Sustainable Life Cycle Design: Circular Economy Case Study in the Automotive Industry

Harun Resit Yazgan , **Derya Eren Akyol** , **and Gulseli Altıntaş**

Abstract While the circular economy (CE) and Sustainability are prominent today, it is realized that network optimization studies should be given importance to ensure the transition to the CE. Overall, this study provides a comprehensive exploration of the CE concept, along with an extensive review and application focused on network optimization aligned with sustainability and CE principles. It primarily includes a comprehensive research and literature review on the CE. A review of the studies that made network optimization using reverse supply chain aiming sustainability and circularity in the production sector was made, and network design and optimization for a sector that has not been studied before was suggested based on the literature. Since transition to circular production has been started by companies in the automotive industry, this study aims to fill the gap in the literature of hybrid circular network design and optimization specific to the automotive industry. For the designed novel network, a linear programming model for profit maximization was developed, solved using optimization software, and the global optimal result was found. Results proved that hybrid network optimization reduced costs. Finally, scenario analyses related to recycling and reuse rates were made, and their effects on the objective function were determined. Scenario analyses showed that recycling rates directly affected profitability. Thus, this study represents a noteworthy contribution to the field.

Keywords Sustainability · Circular economy · Network design · Optimization · Mathematical model · Linear programming

H. R. Yazgan · G. Altıntaş (✉)
Sakarya University, Engineering Faculty, Industrial Engineering, Sakarya, Turkey
e-mail: yazgan@sakarya.edu.tr; gulseliisler@sakarya.edu.tr

D. E. Akyol
Dokuz Eylül University, Engineering Faculty, Industrial Engineering, Izmir, Turkey
e-mail: derya.eren@deu.edu.tr

B. Jarboui et al. (eds.), *Data-Driven Waste Revolution*, Computational Intelligence Methods and Applications, https://doi.org/10.1007/978-981-96-5890-9_2

1 Introduction

Our nature consists of processes that involve an endless cycle, such as a waste-free metabolism. Since an input of one process is the output of another process in nature, no waste is generated. The circular economy model, which is seen as an innovative alternative to the frequently used linear economy model, is a new and advanced system that optimizes natural resources used and keeps them in product life cycles as long as possible. Also, the circular economy model transforms products that have completed their lifecycles into sources of other product and service cycles [44].

There is no clear evidence of the source or originator of the circular economy concept. Most inspirations in the literature are based on the Industrial Ecology Model based on the closed-loop concept originating from Japan and Germany [22, 48]. However, the following names can be seen among those who contributed to the emergence of this concept: [20, 21, 45], and [37].

Circular economy is an important and popular topic today for both companies and academics [11]. At this point, it seems that such network optimization studies should be emphasized in order to effectively ensure the transition to circular economy in the production and consumption sector.

The main motivation of this study is to address the growing need for practical network optimization models that can guide the automotive industry's transition to circular economy, providing both theoretical and real-world benefits. This study is motivated by the need to bridge the gap between circular economy principles and practical applications in the automotive sector, aiming to develop an innovative network optimization model that supports sustainable production.

In the first section, an introduction to the topic is provided, key motivations are highlighted, and the details of the study are briefly discussed. The second section includes detailed explanations of the circular economy and a compilation of studies that have made network optimization in line with sustainability and circular economy. In addition, it covers and aims to develop a new and unique network model in the automotive industry. In the third section, a network model that includes hybrid closed-loop and open-loop supply chain is developed for the life cycle of cars. We also aim to design a mathematical model that will maximize the profit of this network, and we illustrate the results through a case in the fourth section. Additionally, by analyzing 12 scenarios regarding recycling rates, we try to find the types of raw materials with recycling rates that most affect the objective function of the model. Finally, the fourth section addresses the conclusion and future studies, summarizing the key findings and suggesting potential directions for further research. The schematic representation of the general overview of the study is given in Fig. 1.

In the automotive sector, the circular economy model has been put into practice in one of the world's leading car manufacturing companies. However, no practical models beneficial for both theory and real-world applications are encountered. At this point, this study aims to fill the gap in the literature by addressing the circular network design and optimization specifically for the automotive sector.

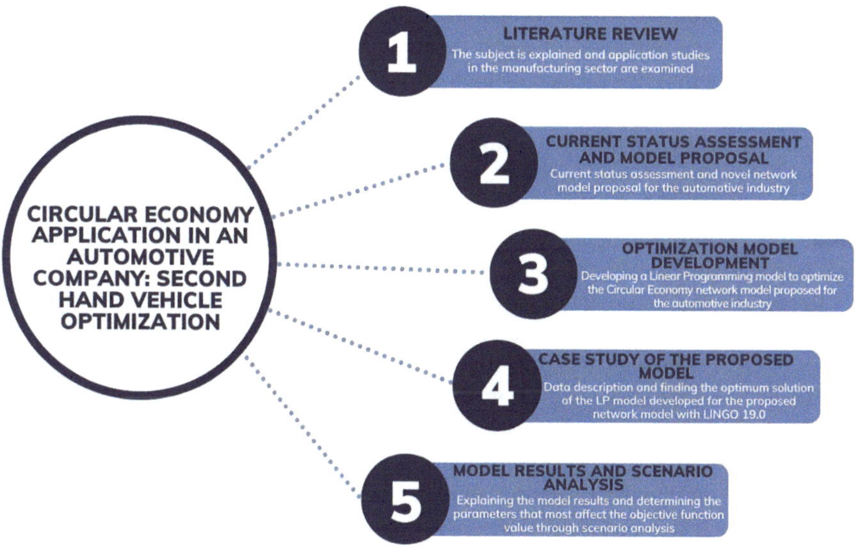

Fig. 1 General overview of the study

2 Literature Review

The circular economy, adopted by the "European Green Deal" [7], is one of the priority policy areas in the new strategy. The European Green Deal, approved in 2020, comprises a set of policy initiatives implemented by the European Commission with the overarching aim of achieving climate neutrality for the European Union (EU) by 2050. With this document, 2050 is targeted for the EU to offer new and better jobs, restore climate neutrality, improve people's well-being, and increase growth.

The 2030 Agenda for Sustainable Development was adopted by the United Nations General Assembly on September 25, 2015, thus taking a step toward implementing the circular economy model. In line with these studies, the EU's objectives for production are to develop some incentives that support cyclical product designs and to create efficient and innovative production-product life cycle processes [43].

It is anticipated that businesses will need support in creating profitable and viable processes using circular economy business models. Accordingly, all stakeholders will have important responsibilities. While the private sector will drive most of the innovation, investment, and implementation, governments must also create new regulatory strategies and benefit from this transformation [32].

Circular economy concept emphasizes designing products with long life cycles and encourages repair, refurbishment, and remanufacturing. Therefore, the circular economy is crucial in the manufacturing sector for several reasons. By minimizing waste and encouraging the reusing and recycling of materials, this concept can contribute to sustainable resource management in the manufacturing sector. It ensures reduced waste generation and encourages recycling of materials. It requires innovation in product design and manufacturing processes. This innovative approach has the potential for more sustainable and resource-efficient production methods. Adopting this concept can ensure cost savings for manufacturers, thanks to the reuse of materials and components. Besides cost savings, it contributes to a reduction in the environmental impact of manufacturing activities and environmental protection. It can help manufacturers comply with legal requirements as environmental regulations become more stringent. Besides all these, embracing circular economy principles can enhance a company's reputation and provide for building a positive corporate image.

Regarding the optimization study of the network model developed within the framework of circular economy, a search is carried out on the Scopus, ResearchGate, ScienceDirect, ProQuest, and National Thesis Center databases, by using the keywords "circular economy," "network," "optimization," "sustainability," "closed loop supply chain," "open loop supply chain," and "automotive." This section of the study includes a review of the studies that propose a network and supply chain optimization on circular economy and sustainability in the manufacturing sector. Studies thought to be relevant to the proposed model are classified, and 33 studies that had the quality to contribute to the study are determined and analyzed.

Some studies propose single-objective linear models. While some of these studies generally cover the production/manufacturing sector [4, 29, 46], there are also some optimization studies in the digital industry, electrical and electronics industry [47], and food and beverage industry [39].

In some studies, we see nonlinear programming approach is suggested and applied in sectors such as the electrical and electronics industry [25] and the network marketing industry [10].

Some studies propose multi-objective linear models. While some of these studies generally covered the production/manufacturing sector, there are also some optimization studies in the public and private sectors [8], energy industry [2], agriculture industry [12, 23], textile industry [6], mining industry [36], electronics industry [19, 38], food industry [31], and battery industry [41].

Studies in the literature propose linear mathematical models, including EoL (end-of-life) vehicles and reverse supply chains in terms of circular economy and sustainability in the automotive industry, and perform optimization studies. The objective functions of some of these studies are maximization [26, 34, 35, 40]. In some studies, the objective function is minimization [3, 14, 28]. In some studies in the literature dealing with the automotive industry, the authors have developed a multi-objective linear model [18, 30]. A study also suggested a multi-objective nonlinear model in the automotive industry [33].

Table 1 complies examined studies published between 2011 and 2023 which focus on the optimization of sustainability and circular economy in the manufacturing sector.

As seen in Table 1, deterministic models were often proposed in studies in the literature. Studies suggesting network models and optimization studies within the framework of circular economy and sustainability were carried out in various sectors. In the automotive industry, it was observed that network designs have been made for end-of-life (scrap) vehicles and especially for the recycling of a part of the

Table 1 Literature review

Authors	Model type	Modeling approach	Objective/s	Application area
Talaei et al. [38]	Fuzzy	MOMILP	Min total cost and min total carbon dioxide emission	Electronics (copier)
Phuc et al. [28]	Fuzzy	MILP	Min total cost	Manufacturing (vehicles)
Wu et al. [46]	Fuzzy	MILP	Min total cost	Manufacturing (general)
Kuşakçı et al. [18]	Fuzzy	MOMILP	Min total cost and max aspiration level of fuzzy constraints	Manufacturing (vehicles)
Nayeri et al. [23]	Fuzzy	MOMILP	Min total cost, reduce and control total carbon emissions, and max social impacts	Agriculture (water tanker)
Çınar [2]	Fuzzy	MOMILP	Min total reverse logistics network cost and min total carbon emissions	Energy (wind energy)
Torgul and Paksoy [40]	Fuzzy	MILP	Max total profit	Manufacturing (automotive steel)
Tosarkani and Amin [41]	Fuzzy-Stochastic	MOLP	Max profit and max the environmental compliance	Battery (lead acid batteries)
Hajipour et al. [10]	Stochastic	MINLP	Max total profit	Network marketing (general)
Simic [35]	Stochastic	MILP	Max expected profit	Manufacturing (vehicles)
Karagöz et al. [14]	Stochastic	MILP	Min total cost	Manufacturing (vehicles)
Shahedi et al. [33]	Stochastic	MOMINLP	Max total profit, min effects of pollutants, and max number of employment opportunities	Automotive (general)
Ene and Öztürk (2014)	Probabilistic	MILP	Max total profit of reverse supply chain network	Manufacturing (general)
Shankar et al. [34]	Probabilistic	MILP	Max contribution margin or profit	Automotive (general)
Öztürk and Aksoy [27]	Probabilistic	MOMILP	Max total profit, max recovery rate, and min total recovery time	Manufacturing (general)

(continued)

Table 1 (continued)

Authors	Model type	Modeling approach	Objective/s	Application area
Pishvaee et al. [29]	Deterministic	MILP	Min total cost	Manufacturing (general)
Fahimnia et al. [6]	Deterministic	MOMILP	Min total cost and min carbon emission	Manufacturing (textile)
Özceylan et al. [25]	Deterministic	MINLP	Min total cost	Electrical and electronics (hand light)
Demirel et al. [3]	Deterministic	MILP	Min total cost	Manufacturing (vehicles)
Özceylan [24]	Deterministic	MILP	Min total cost of the closed-loop supply chain network	Digital (camera)
Kaya and Alumur Alev [15]	Deterministic	MILP	Max total profit	Manufacturing (general)
Özceylan et al. [26]	Deterministic	LP	Max revenue	Manufacturing (vehicles)
Kır et al. [17]	Deterministic	MINLP	Min total cost	Recycling (tire)
Rohmer et al. [31]	Deterministic	MOMILP	Min total cost and min total environmental impact	Food (nutrient)
Jabarzadeh et al. [12]	Deterministic	MOMILP	Max profit, customer responsiveness, and quality	Agriculture (fruit)
Golpira and Javanmardan [8]	Deterministic	MOMILP	Min cost, max job opportunities, and min work-related damages	Public sector and private sector
Lu et al. [19]	Deterministic	MOMILP	Min total cost and min pollution by heavy metals	Electronics (general)
Soleimani [36]	Deterministic	MOLP	Min energy consumption and max profit	Mining
Torgul et al. [39]	Deterministic	MILP	Max total profit	Food (juice)
Yıldızbaşı and Arıöz [47]	Deterministic	MILP	Max total profit from the flow in the network	Electrical and electronics (printed circuit boards)

(continued)

Table 1 (continued)

Authors	Model type	Modeling approach	Objective/s	Application area
Kabadurmuş et al. [13]	Deterministic	MILP	Max total distributed food waste and min total cost	Food (general)
Kazançoğlu et al. [16]	Deterministic	MILP	Min total cost, CO_2 emissions, and particulate matter concentrations	Manufacturing (home appliances)
Rajak et al. [30]	Deterministic	MOMILP	Min environmental effects and max net profit	Automotive (steering column)
This Study	Deterministic	MILP	Max total profit of network	Automotive

car, but no study has been found that takes into account all parts and raw materials inside the car. Unlike the studies in the literature, this study proposes a network model for all parts and raw materials from cars that have not yet become scrap (the parts of which are also suitable for reuse). In addition, reverse logistics or closed-loop approaches are seen in most of the studies in the literature, while a hybrid approach is found in the automotive industry. At this point, the difference in this study is that it proposes a network model that includes a hybrid supply chain specific to the automotive industry.

3 Problem Definition and Proposed Model

Circular economy ensures the efficient and sustainable use of resources, thus reducing waste in a company. It allows cost savings and minimizes environmental impacts. Additionally, by focusing on sustainable production and design principles, reputation and recognition can be strengthened by gaining a competitive advantage. Considering all these benefits, the world's leading companies have become interested in the circular economy.

In this chapter, the production system that a well-known automotive company, one of the companies interested in the subject, plans to implement is discussed. In a facility of this company, works have started to add a new and circular line in addition to the existing production line. By examining the shared information and data regarding this company's new production plan, the problem definition is made throughout the automotive industry. According to this definition, a novel network model that includes circularity is recommended. This network includes hybrid closed-loop and open-loop supply chains for the life cycle of cars. A mathematical model with a maximum profit objective that determines the product flows of this model is developed.

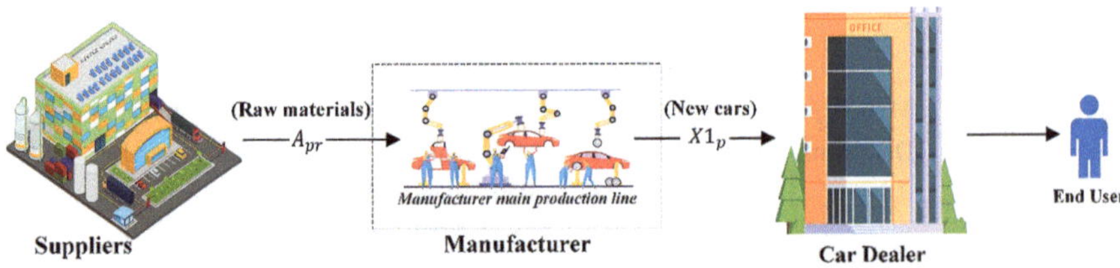

Fig. 2 The current forward production system in the automotive industry

3.1 Problem Definition

Unlike the studies given in the previous chapter, this study proposes a general network model specific to the automotive industry. A novel and hybrid network model for the automotive industry is proposed by examining the information shared by the company operating in this industry that aims to make its production compatible with the circular economy. According to the information shared by the examined company, the company plans to restructure the automobile production process in the factory layout to increase circularity.

For the automobile type under consideration, there are both renew (reconditioned) car and new car demands from the customer in the sector. Therefore, the automobile manufacturer plans to dedicate one production line to renew secondhand car production while continuing to produce new cars on the main line. The following three centers are included in the system to provide reverse flow: Collection Center, Disassembly/Dismantling/Rig Testing Center, and Reuse/Repair Center. In addition to these centers, recycling facilities (stakeholders) and disposal centers are also included in the system as outsource.

In the automotive sector, the forward material flow network in the system is shown in Fig. 2.

Currently, car production only includes the forward supply chain system. As can be seen in Fig. 2, the manufacturer supplies all of the raw materials from the supplying companies. The manufacturer passes the raw materials through the production process, produces cars, and sells them to the car dealer. The car dealer delivers the cars to the end users.

The end user drives the car but does not create a reverse supply chain. Used cars become scrap in this system when they reach the end of their useful life. Then, as companies are obligatory, cars and car parts collected from scrap vehicle warehouses are sent directly to the disposal centers in accordance with Environment Law No. 2872 of the Republic of Türkiye.

Assumptions regarding the problem are given below:

- The manufacturer, collection center, disassembly/dismantling center, and reuse center are controlling from a single center.
- End user demands are exactly met for each period.
- No waste is created during processes in the facilities and transportation.

- The disassembly/dismantling center is a unit that transforms used cars into raw materials.
- Facilities' capacities in the network are certain and constant for each period.
- Transportation costs do not vary based on the type or dimensions of the parts determined in kilograms.
- The recycling rates are known beforehand.
- The raw materials obtained from the reuse center and recycling facility are not of poor quality from the original raw materials.
- The sum of the percentages of the transportations made from the disassembly/dismantling center to different centers is 1.
- There is no holding cost.
- For the production of new cars, only zero raw material from the supplier is used.
- In renew secondhand car manufacturing, after the raw materials are supplied from the reuse and recycling center, the rest is purchased from the suppliers as zero raw material to meet end user demand.
- Raw material transportation to the suppliers' and recycling facilities' manufacturing facility is their responsibility, and they bear the cost. This has delegated to the car dealer to endure the difficulties of collecting used cars from the end user so that only the purchasing cost is covered. And since the disposal center has its own transportation methods for harmful materials, transportation of the raw materials to be disposed of from the disassembly/dismantling center is the disposal center's responsibility and bears the cost.
- The cost of destroying the materials is legally covered by the automotive companies.
- Product collection from the end user is the responsibility of the car dealer. Used cars are purchased from car dealer and bear the cost of transportation.
- Since production line 2 has just started production, in the first period, the raw materials are used from the warehouse.
- In the model, one period is increased each time the manufacturer is reached.

3.2 Proposed Model

This study proposes a new network model for the evolution of a linear material flow network to a circular material flow network. The proposed model also considers the backward supply chain, in addition to the forward supply chain. The company, whose information is examined during the proposal of this network model, plans to add a new production line in which raw materials supplied from reuse and recycling centers will be used, while new cars continue to be produced on the current production line. The company plans to produce two types of vehicles from these two lines: new and renew cars. In the current production line, raw materials supplied from reuse and recycling centers are not used, only raw materials supplied from suppliers are used, and new car is produced as the final product. In the production line that the company plans to add to the system, both raw materials supplied from reuse and recycling centers and raw materials supplied from suppliers selling new raw materials are used, and a renew car is produced as the final product. In this way, while the users' demand for cars produced from completely zero raw materials will continue to be met, it will also meet the demand for cars produced using repaired and recycled raw materials (with low sales prices).

Figure 3 shows the proposed material flow network with circularity, including the reverse supply chain for recycled automobile manufacturing.

As we see from Fig. 3, there are two production lines. Raw materials ($A1_{pr}$) are supplied from the suppliers to the first production line of the manufacturer, and new cars are produced with only these raw materials. The materials obtained from used cars such as metal, plastic, rubber, glass, oil, and other materials are converted into secondary raw materials (U_{pr}, S_{pr}) in the reuse/repair center and recycling facilities. Then, they are used as the raw materials of the second production line together with raw materials supplied by the suppliers.

Two different cars (new and renew cars) are produced from two production lines. Pressing, welding, assembly, painting, and quality control operations are carried out in these production lines. In the first production line, new cars are produced using only raw materials supplied by the suppliers. In the new second production line, renew cars are produced using raw materials supplied by suppliers, reuse (the flow is given by U_{pr}) and recycling centers (the flow is given by S_{pr}). These two types of cars are delivered to end users as new ($X1_p$) and renew ($X2_p$) cars through the car dealer (as a retailer) with different sale prices. The car dealer also purchases old cars, at most 20 years old, i.e., used cars, from end users, with campaigns frequently used in product sales, such as "give the old one, get the new one" and "we buy your cars for their value." The term used cars includes cars that are no more than 20 years old, have not reached the end of their useful life, and have not been scrapped.

The identified company sells more than half of its European sales in the retail market; therefore car dealer is a retailer in this system. Car dealer sends the used cars to the Collection Center (the flow is denoted by Y_p). Used cars can be easily stored in the Collection Center and sent to the Disassembly Center (the flow is given by Z_p) when user demand needs to be met.

In the Disassembly Center, separation, cleaning, repair, and inspection operations are carried out. In this center, disassembly and dismantling parts from cars are separated according to six types of materials: metal, plastic, rubber, glass, oil, and other materials. It is not possible in real life for all of these six types of materials to be reusable or recyclable. For this reason, reuse and recycling rates are defined in the model.

The materials that can be reused (secondary materials) (Q_{pr}) in car production are sent to the Reuse/Repair Center. Six types of materials sent to the Reuse Center are processed in this facility in such a way that they can be used in car production again with the tests, processes, and controls. The reusable raw materials (secondary raw materials) (U_{pr}) are sent to the manufacturer to be used in the second production line of the manufacturer and to produce renew cars.

The secondary materials that can be recycled and used in car production are sent (sold) to Recycling Facilities (the flow is denoted by W_{pr}). The six types of materials

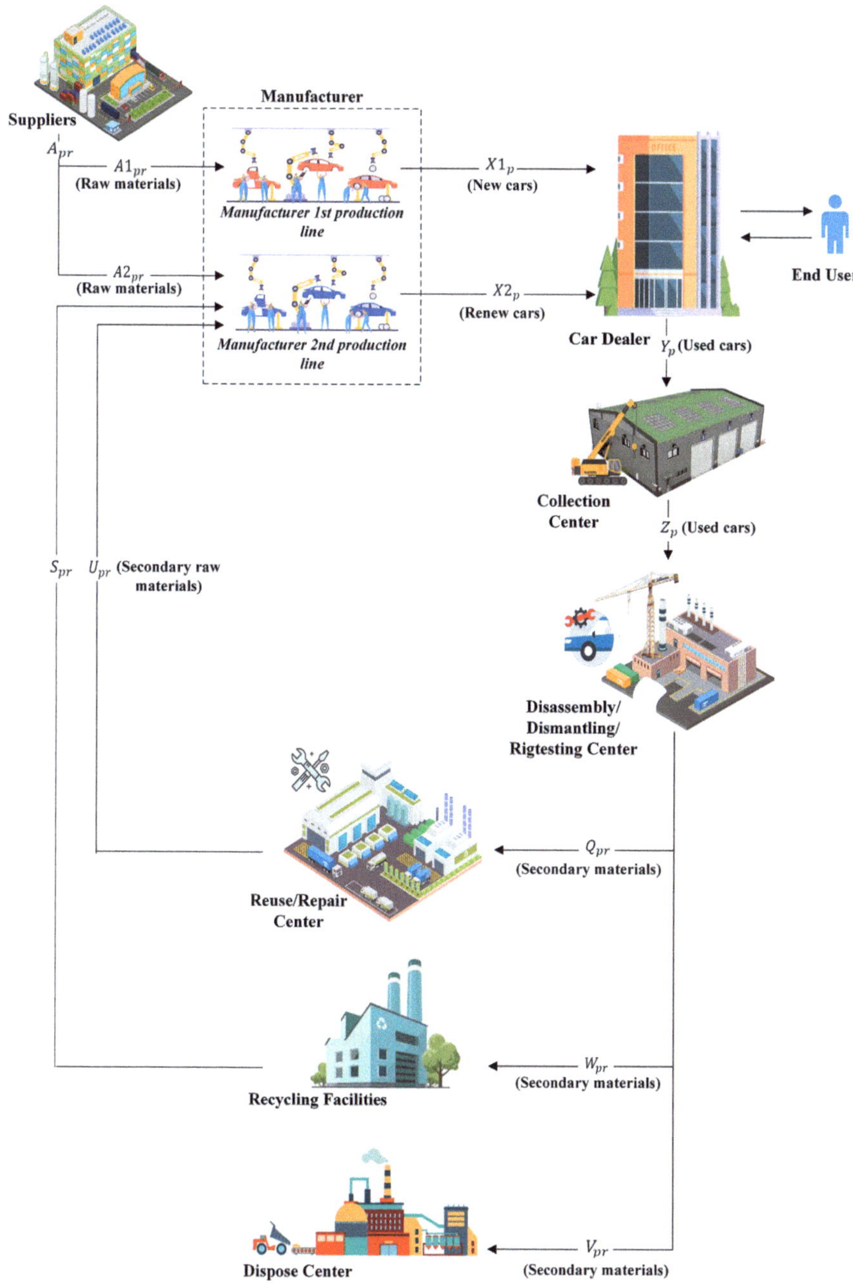

Fig. 3 Schematic representation of the proposed circular economy model in the automotive industry

sent to the recycling facilities are turned into raw materials again with the recycling processes carried out in the facilities specialized in their fields. The reusable raw materials are procured (purchased) by the manufacturer from the recycling facilities to be used in the second production line and to produce renew cars (the flow is given by S_{pr}).

A certain number of cars can be manufactured with raw materials from the Reuse Center and the Recycling Facility. The remaining part of the raw material required to meet the end user demand is purchased from the supplier companies.

The materials that cannot be used in the production of new cars are sent to the Dispose Center for disposal in accordance with the Republic of Türkiye Environmental Law No. 2872.

3.3 *Formulation of Mathematical Model*

In this section, a mixed-integer linear programming (MILP) model is developed to optimize the proposed circular economy model for the automotive industry. The objective of the model is to maximize the profit of the proposed network model.

Notations, parameters, and decision variables are presented in the following:

Indices:

$p = 1, 2, \ldots, P$ set of periods

$r = 1, 2, \ldots, R$ set of raw materials

Decision variables:

$X1_p$: amount of new car transported from manufacturer's production line 1 to end user in period p (kg)

$X2_p$: amount of renew car transported from manufacturer's production line 2 to end user in period p (kg)

Y_p: amount of used car transported from car dealer to collection center in period p (kg)

Z_p: amount of used car transported from collection center to disassembly/ dismantling center in period p (kg)

Q_{pr}: amount of secondary material r transported from disassembly/ dismantling center to reuse center in period p (kg)

W_{pr}: amount of secondary material r transported from reuse center to recycling facility in period p (kg)

V_{pr}: amount of secondary material r transported from recycling facility to dispose center in period p (kg)

U_{pr}: amount of secondary raw material r transported from reuse center to manufacturer in period p (kg)

S_{pr}: amount of secondary raw material r transported from recycling facility to manufacturer in period p (kg)

A_{pr}: amount of total raw material r transported from suppliers to manufacturer in period p (kg)

$A1_{pr}$: amount of raw material r transported from suppliers to manufacturer's production line 1 in period p (kg)

$A2_{pr}$: amount of raw material r transported from suppliers to manufacturer's production line 2 in period p (kg)

I_{pr}: amount of raw material r in the manufacturer's warehouse in period p (kg)

Parameters:

$SP1_p$: sales price of new car to car dealer in period p (€/kg)

$SP2_p$: sales price of renew car to car dealer in period p (€/kg)

IP_r: sales price of material r to recycling facilities (€/kg)

CH_r: purchasing cost of raw material r from suppliers (€/kg)

fi_p: fixed facility cost of manufacturer in period p (€)

fc_p: fixed facility cost of collection center in period p (€)

fd_p: fixed facility cost of disassembly/dismantling center in period p (€)

fk_p: fixed facility cost of reuse center in period p (€)

$OP1$: operation cost of new car manufacturing in manufacturer (€/kg)

$OP2$: operation cost of renew car manufacturing in manufacturer (€/kg)

OPC: operation cost of used car in collection center (€/kg)

OPD: operation cost of used car in disassembly/dismantling center (€/kg)

OPK: operation cost of materials in reuse center (€/kg)

$TCIM1$: transportation cost of new cars transported from manufacturer's production line 1 to car dealer (€/kg)

$TCIM2$: transportation cost of renew cars transported from manufacturer's production line 2 to car dealer (€/kg)

$TCNC$: transportation cost of used cars transported from car dealer to collection center (€/kg)

$TCCD$: transportation cost of used cars transported from collection center to disassembly/dismantling center (€/kg)

$TCDK$: transportation cost of secondary materials transported from disassembly/dismantling center to reuse center (€/kg)

$TCDG$: transportation cost of secondary materials transported from disassembly/dismantling center to recycling facility (€/kg)

$TCKI$: transportation cost of secondary raw materials transported from reuse center to manufacturer (€/kg)

EX_r: disposal cost of material r (€/kg)

$PriceG_r$: purchasing cost of raw material r from recycling facility (€/kg)

$PriceJAL$: purchasing cost of old vehicle from car dealer (€/kg)

$DM1_p$: new car demand of car dealer in period p

$DM2_p$: renew car demand of car dealer in period p

$CAPI$: material capacity of manufacturer (kg)

$CAPC$: material capacity of collection center (kg)

$CAPD$: material capacity of disassembly/dismantling center (kg)

$CAPK$: material capacity of reuse center (kg)

Car_r: ratio of material r in a car to total weight of the car

$k1_r$: ratio of material r transported to reuse center

$k2_r$: ratio of material r transported to recycling facility

$k3_r$: ratio of material r transported to dispose center

Objective Function:

The objective function is to maximize profit. Maximization of profit consists of two components: total revenue and total cost.

$$maxprofit = totalrevenue(TR) - totalcost(TC) \tag{1}$$

$maximize\ p =$

$$\sum_{p=1}^{P} SP1_p X1_p + \sum_{p=1}^{P} SP2_p X2_p + \sum_{p=1}^{P}\sum_{r=1}^{R} IP_r W_{pr} + \sum_{p=1}^{P}\sum_{r=1}^{R} CH_r U_{pr} \tag{2}$$

$$-\sum_{p=1}^{P} fi_p + fc_p + fd_p + fk_p \tag{3}$$

$$-\sum_{p=1}^{P}[OP1(X1_p) + OP2(X2_p) + OPC(Z_p)]; \tag{4}$$

$$-\sum_{p=1}^{P}\sum_{r=1}^{R}[OPD(Q_{pr} + W_{pr} + V_{pr}) + OPK(U_{pr})] \tag{5}$$

$$-\sum_{p=1}^{P}[TCIM1(X1_p) + TCIM2(X2_p) + TCNC(Y_p) + TCCD(Z_p)] \tag{6}$$

$$-\sum_{p=1}^{P}\sum_{r=1}^{R}[TCDK(Q_{pr}) + TCDG(W_{pr}) + TCKI(U_{pr})] \tag{7}$$

$$-\sum_{p=1}^{P}\sum_{r=1}^{R} EX_r V_{pr} \tag{8}$$

$$-\sum_{p=1}^{P}\sum_{r=1}^{R} PriceG_r S_{pr} - \sum_{p=1}^{P}\sum_{r=1}^{R} CH_r A_{pr} - \sum_{p=1}^{P} PriceJAL(Y_p) \tag{9}$$

Equation 2, which is one of the subequations of the objective function, includes the revenue from the sale of new cars and renew cars to the car dealer and revenue from the sale of all secondary materials to recycling facilities. Also it includes the purchasing cost of each raw material from the suppliers multiplied by the total raw material transported from the reuse center to the manufacturer for all periods. In

this way, the profit of the manufacturer from supplying secondary raw materials from the reuse center instead of supplying zero raw materials from the suppliers is taken into account. Equation 3 includes the total fixed facility costs of the manufacturer, collection center, disassembly/dismantling center, and reuse center for all periods. Fixed facility costs include items such as facility rents, insurance premiums, cleaning expenses, and office expenses. Equations 4 and 5 include the total operating costs for the processing of all products and raw materials at the facilities of all periods. Equations 6 and 7 include the total transportation cost of all products and raw material transportation between internal and external source facilities of all periods. Equation 8 includes the total cost of disposal of all materials sent to the disposal center of all periods. Equation 9 includes total purchasing costs from external sources for all periods. This equation includes the following costs: purchasing cost of all raw materials from the recycling facilities, used cars from the car dealer, and all zero raw materials from the suppliers.
Constraints:

$$X1_p - DM1_p = 0 \quad \forall p \tag{10}$$

$$X2_p - DM2_p = 0 \quad \forall p \tag{11}$$

Demand constraints (Eqs. 10 and 11) ensure that all end user demand is met, but no product is produced in excess of demand.

The constraints between (12) and (22) are equilibrium constraints.

$$A1_{pr} = Car_r\, X1_{p+1} \quad \forall p, r \tag{12}$$

$$A2_{pr} + S_{pr} + U_{pr} = Car_r\, X2_{p+1} + I_{(p+1)r} \quad \forall p, r \tag{13}$$

$$Y_p - X1_p - X2_p = 0 \quad \forall p \tag{14}$$

$$Z_p - Y_p = 0 \quad \forall p \tag{15}$$

$$\sum_{r=1}^{R} \left(Q_{pr} + W_{pr} + V_{pr}\right) - Z_p = 0 \quad \forall p \tag{16}$$

$$Z_p\, Car_r = Q_{pr} + W_{pr} + V_{pr} \quad \forall p, r \tag{17}$$

$$Z_p\, Car_r\, k1_r = Q_{pr} \quad \forall p, r \tag{18}$$

$$Z_p\, Car_r\, k2_r = W_{pr} \quad \forall p, r \tag{19}$$

$$Z_p\, Car_r\, k3_r = V_{pr} \quad \forall p, r \tag{20}$$

$$U_{pr} = Q_{pr} \quad \forall p, r \tag{21}$$

$$S_{pr} = W_{pr} \quad \forall p, r \tag{22}$$

Constraints between (23) and (26) stipulate the capacity constraints of the relevant center.

$$\sum_{r=1}^{R} U_{pr} + S_{pr} + A_{pr} + I_{pr} \leq CAPI \quad \forall p \tag{23}$$

$$Y_p \leq CAPC \quad \forall p \tag{24}$$

$$Z_p \leq CAPD \quad \forall p \tag{25}$$

$$\sum_{r=1}^{R} Q_{pr} \leq CAPK \quad \forall p \tag{26}$$

$$A_{pr} = A1_{pr} + A2_{pr} \quad \forall p, r \tag{27}$$

Constraint 27 specifies the sum of the raw materials supplied to two different production lines from the suppliers for ease of formulation and calculation.

$$X1_p, X2_p, Y_p, Z_p, Q_{pr}, W_{pr}, V_{pr}, U_{pr}, S_{pr}, A_{pr}, A1_{pr}, A2_{pr}, I_{pr} \geq 0 \tag{28}$$

Constraint (28) means that the values of the decision variables must be at least 0.

4 Case Study and Computational Results

Conducting a case study of the circular economy in the automotive industry provides the opportunity to demonstrate practical applications of the principle of sustainability to the literature and industry professionals. Case studies can make a significant contribution to manufacturers in developing their strategies and making strategic decisions. Thus, they can play an important role in shaping future industry leaders.

In this part of the study, the data description of the model developed specifically for the automotive industry is given, and the model is illustrated with a case study. Results obtained using a commercial software are given. Finally, with changes in certain parameters that ensure circularity in the model, the parameters that have the highest impact on the objective function of the model are clarified.

4.1 Description of Data

In the proposed model, we consider two production types (new car and renew car) of a particular car model for circular economy application in the automotive sector. As we see in Table 1, six types of raw materials are considered. The model is run with the data that cover 12 periods. Each period is 1 month.

In the case study, data from one of the most popular car models in Europe, which has been produced for years by a car manufacturer with a strong presence in various European markets, are used. The company, which operates production facilities in many countries and has a widespread dealer network in Türkiye, offers sales, service, and spare parts support to its customers. Overall, the company is a prominent automotive brand in both Europe and Türkiye, known for its diverse vehicle lineup, focus on electric mobility, and strong manufacturing and sales presence in the region.

The material capacity of the manufacturer denoted by *CAPI* corresponds to the maximum amount of car manufacturing per period. This parameter is found by monthly calculation of the total car production data in 2022 at the main factory of the specified vehicle manufacturer which is 188,537,076.83 (kg). The material capacity of the collection center is given by *CAPC* which refers to the maximum amount of material storage per period. The material capacity of the disassembly/dismantling center is indicated by *CAPD* that corresponds to the maximum amount of car disassembly, dismantling, and rig testing per period. The material capacity of the reuse center is given by *CAPK* which corresponds to the maximum amount of material that can be reused per period. Since there are no factories other than the manufacturer in real life yet, restricting their capacities would not be realistic at this point. Hence, although there are capacity constraints in the proposed model, capacities other than the material capacity of the manufacturer are assigned as unlimited.

In the problem under consideration, fixed facility costs include rent for the factory, depreciation on equipment, property taxes, salaries, and insurance premiums. The fixed facility cost of the manufacturer is given by considering the fixed facility costs shared by an automobile manufacturer for the total car production in Türkiye in 2022. The fixed facility cost of the manufacturer is denoted by fi_p. Since there are no collection, disassembly, and reuse centers in real life yet, their data are obtained from a government-authorized recycling facility operating in the automotive industry in Düzce/Türkiye. This company performs automotive recycling and waste management operations such as hazardous/nonhazardous waste, storage, recovery, and scrap metal/end-of-life vehicle processing. Since the processes in the company are similar to the processes in the proposed circular model, the data shared by this factory are important at this point. Because these operations are currently performed on a single facility at the company, fixed costs are separated by operations. Due to the data security policy of the company, cost data close to real life are obtained. The fixed facility cost of the collection center is given by fc_p, of the disassembly/dismantling center is denoted by fd_p, and of the reuse center is given by fk_p, and they get the same value for each period.

The car dealer, which is an external source, collects used cars from customers through some campaigns. This collection cost and the service cost of the car dealer are reflected in this system as the purchasing cost of used vehicles from the car dealer. The purchasing cost of used cars from the car dealer is denoted by *PriceJAL*, obtained from the specified recycling facility.

The operation cost of new car manufacturing by the manufacturer is given by considering the operation costs shared by an automobile manufacturer for the total car production in Türkiye in 2022. The operation cost of new car manufacturing by the manufacturer is denoted by *OP1*. Assuming that the operation cost of renew car production will be 25% more than the operation cost of new car production, the operation cost of renew car manufacturing in the manufacturer is denoted by *OP2*. Collection center, disassembly center, and reuse centers' operation costs data are obtained from the specific recycling facility operating in the automotive industry in Düzce/Türkiye. According to data obtained from this company close to real life, the operation cost of used cars in the collection center is denoted by *OPC*, of used cars in disassembly/dismantling center is given by *OPD*, and of materials in the reuse center is denoted by *OPK*.

Inter-facility material transportation costs represent the freight costs of land transportation and include distance, fuel costs, tolls of highways and bridges along the route, and driver fees. Values of the transportation cost parameters are obtained close to the real value from the recycling company. The transportation cost of new cars transported from the manufacturer's production line 1 to the car dealer is denoted by *TCIM1* and of renew cars transported from the manufacturer's production line 2 to the car dealer is given by *TCIM2*. The transportation cost of used cars transported from car dealer to the collection center is given by *TCNC*, of used cars transported from the collection to the disassembly/dismantling center is given by *TCCD*, of secondary materials transported from the disassembly/dismantling center to the reuse center is given by *TCDK*, of secondary materials transported from the disassembly/dismantling center to the recycling facility is denoted by *TCDG*, and of secondary raw materials transported from the reuse center to the manufacturer is denoted by *TCKI*.

Table 2 presents the values of the parameters whose descriptions are given.

In the Türkiye Automotive Plastics Sector Monitoring Report [42]. The expected ratio of the main materials to be used in automobile production in Türkiye for the years 2017–2020, as provided in the Türkiye Automotive Plastics Sector Monitoring Report (Türk Plastik Sanayicileri Araştıırma Geliştirme ve Eğitim Vakfı [PAGEV], 2017), is used in this study. Table 3 shows forecasted rates of raw materials used in automobile production in Türkiye in 2020 (%).

In the mathematical model solution, the forecasted rate data for 2020 in the report published by PAGEV (2017) are used. For the mathematical model solution, raw materials with a ratio of more than 2% in total consumption of materials to be used in automobile production are considered. These raw materials and their ratios are given in Table 4, and the parameter is Car_r.

Table 2 Cost related parameters

Parameter	Value
OP1	€2.00/kg
OP2	€2.50/kg
OPC	€1.20/kg
OPD	€1.95/kg
OPK	€1.30/kg
TCIM1	€0.5
TCIM2	€0.5
TCNC	€0.4
TCCD	€0.3
TCDK	€0.6
TCDG	€0.6
TCKI	€0.75
$fi_p(\forall p)$	€7855
$fc_p(\forall p)$	€6646
$fd_p(\forall p)$	€7250
$fk_p(\forall p)$	€6042
PriceJAL	€2.674/kg

The purchasing cost of the raw material r from suppliers denoted by CH_r refers to the cost of purchasing primary (i.e., zero) raw materials from relevant producers. Producers of raw materials (suppliers) price them including transportation costs, that is, this price includes the cost of transportation of raw materials from suppliers to the manufacturer. These values are based on the sales prices of relevant suppliers producing raw materials and original products for the automotive industry. Table 4 shows the values of the parameters that take values according to the material types of the model.

Secondary materials are sold to recycling facilities. Recycling facilities add value to the materials, making them reusable in car production, and sell them to the manufacturer with recycling costs included. These facilities do not undertake the cost of product transportation from the disassembly/dismantling center to recycling facilities but undertake the transportation cost of secondary raw materials to the manufacturer after sale. The sales price of the material r to recycling facilities is given by IP_r, and the purchasing cost of the raw material r from recycling facilities is denoted by $PriceG_r$.

When cars are dismantled, hazardous and nonhazardous waste is produced. These wastes are sent to authorized disposal facilities that deal with vehicle destruction and preparation of destruction reports by legal legislation. The cost of disposing of waste secondary materials in disposal facilities is indicated by EX_r and obtained from an authorized facility operating in Kocaeli/Gebze.

For application, data from one of the most popular car models in Europe, which has been produced for years by a car manufacturer with a strong presence in various European markets, are used. The company, which operates production facilities in many countries and has a widespread dealer network in Türkiye, offers

Table 3 Forecasted rates of raw materials used in automobile production in Türkiye in 2020

Raw material	Forecasted usage rates for 2020 (%)
Plain steel	34.1
High and medium strength steel	16.2
Stainless steel	1.8
Other steels	0.6
Total steel	52.7
Iron casting	2.2
Aluminum	9.8
Magnesium	0.3
Copper and brass	1.3
Lead	2.1
Zinc casting	0.2
Powder metal	1.0
Other metals	0.1
Metals other than steel	16.9
Total metals	68.5
Plastics	14.1
Rubber	7.8
Covering	1.1
Textile	1.4
Fluids and oils	5.2
Glass	2.1
Other	2.3
Total	100

Table 4 Parameters according to material types

Parameter	Metal	Plastic	Rubber	Glass	Oil	Other
CH_r	1.510	0.700	0.667	1.000	0.500	0.600
$PriceG_r$	0.605	0.350	0.233	0.280	0.250	0.400
IP_r	0.423	0.272	0.121	0.146	0.105	0.180
EX_r	0.098	0.090	0.055	0.050	0.030	0.020
Car_r	0.685	0.141	0.078	0.021	0.052	0.023
$k1_r$	0.050	0.080	0.070	0.060	0.020	0.070
$k2_r$	0.080	0.090	0.080	0.070	0.050	0.090
$k3_r$	0.870	0.830	0.850	0.870	0.930	0.840

sales, service, and spare parts support to its customers. Overall, the company is a prominent automotive brand in both Europe and Türkiye, known for its diverse vehicle lineup, focus on electric mobility, and strong manufacturing and sales presence in the region. For the data in Table 5, the specified car type is determined as the source. Table 5 shows the values of the parameters that take values according to the periods of the model.

Table 5 Parameters according to periods

Parameter	1	2	3	4	5	6	7	8	9	10	11	12
$SP1_p$	21934	20006	21561	23680	22993	24327	26228	27012	26747	26576	26852	23902
$SP2_p$	17547	16005	17249	18944	18394	19462	20982	21610	21398	21261	21482	19122
$DM1_p$	21420	22681	30523	22733	26573	34291	20193	15983	23597	21385	22832	25691
$DM2_p$	12852	13608	18314	13640	15945	20575	12116	9590	14158	12831	13699	15415

All cost and price data used in the application of the model are calculated by taking into account the average [1].

The sales price of a new car to car dealer data is determined as the average of the sales price values of all packages (five different packages are considered, from basic to advanced) of the specified car model for each month of 2022, based on the reports shared by the company. The sales price of renew car data, on the other hand, is determined as 80% of the new car sales price, based on the ratio of new and secondhand car sales price data in Türkiye in 2022 [9].

End user demand for new car data, based on reports shared by the company, is averaged over the months of sales number data for the specified car model in Europe between 2016 and 2021, and the average values of these 6 years are used for manufacturing in 2022. Car sales data are from the following countries: Austria, Belgium, Cyprus, Czech Republic, Denmark, Estonia, Finland, France, Germany, Great Britain, Greece, Hungary, Iceland, Ireland, Italy, Latvia, Lithuania, Luxembourg, Netherlands, Norway, Poland, Portugal, Romania, Slovakia, Slovenia, Spain, Sweden, and Switzerland. End user demand for renew car data, on the other hand, is determined as 60% of new vehicle demand, based on the ratio of new and secondhand car sales in Türkiye in 2022 [9].

The curb weight of the car, which is needed to use all the data in Table 3 in the model application, is taken as 1103 kg from the defined car model. From the parameters whose definitions are given, in the model, the demand data are multiplied by the curb weight of the car to find the demand in kilograms, and the sales price data are divided by the curb weight of the car to find the amount per kilogram.

4.2 Computational Results

Using the data given in Sect. 4.1, the mixed-integer linear programming model is illustrated with a case study. Commercial software is used to find the optimal solution of the proposed model. A linear programming model is developed for the proposed network model.

The program is run for 12 periods, and the global optimum result is found. The objective value is €4,992,094,000. It should be noted that the variables A_{pr}, $A1_{pr}$ and $A2_{pr}$ are calculated according to the end user demand of the next period.

In the case study of the circular economy model proposed in the automotive industry, the values of the decision variables are schematically represented for the first period in Fig. 4.

4.3 Scenario Designs

In the scenario designs, the effect of the change in the reuse and recycling rate parameters on the value of the objective function is measured, and the most

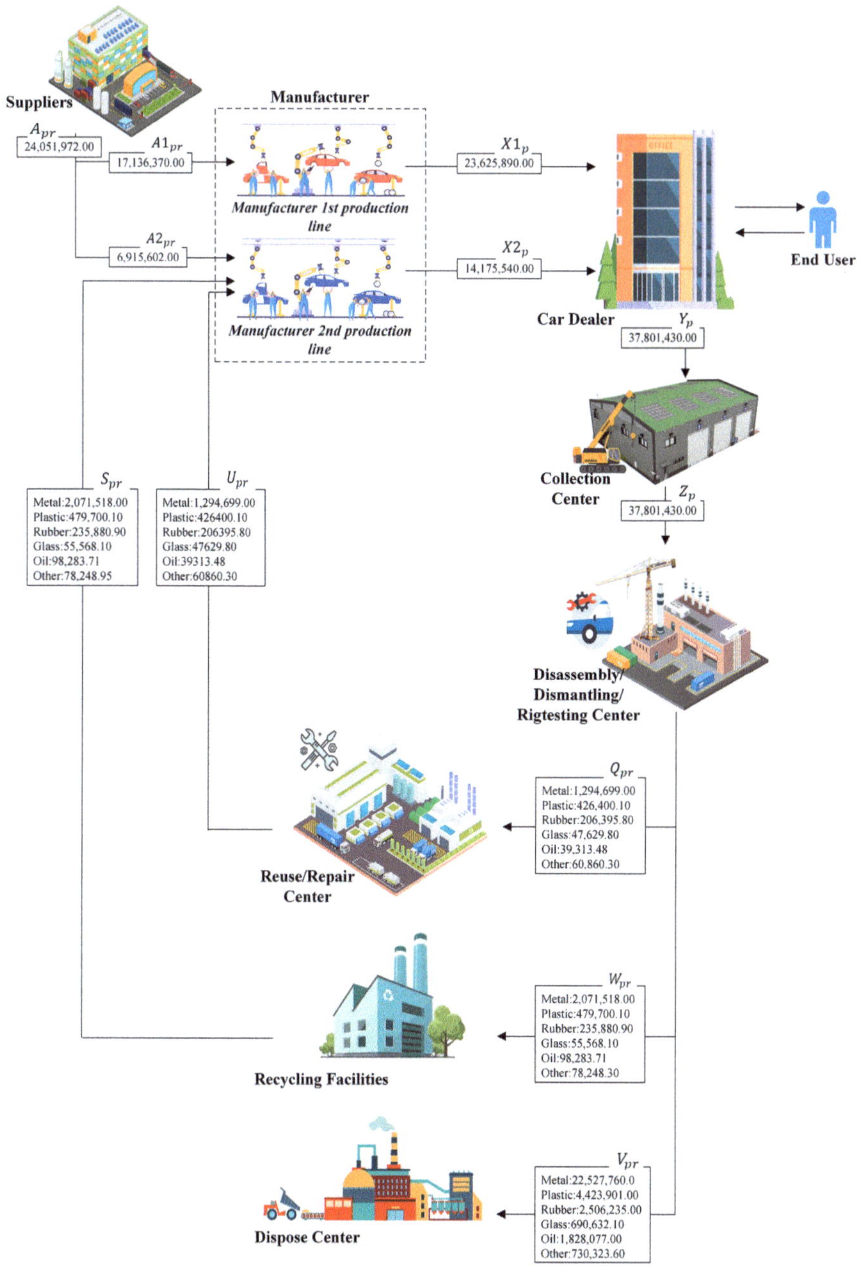

Fig. 4 Schematic representation of case study results for the first period

influencing parameters are determined. In the model, recycling and reuse are also taken into account in line with circular economy goals and strategies in a classic (linear) car product life cycle; recycling and reuse rate parameters $k1_r$ and $k2_r$ can be assumed to be important parameters of the model.

A 10% increase in the recycling rate of a raw material type means that 10% of the amount of raw material that is transported to the dispose center in the case study will now be transported to the recycling center. Thus, a certain amount of raw material that would normally be waste will be transported to the recycling center and reused in the system.

Likewise, a 10% increase in the reuse rate of a raw material type means that 10% of the amount of raw material that is transported to the dispose center in the case study will now be transported to the reuse center. Thus, a certain amount of raw material, which would normally be waste, will be sent to the reuse center and reused in the system.

In order to measure the effect of a 10% increase in recycling and reuse rate parameters separately on the value of the objective function, 12 scenarios are conducted. These scenarios are:

- Scenario 1: Change in the Ratio of the Material Transported to Reuse Center $k1_r$: Six scenarios are considered by changing the ratio of the material transported to the reuse center. It is aimed to determine the effect of this change in the parameter on the objective function.

- Scenario 2: Change in the Ratio of the Material Transported to Recycling Facility $k2_r$: Six scenarios are considered by changing the ratio of the material transported to the recycling facility. It is aimed to determine the effect of this change in the parameter on the objective function.

The results of 12 scenario designs are determined. The objective value, model class, and solution type of the 12 scenario designs are given in Table 6. Also, the ratio of difference in the objective function is given for each scenario.

Table 6 Scenario designs results

Scenario	Objective Value (€)	Difference (%)
Scenario 1	5,003,118,000.00	0.2208
Scenario 2	4,983,043,000.00	−0.1813
Scenario 3	4,986,769,000.00	−0.1067
Scenario 4	4,991,347,000.00	−0.0150
Scenario 5	4,987,696,000.00	−0.0881
Scenario 6	4,990,328,000.00	−0.0354
Scenario 7	5,015,578,000.00	0.4704
Scenario 8	4,992,155,000.00	0.0012
Scenario 9	4,991,806,000.00	−0.0058
Scenario 10	4,992,325,000.00	0.0046
Scenario 11	4,991,408,000.00	−0.0137
Scenario 12	4,991,765,000.00	−0.0066

Commercial software is used to find the global optimal solution of the mixed-integer linear programming model developed for the proposed network model for scenario designs. The value of the objective function varies for each scenario. The difference in objective function values is compared with the main model; the percent difference is calculated with Eq. 29.

$$\% \text{ difference}_n = \frac{(B_n) - (A)}{(A)} \times 100 \tag{29}$$

n: scenario number
A: value of the objective function in the main model
B_n: value of the objective function in scenario n

The difference in the objective function when we change the values of the parameters also depends on the other parameters, variables, and multipliers of the changed parameter in the objective function. It is seen that the solution model of each scenario is in the mixed-integer linear programming class, and the developed model found the global optimal solution for each scenario.

As we see in Table 6, the parameter that causes the most positive difference as a result of a 10% increase in the value of the objective function is the ratio $k2_1$ of metal transported to the recycling facility. The second and third parameters that have the strongest positive effect on the objective function are the ratio $k1_1$ of the metal materials transported to the reuse center and the ratio $k1_2$ of the plastic materials transported to the reuse center. According to the scenario designs, it is proven that network profit will increase by increasing these parameters, which have a high positive effect on the objective function. In order to increase these parameters, it is recommended to reduce the operational costs in the relevant companies in real life. Transportation costs can be reduced. These recycling and reuse rates can be increased by improving raw materials through R&D studies and making their recyclability properties higher. When these rates are increased, the circularity of the system will increase, and the circular economy (ideal system) will be approached. It is recommended to increase the values of these parameters for businesses to increase their profits.

As we see in Table 6, the parameter that causes the most negative difference as a result of a 10% increase in the value of the objective function is the ratio $k2_2$ of the plastic transported to the recycling facility. The second and third parameters that have the strongest negative effect on the objective function are the ratio $k2_4$ of the glass transported to the recycling facility and the ratio $k2_3$ of the rubber transported to the recycling facility.

According to the scenario designs, it is proven that by increasing these parameters, which have a negative effect on the objective function, the network profit will decrease. In order to reduce the negative impact of these parameters on the objective function, it is recommended to reduce the relevant operation and transportation costs in real life. Raw materials can be improved through R&D studies, and their recyclable properties can be increased. In fact, when these rates are increased, the circularity of the system will increase, and the circular economy, which is an ideal system, will be approached. Since many factors such as sustainability, resource efficiency, waste management, cooperation, innovation, and environmental responsibility are important in the automotive industry, in addition to the cost factor, it is recommended that strategic decision-making should not be made only by considering the costs.

5 Conclusion and Future Studies

In the manufacturing sector, improving recycling and reuse processes at the end of a product's life is emerging as a viable solution to ensure the sustainable use of critical materials. In particular, for end-of-life vehicles [5], sets strict targets to promote the reuse, recycling, and recovery of materials.

Network models developed in accordance with the circular economy in different sectors are one of the focal points of efforts to achieve economic, environmental, and social sustainability goals. Therefore, it seems that network optimization studies should be given importance to ensure the transition to circular production.

In the future, there will be a potential for a significant customer demand for renew secondhand cars, particularly in countries where cars are expensive. Also, in the developed countries of the world, the demand for renew car production can be high to reduce the environmental effects in production, regardless of profit and cost. Although some companies in the automotive sector have started practicing, it is seen that there is a gap in the literature in the field of case studies in this sector.

This study provides a comprehensive exploration of the circular economy concept, along with a review and application focused on network optimization aligned with sustainability and circular economy principles.

A literature review is presented, focusing on the origins, logic, structure, and core principles of circular economy and providing detailed explanations of these aspects. A clear distinction between circular and linear economy is outlined, and a collection of relevant global studies related to the concept is incorporated. Furthermore, the principles, processes, potential benefits, and potential barriers are comprehensively listed. The study also examines the relationship between the circular economy and the product life cycle.

A compilation and review of network designs and optimization studies that integrate reverse logistics and sustainability concepts into the product life cycle is presented.

Moreover, this study aimed to pioneer a new network model specific to the automotive industry and develop a mathematical model that optimizes profits within this network. An innovative network design is proposed that addresses a significant gap in this framework in the existing literature. This network model includes hybrid closed-loop and open-loop supply chains for the life cycle of cars. A mathematical model specific to the proposed network is formulated to maximize profit and represent a noteworthy contribution to the field.

The proposed MILP model is illustrated with a case study to apply and validate the model. The case study involving the developed linear programming model is effectively solved using commercial software and found the global optimal solution.

Finally, 12 scenarios are examined to determine the effects of the parameters that ensure circularity in the model (recycling and reuse rates) on the objective function. Through these analyses, crucial ratios that have the most impact on the objective function are successfully determined. As a result of the scenario designs, it is seen that the total profit of the companies will increase by increasing the amount of the determined type of raw materials sent to one of the reuse and recycling centers. In addition, a new perspective is added to the literature: studies to increase these parameters that positively affect the total profit.

The findings of this study offer valuable insights for managers in the automotive sector, emphasizing the economic and environmental benefits of implementing circular economy practices. The developed optimization model shows that integrating closed-loop supply chains can reduce waste and enhance resource efficiency, leading to good financial outcomes. Managers should prioritize investments in recycling technologies and supply chain optimization to achieve a competitive edge in a sustainability-focused market. The scenario analyses further illustrate how varying recycling rates can alter profitability, underscoring the need for flexible and adaptive strategies in circular network management.

The practical implications of this study highlight the importance of adopting circular network designs in the automotive sector. This study emphasizes the value of integrating circular economy principles into the automotive supply chain. The model developed provides actionable insights on how companies can optimize their network to improve both environmental and economic outcomes and serves as a guide for enhancing supply chain performance through effective reuse and recycling strategies. The practical application of this model can help companies minimize costs, reduce environmental impact, and improve overall operational efficiency, making circular economy not just a theoretical concept but a viable business strategy.

For future studies, a promising opportunity is seen to improve the model in this study by adding additional targets and constraints with a particular focus on environmental parameters. In addition, original network designs can be developed for different subsectors, and optimization studies with different objectives can be made for the application of this concept to the manufacturing sector. Thus, various optimization studies that appeal to different targets and companies can be enabled.

Competing Interests The authors have no conflicts of interest to declare that are relevant to the content of this chapter.

References

1. Central Bank of the Republic of Turkey (2022) Exchange rates. Available via TCMB. https://www.tcmb.gov.tr/. Accessed 25 July 2023
2. Çınar S (2022) A fuzzy multi-objective mixed integer linear programming model for end of life use. Duzce Univ J Sci Technol 10(2):872–892
3. Demirel E, Demirel N, Gökçen H (2016) A mixed integer linear programming model to optimize reverse logistics activities of end-of-life vehicles in Turkey. J Clean Product 112:2101–2113
4. Ene S, Öztürk N (2014) Open loop reverse supply chain network design. Proc Soc Behav Sci 109:1110–1115
5. European Parliament and of the Council (2000) Directive 2000/53/EC of the European parliament and of the council of 18 September 2000 on end-of life vehicles. Available via EUMonitor. https://www.eumonitor.eu/9353000/1/j9vvik7m1c3gyxp/vhckn6df9zzn. Accessed 11 Nov 2023
6. Fahimnia B, Sarkis J, Dehghanian F, Banihashemi N, Rahman S (2013) The impact of carbon pricing on a closed-loop supply chain: an Australian case study. J Clean Product 59:210–225
7. Fetting C (2020) The European green deal. ESDN Report, ESDN Office, Vienna
8. Golpira H, Javanmardan A (2021) Decentralized decision system for closed-loop supply chain: a bi-level multi-objective risk-based robust optimization approach. Comput Chem Eng 154:107472
9. Good Car Bad Car Automotive Sales Data (2022) Renault clio sales figures. Available via Car Sales Base. https://carsalesbase.com/europe-renault-clio/. Accessed 11 Nov 2023
10. Hajipour V, Tavana M, Di Caprio D, Akhgar M, Jabbari Y (2019) An optimization model for traceable closed-loop supply chain networks. Appl Math Model 71:673–699
11. İşler G, Akyol DE, Yazgan HR (2023) Theory and research concerning the circular economy model and future trend. In: International symposium on intelligent manufacturing and service systems. Springer, Singapore, pp 769–782
12. Jabarzadeh Y, Reyhani Yamchi H, Kumar V, Ghaffarinasab N (2020) A multi-objective mixed-integer linear model for sustainable fruit closed-loop supply chain network. Manage Environ Qual Int J 31(5):1351–1373
13. Kabadurmuş O, Kazançoğlu Y, Yüksel D, Pala MÖ (2022) A circular food supply chain network model to reduce food waste. Ann Oper Res 1–31. https://doi.org/10.1007/s10479-022-04728-x
14. Karagöz S, Aydın N, Simic V (2022) A novel stochastic optimization model for reverse logistics network design of end-of-life vehicles: a case study of Istanbul. Environ Model Assessment 27(4):599–619
15. Kaya A, Alumur Alev S (2017) Closed loop supply chain network design. Ind Eng 28(2):2–18
16. Kazançoğlu Y, Yüksel D, Sezer MD, Mangla SK, Hua L (2022) A green dual-channel closed-loop supply chain network design model. J Clean Product 332(1):130062
17. Kır S, Ercan Cömert S, Yener F, Yazgan HR, Candan G (2019) Hazardous waste recycling: end of life tires case. Polish Acad Sci Inst Phys 135(4):681–683
18. Kuşakçı AO, Ayvaz B, Cin E, Aydın N (2019) Optimization of reverse logistics network of end of life vehicles under fuzzy supply: a case study for istanbul metropolitan area. J Clean Product 215:1036–1051
19. Lu S, Hou W, Li Z, Xia J, Xie L, Su H (2022) Design and optimisation of a green manufacturing-recycling network considering heavy metal pollutants–an electronic assembly case. Int J Product Res 60(9):2830–2849
20. Lyle JT (1994) Regenerative design for sustainable development, rethinking the mind in nature. Wiley, Hoboken
21. McDonough W, Braungart M, Clinton B (2013) The upcycle: beyond sustainability-designing for abundance. North Point Press, Berkeley

22. Murray A, Skene K, Haynes K (2017) The circular economy: an interdisciplinary exploration of the concept and its application in a global context. J Bus Ethics 140(3):369–380
23. Nayeri S, Paydar MM, Asadi-Gangraj E, Emami S (2020) Multi-objective fuzzy robust optimization approach to sustainable closed-loop supply chain network design. Comput Ind Eng 148:106716
24. Özceylan E (2016) Simultaneous optimization of closed-and open-loop supply chain networks with common components. J Manufact Syst 41:143–156
25. Özceylan E, Paksoy T, Bektaş T (2014) Modeling and optimizing the integrated problem of closed-loop supply chain network design and disassembly line balancing. Transp Res E Logist Transp Rev 61:142–164
26. Özceylan E, Demirel N, Çetinkaya C, Demirel E (2017) A closed-loop supply chain network design for automotive industry in Turkey. Comput Ind Eng 113:727–745
27. Özturk G, Aksoy HK (2021) Closed loop supply chain optimization with linear physical programming approach. J Stat Appl Sci 2(2):14–28
28. Phuc PNK, Vincent FY, Tsao YC (2017) Optimizing fuzzy reverse supply chain for end-of-life vehicles. Comput Ind Eng 113:757–765
29. Pishvaee MS, Rabbani M, Torabi SA (2011) A robust optimization approach to closed-loop supply chain network design under uncertainty. Appl Math Model 35(2):637–649
30. Rajak S, Vimal KEK, Arumugam S, Parthiban J, Sivaraman SK, Kandasamy J, et al (2022) Multi-objective mixed-integer linear optimization model for sustainable closed-loop supply chain network: a case study on remanufacturing steering column. Environ Develop Sustain 24(5):6481–6507
31. Rohmer SUK, Gerdessen JC, Claassen GDH (2019) Sustainable supply chain design in the food system with dietary considerations: a multi-objective analysis. Eur J Oper Res 273(3):1149–1164
32. Sapmaz Veral E (2018) The recent measures and the strategies of The EU member states towards circular economy transition. Ankara Univ J Environ Sci 17(2):463–488
33. Shahedi A, Nasiri MM, Sangari MS, Werner F, Jolai F (2022) A stochastic multi-objective model for a sustainable closed-loop supply chain network design in the automotive industry. Process Integr Optim Sustain 6(1):189–209
34. Shankar R, Bhattacharyya S, Choudhary A (2018) A decision model for a strategic closed-loop supply chain to reclaim end-of-life vehicles. Int J Product Econ 195:273–286
35. Simic V (2019) Interval-parameter conditional value-at-risk two-stage stochastic programming model for management of end-of-life vehicles. Environ Model Assess 24(5):547–567
36. Soleimani H (2021) A new sustainable closed-loop supply chain model for mining industry considering fixed-charged transportation: a case study in a travertine quarry. Resour Policy 74:101230
37. Stahel WR (2016) The circular economy. Nature 531(7595):435–438
38. Talaei M, Moghaddam BF, Pishvaee MS, Bozorgi-Amiri A, Gholamnejad S (2016) A robust fuzzy optimization model for carbon-efficient closed-loop supply chain network design problem: a numerical illustration in electronics industry. J Clean Product 113:662–673
39. Torgul B, Demiralay E, Paksoy T (2021) Network design and optimization in the agri-food supply chain: a juice business example. J Transp Logist 6(1):49–70
40. Torğul B, Paksoy T (2024) Multi-level competitive closed-loop supply chain network design: a comparative analysis. J Polytech 27(1):237–253
41. Tosarkani BM, Amin SH (2019) An environmental optimization model to configure a hybrid forward and reverse supply chain network under uncertainty. Comput Chem Eng 121:540–555
42. Türk Plastik Sanayicileri Araştırma Geliştirme ve Eğitim Vakfı [PAGEV] (2017) Türkiye automotive plastics sector monitoring report. Available via PAGEV. https://pagev.org/turkiye-plastik-sektor-izleme-raporu-2017-5a785e4022b85. Accessed 25 Jul 2023
43. United Nations (2023) Sustainable development goals. https://sdgs.un.org/goals. Accessed 25 Jul 2023
44. Ünlütürk Ş, Kihtir Öztürk P, Kardeş S, Birdal M, Saral B, Aşıroğlu B, et al (n.d.) Circular economy guide for businesses. https://business4goals.org/PDF/Dongusel_Ekonomi_Rehberi.pdf. Accessed 25 Jul 2023

45. Wise C, Pawlyn M, Braungart M (2013) Eco-engineering: living in a materials world. Nature 494:172–175
46. Wu GH, Chang CK, Hsu LM (2018) Comparisons of interactive fuzzy programming approaches for closed-loop supply chain network design under uncertainty. Comput Ind Eng 125:500–513
47. Yıldızbaşı A, Arıöz Y (2021) A novel hybrid network optimization model for printed circuit boards recycling: a circular economy perspective. Clean Technol Environ Policy 23:2989–3013
48. Yuan Z, Bi J, Moriguichi Y (2006) The circular economy: a new development strategy in China. J Ind Ecol 10(1–2):4–8

Exploring Options for Designing an Optimal Circular Spent Coffee Grounds Reverse Logistics Network: An Optimization-Based Approach

Amin Chaabane, Hanieh Zohourfazeli, Ali Sabaghpourfard, and Armin Jabbarzadeh

Abstract In this chapter, we explore the development of new business models aimed at creating smart and sustainable systems for collecting and treating spent coffee grounds (SCGs). The focus is on evaluating two different models: one involving circular coffee shops (CCs) and the other without. The evaluation compares the efficiency, environmental impact, and economic viability of both models. The research proposes a smart and sustainable system leveraging new technology to streamline the coffee value chain while minimizing environmental and social impacts. The core of this research lies in designing an optimal SCG collection network. Circular coffee shops serve as local depots equipped with pre-drying technologies, which reduce contamination risks and improve SCG collection efficiency. The findings indicate that the model involving circular coffee shops significantly reduces contamination, improves product quality, and decreases economic and environmental impacts compared to the model without circular coffee shops. Incorporating new technologies into coffee management practices aligns with the latest advancements in sustainable and circular practices, offering a viable path toward circularity in coffee waste management.

Keywords Circular economy · Industry 4.0 technologies · Smart spent coffee ground sustainability

A. Chaabane · H. Zohourfazeli (✉) · A. Jabbarzadeh
École de Technologie Supérieure (ÉTS), Montréal, QC, Canada
e-mail: Amin.chaabane@etsmtl.ca; Hanieh.zohourfazeli.1@ens.etsmtl.ca;
Armin.jabbarzadeh@etsmtl.ca

A. Sabaghpourfard
Air Canada Vacation, Montréal, QC, Canada
e-mail: Asabaghpourfard@vacv.com

B. Jarboui et al. (eds.), *Data-Driven Waste Revolution*, Computational Intelligence Methods and Applications, https://doi.org/10.1007/978-981-96-5890-9_3

1 Introduction

1.1 Background

In recent years, global consumption of coffee has experienced a steady rise. However, this industry is concerned about the large amount of unused waste streams [34]. Coffee shops (CSs) are one of the primary sources of spent coffee ground (SCG) generation that can collaborate with treatment facilities to help with coffee valorization issues and produce various products that can be sold in the target market, creating economic prospects. Recognizing the ecological effects of SCG and the potential for its value recovery, there is a growing need to move toward a circular economy (CE) and implement sustainable SCG management strategies to help reduce the negative impacts.

The current scheme of waste management systems follows an old model that consists of large fleets of trucks that travel daily for long distances and often by unnecessary routes [58]. The traditional waste management systems often become costly and exhaust the municipal budgets [30]. Uncertainty further complicates waste management and makes designing an efficient collection process challenging. If the waste collection system is not streamlined adequately under uncertainty, it can lead to a decline in the quality of recovered products and result in a costly collection process. Addressing these issues requires a paradigm shift toward more intelligent and sustainable waste management practices. Therefore, mitigating the impact of uncertainties by adopting Industry 4.0 (I4.0) technologies, which represent the Fourth Industrial Revolution and involve advanced digital technologies, into reverse logistics (RL), the process of moving goods back to the manufacturer for returns or recycling becomes crucial to optimizing operations.

By strategically locating collection points, optimizing transportation routes, and incorporating advanced equipment, it becomes possible to streamline the collection and utilization of SCG while minimizing costs and environmental impact. This chapter proposes a novel, sustainable, and intelligent approach to designing the RL for collecting SCG, incorporating disruptive technologies, and integrating location, allocation, capacity, and routing decisions. A real-life case study is also provided to illustrate the practical relevance of the problem.

1.2 Motivation, Research Gaps, and Scope of Research

The global coffee industry is experiencing a significant rise in consumption, bringing opportunities and challenges. The coffee value chain, which spans production to consumption, must be managed sustainably to mitigate environmental and societal impacts. One critical issue within this chain is the disposal of SCG, which, when treated as mere organic waste, can negatively affect the environment.

The timely collection of SCG is crucial to prevent contamination and associated health hazards. However, the SCG collection network is fraught with complexities and uncertainties. Given these challenges, we must develop a cost-effective, environmentally friendly, intelligent SCG management system.

This research is motivated by the potential to transform SCG from waste into valuable products such as biofuel, activated carbon, fertilizers, and energy. Such an approach mitigates environmental pollution, promotes resource efficiency, and creates economic opportunities. This research aims to design and implement a sustainable SCG management system that aligns with the principles of the CE, ultimately contributing to a more sustainable and resilient coffee industry.

Montreal serves as an ideal case study for this research. Previous findings, such as those by Carassou [11], suggest that implementing a dedicated SCG recovery system in Montreal could significantly enhance waste management, sustainability, and economic viability within the local coffee industry. This research is driven by the need to address the specific challenges faced by Montreal's coffee sector and explore innovative solutions for SCG valorization.

Amidst the growing interest in sustainable coffee value chains, it is crucial to acknowledge the significant research gaps that persist in effectively managing SCG, especially under uncertain conditions. The absence of a properly streamlined waste collection system during such circumstances can lead to a decline in the quality of recovered products, resulting in a costly collection process. Current waste collection systems often lack efficiency, leading to compromised quality of recovered products and increased costs. To optimize operations, it is imperative to mitigate the impact of uncertainties by adopting new technologies. This research aims to bridge these gaps by exploring the integration of I4.0 technologies in the coffee value chain, underlining the urgency and relevance of this work.

Our research primarily focuses on collecting and treating SCG within the RL segment of the coffee value chain. We aim to address sustainability challenges and operational uncertainties. This specific approach allows for a detailed exploration of how I4.0 technologies can improve these crucial aspects, ultimately contributing to the development of an innovative and sustainable coffee value chain.

1.3 Problem Statement

The current waste management systems could be more efficient and rely on outdated models that involve large fleets of trucks traveling long distances, resulting in high costs and significant environmental impact. This is particularly problematic in collecting and managing SCG, where uncertainties related to collection processes pose substantial challenges to developing cost-effective and sustainable RL networks. These inefficiencies hinder long-term operational efficiency and economic viability.

Furthermore, the coffee industry's absence of intelligent and sustainable waste management practices exacerbates environmental concerns and places additional strain on municipal budgets. Therefore, there is a critical need for innovative solutions that optimize the collection and utilization of SCG. To address these challenges, the following research questions are posed:

Q1: How can we optimize the design and operation of an intelligent, sustainable SCG RL network and develop an effective business model?

Q2: How can uncertainty affect the selection and viability of business models for managing SCG, and how can these uncertainties be mitigated to ensure long-term success?

Q3: How do we evaluate the impact of new technologies on operations and performance of business models for SCG management, and what strategies can be employed to leverage these technologies for improved efficiency and sustainability?

1.4 Objectives of the Research

This research evaluates the economic feasibility and advantages of incorporating I4.0 technologies into the SCG value chain to improve waste management practices. It involves examining strategic and operational decisions related to the optimal location, allocation, and routing among facilities. Additionally, the research seeks to demonstrate how disruptive technologies, such as pre-drying methods and smart bins, can improve the quality and efficiency of the SCG collection network, thereby reducing transportation frequency and minimizing contamination risks.

The main objective of this research is to create a comprehensive framework for developing a smart and sustainable SCG management system and propose a viable business model for its implementation. To achieve this, we have identified several specific objectives. These include proposing optimization models for designing an RL network, evaluating various business models, and estimating the transportation costs associated with the collection of SCG.

We also aim to assess different scenarios to analyze the robustness of the model under varying conditions. Furthermore, we provide a detailed financial outlook for the project, offering insights into the economic feasibility and viability of the proposed management system. Therefore, this work significantly contributes to addressing the crucial challenge of efficiently collecting and treating SCG.

Concentrating on Montreal, we aim to demonstrate the practical benefits of a sustainable SCG management system. By valorizing SCG into activated carbon, we aim to reduce environmental pollution, promote resource efficiency, and create economic opportunities.

1.5 Contribution

One primary contribution is introducing two novel business models specifically designed for analyzing SCG management. These models and their variants are meticulously developed to address the unique challenges and opportunities associated with SCG management, filling a notable literature gap.

Another key contribution of the study involves conducting a thorough assessment and comparison of the effectiveness and feasibility of the proposed business models for SCG management. By systematically evaluating the performance of these models under various scenarios and conditions, the research provides valuable insights into their practical applicability and potential impact on waste management practices.

A third significant contribution revolves around delving into the impact analysis of uncertainty on SCG management. With a specific focus on elucidating the system's resilience in confronting unforeseen challenges, the research sheds light on the adaptive capacity and robustness of the proposed business models in the face of changing environmental, economic, and operational conditions. This analysis enhances our understanding of the dynamic nature of SCG management and informs strategies for mitigating risks and enhancing system resilience.

1.6 Organization of This Chapter

The remainder of the chapter is structured as follows. In Sect. 2, an extensive examination is conducted on the pertinent literature concerning I4.0 technologies applied to restaurant food waste management and coffee value chain, alongside a thorough exploration of existing vehicle routing problem (VRP) models addressing the challenges of RL design within the waste management sector. This chapter also encompasses a review of the literature concerning the generation, collection, and treatment processes of SCG. Section 3 elucidates the overarching research methodology employed in this study.

Following this, Sect. 4 delineates the problem description, underlying assumptions, and notation employed and presents the formulation of the mathematical models. Subsequently, Sect. 5 presents an illustrative case study and a detailed exposition of the obtained results and managerial insights. This chapter also encompasses an analysis of various scenarios aimed at discerning the network's economic viability and examining the project's financial prospects. Finally, conclusive remarks, the research limitations, and potential avenues for future exploration are discussed.

This research led to several publications and oral presentations at the following conferences:

1. International Conference on Computers and Industrial Engineering (CIE) (Zohourfazeli et al. [78])
2. International Conference on Control, Decision and Information Technologies (CoDIT) (Zohourfazeli et al. [80])
3. International Conference on Connected Innovation and Technology X.0 (ICC-ITX.0) (Zohourfazeli et al. [79])
4. Optimization Days conference (OD) (Zohourfazeli et al. [81])

2 Literature Review

2.1 Restaurant Food Waste and New Technologies

The food industry is a significant contributor to global waste, and a considerable portion of this waste is restaurant food waste (RFW), which can lead to environmental and social problems and food safety risks [47, 63, 74]. For instance, RFW disposal contributes to generating gases and causing climate change. Additionally, RFW eliminates using natural resources to produce, transport, and prepare food. Also, RFW can cause water, soil, visual pollution, and strong odors. In addition, it can attract pests and create unsanitary conditions, spreading disease, and contamination. This can result in serious health issues if not appropriately addressed [73].

Moreover, various uncertainties and the COVID-19 pandemic have highlighted the importance of efficient and sustainable RFW management [37]. Consequently, in recent years, there has been a growing trend of implementing I4.0 technologies to encourage sustainable and intelligent RFW (SSRFW) management practices [5, 8, 10, 15, 17, 18, 20, 25, 28, 29].

A part of the literature review in this study led to a publication and presentation at International Conference on Computers and Industrial Engineering (CIE) [78]. We provided a holistic view of the entities in SSRFW management practices and the trend of publications on I4.0 technologies in RFW management. In addition, we offered an overview of the reviewed papers that consider I4.0 technologies such as the Internet of Things (IoT), radio frequency identification (RFID), big data (BD), artificial intelligence (AI), augmented reality (AR), machine learning (ML), and virtual reality (VR) in the context of RFW. For instance, IoT and RFID are used for real-time tracking and monitoring of food waste throughout the supply chain [4, 49–51, 62]. Big Data and AI analyze patterns and predict waste generation to optimize inventory and reduce surplus [7, 23, 35, 52, 54–56]. AR and VR provide immersive training for staff on best practices in waste reduction and resource management [6, 13, 36]. Machine learning automates sorting and categorization of waste to enhance recycling processes [2, 3, 9, 16, 19, 40, 44, 45, 66, 69, 77]. The findings underscore the critical role of implementing a CE framework to reduce waste and foster a more sustainable system [14]. Transitioning toward a CE involves strategies for managing RFW by decreasing the volume of discarded food and exploring methods to repurpose RFW as a valuable resource, thereby minimizing disposal.

The reviewed literature highlights that incorporating I4.0 technologies into RFW management practices can significantly facilitate the establishment of a CE, enhancing sustainability through innovative and efficient waste management solutions. These technologies enable more intelligent tracking, processing, and utilization of waste, aligning with the principles of a CE. Upon reviewing the literature, we have determined that there is a need to focus on SSRFW management practices using I4.0 technologies, particularly on the growing significance of the CE and the uncertainties. Primary research aspects of SSRFW and the contribution of I4.0 technologies are traceability, monitoring, and decision-making [71]. However, several possible challenges must be addressed, such as limited adoption and scalability, cost-effectiveness, lack of data, cybersecurity risks, interoperability, limited public awareness, resistance to change, and environmental concerns [32, 33, 38, 39, 43, 46, 53, 59, 72].

Based on this review, one of the potential areas for future research is exploring the interoperability of I4.0 technologies in SSRFW and addressing integration issues. Also, finding solutions to reduce the impact of uncertainties in SSRFW beyond the pandemic and studying the relationship between I4.0 technologies and performance indicators should be considered. In addition, examining stakeholders' roles, objectives, and interactions in SSRFW management is necessary. Moreover, concerns regarding data privacy and security and the possibility of I5.0 technologies in the SSRFW domain can be addressed. These insights are significant for SCG management, as SCG is a type of RFW commonly produced in CS. SCG should be treated separately from other organic waste because its content can disrupt treatment processes when mixed in bulk. Additionally, SCGs have specific applications which are best utilized when they are processed separately. This targeted treatment maximizes their environmental and economic benefits and reduces the risk of methane emissions from landfills. Efficiently managing SCG through SSRFW practices and I4.0 technologies can enhance sustainability within the coffee industry and contribute to broader efforts in waste reduction and resource optimization.

2.2 *Coffee Value Chain, SCG, and I4.0 Technologies*

SCG, a natural by-product of the brewing process, is rich in sugar, oil, and other energy-dense compounds, which gives it inherent value [34]. SCG is often produced in CS and households, and the quantity generated depends on various factors, such as coffee consumption patterns, brewing methods, and the scale of coffee production. SCG, as wet organic waste, poses a significant threat, creating numerous negative environmental and social impacts [31, 48, 67]. Typically, SCGs are disposed of in general waste and sent to landfills, where they release methane. This potent greenhouse gas is a major contributor to global warming. The disposal of organic waste in landfills can lead to the production of hazardous gases, significantly contributing to greenhouse gas emissions and polluting soil and water bodies.

Efficient collection of SCG is essential to prevent its accumulation in landfills and harness its potential value. SCG can undergo various treatment processes to extract value or mitigate environmental impacts. Comparing several practices of SCG treatment, composting is a standard method in which SCG is mixed with other organic waste to produce nutrient-rich compost for soil amendment. Additionally, SCG can generate energy through anaerobic digestion or combustion, contributing to renewable energy production. Producing activated carbon from SCG is one of several products that can be obtained after treatment. The process involves pretreating the SCG, activating them, washing and drying the resulting activated carbon, and sizing and packaging it commercially. Activated carbon from SCG can be applied in water treatment, air purification, environmental remediation, and

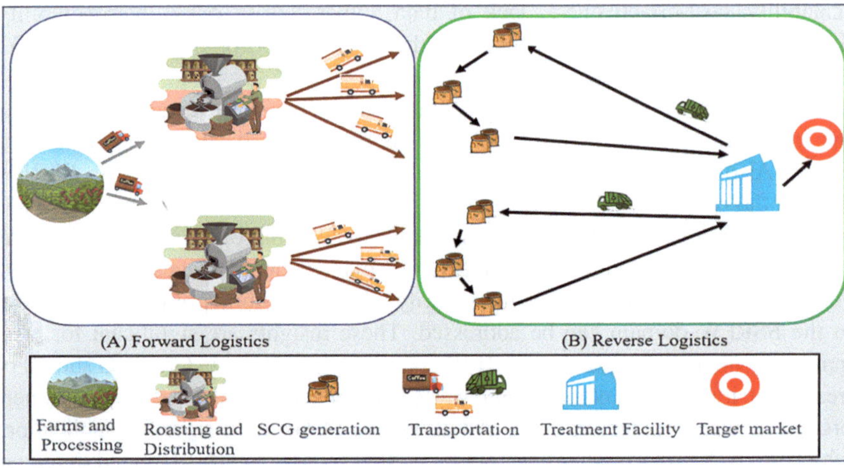

Fig. 1 Schematic of the coffee value chain

industrial processes, offering a sustainable solution for waste management and creating valuable products with diverse uses.

During the SCG collection phase, pre-drying can be carried out at the local depot using new technologies to reduce the moisture content before further treatment. The equipment comes in various sizes, ranging from ultrasmall to medium scale. Once pre-dried, the SCG can be transported to a treatment facility for additional drying processes.

Given the increasing awareness and need for sustainable practices, there is a compelling demand to adopt innovative SCG management strategies that align with the principles of a CE [22]. Circular coffee involves forward logistics, including growing, processing, and distributing coffee, and RL, where SCG is collected, treated, and repurposed, as shown in Fig. 1.

Initially, coffee is grown on farms, sorted, distributed, and sold to the intended markets. Subsequently, SCG is gathered from waste generation sites, with CS being a significant source of such waste. Due to the realization of the effects of SCG, these CSs utilize diverse approaches in line with RL network designs and regulations. As an illustration, specific CS can create compost from SCG on their premises, while others seek to partner with treatment facilities to address coffee valorization challenges [24, 34]. However, various factors, including environmental conditions, market demand fluctuations, and disruptions like the COVID-19 pandemic, contribute to these uncertainties. The coffee industry faces additional challenges, such as climate change, overproduction, price volatility, and the need for sustainable practices [1].

The literature highlights the importance of efficient management strategies for SCG to minimize environmental pollution and maximize resource recovery. Integrating new technologies and optimization strategies into the design of RL networks presents a promising solution. Such a system would ensure efficient collection, treatment, and valorization processes, thereby reducing environmental impact and enhancing resource efficiency. The system will prioritize efficiency by minimizing unnecessary travel routes, meeting customer demands for optimal collection methods, preventing overflow and contamination in waste bins, and reducing routing time.

To our knowledge, while there are studies on coffee valorization, there is a lack of comprehensive research exploring integrating I4.0 technologies into coffee production and SCG treatment. We aim to bridge this gap by analyzing and discussing the potential benefits, challenges, and opportunities associated with adopting I4.0 technologies throughout the coffee value chain. It provides insights into implementing smart features in CS, intelligent logistics for SCG collection, and the establishment of smart treatment facilities. By addressing this research gap, we provide valuable perspectives for researchers and stakeholders interested in advancing circular coffee management practices by leveraging I4.0 technologies.

A part of the literature review in this study led to a publication and presentation at International Conference on Connected Innovation and Technology X.0 (ICCITX.0) [79]. We provided a comprehensive overview of the components involved in circular coffee management practices. Specifically, it focuses on three key areas: (1) incorporating smart features in SCG generation points, (2) utilizing advanced logistics systems, and (3) establishing intelligent SCG treatment facilities. We presented an overview of the diverse approaches and technological solutions proposed to address the sustainable management of coffee value chain.

Research demonstrates the vast potential of integrating I4.0 technologies throughout the coffee value chain, offering enhanced efficiency and environmental stewardship opportunities. By embracing the transformative potential of I4.0 technologies, the coffee industry can move closer to achieving its sustainability and circularity goals, ensuring a more resilient future.

Future research in the field of circular coffee and the application of I4.0 technologies should address uncertainties in supply chain management, optimize SCG treatment processes, assess social and environmental impacts, integrate stakeholder perspectives, conduct cost-benefit analyses, and consider regulatory and policy frameworks to drive sustainability and circularity in the coffee value chain.

2.3 RL and VRP Models for Waste Management

The concept of a CE directly relates to organic waste management by emphasizing the reduction of waste and the optimal use of resources. This approach is particularly relevant to coffee waste management and significantly contributes to organic waste streams. By integrating these CE principles into coffee waste management, we can significantly reduce waste, lower environmental impact, and create value from materials that would otherwise be discarded. Waste management is a critical facet of modern urban living, posing significant challenges to sustainability, environmental preservation, and resource optimization.

The design of networks is crucial in waste management systems as it directly impacts their efficiency, sustainability, and environmental footprint. Van Engeland et al. [70] conducted a review aiming to integrate the fields of RL and waste management, focusing on strategic network design in waste RL chains. They provided an overview of existing efforts in this area, highlighting the importance of environmental, social, and performance indicators in multi-objective models. Additionally, they emphasized the potential of incorporating various stakeholders into network design, addressing future developments.

RL and VRP models offer a promising framework for addressing these challenges by streamlining waste collection and disposal processes [26]. RL, originating from traditional logistics, involves the planning, implementing, and controlling of the reverse flow of products or materials from consumers to manufacturers or designated disposal sites. In waste management, RL pertains to collecting, sorting, and recycling or disposing of waste materials [68].

The emergence of I4.0 has brought about opportunities for digital transformation and significant changes in various industries and sectors. I4.0 technologies boost efficiency and customization in manufacturing, enhance personalized medicine and telehealth in healthcare, optimize precision farming in agriculture, improve customer experiences and inventory in retail, enable smart grids in energy, and transform logistics with autonomous vehicles and real-time tracking. Companies and governments globally are increasingly focused on RL, driven by the increasing pressure to embrace CE principles. The gradual adoption of disruptive technologies, such as IoT, AI, DT, and robotics, promises significant enhancements to traditional RL operations. Sun [55] highlights a notable absence of systematic analysis regarding how these I4.0 technologies impact RL. The author begins with a comprehensive literature review and conceptualization of RL4.0, aiming to thoroughly analyze the implications of I4.0 technologies for smart and sustainable RL. Drawing from this conceptualization, an enhanced two-level decision-support framework is proposed, integrating multi-objective optimization and dynamic simulation to facilitate robust strategic decisions amid dynamic and uncertain conditions. Finally, several managerial and research implications for industrial practitioners and academic researchers are presented.

Smart features, like intelligent bins, offer real-time data and advanced functionality in waste management systems. Ramos et al. [61] emphasize that sensor data must be combined with optimization procedures to determine the most efficient collection routes, maximizing waste collection while minimizing transportation costs and enhancing efficiency. Their study investigates three operational management approaches to define dynamic optimal routes using real-time bin fill-level information. They suggest integrated decision-making for future research that considers location and routing decisions. They argue that these findings highlight the necessity for a model capable of selecting bins based on their fill levels and locations.

LVRP models, as a variant of VRP, are mathematical optimization techniques used to determine efficient routes for vehicles to service a set of geographically dispersed locations, considering various constraints such as vehicle capacity, time windows, and environmental factors. Sar and Ghadimi [64] reviewed recent VRP in RL studies from 2000 to 2022, analyzing publication trends, problem characteristics, solution methods, and applications. They identified emerging issues and suggested future research directions.

We reviewed RL and VRP studies from 2017 to the present, encompassing the coffee industry SCG and other sectors to stay informed about the latest developments. From these studies, it is recommended for future studies to consider the distinctive attributes of each network, influenced by the diversity of its stakeholders and their objectives, as well as the pertinent environmental, social, and economic considerations. These suggestions emphasized that each network possesses unique specifications that set it apart from others, demanding a more profound understanding of the issues and the identification of fitting solutions. Additionally, the authors highlighted technological progress and the imperative shift toward sustainability through intelligent solutions. Recent literature emphasizes the benefits of technology in waste management, highlighting the importance of integrating intelligent solutions for more efficient waste collection. This suggests that future research should explore the potential of intelligent technologies in this field.

The reviewed papers considered routing among the waste generation points and addressed the importance of innovative methods to enhance network design and routing efficiency [12, 21, 27, 41, 42, 60, 65, 75, 76]. The literature review highlights the importance of establishing a smart and sustainable system within the coffee sector. With growing concerns about environmental sustainability and resource optimization, there is an increasing need to leverage innovative solutions to address the challenges faced by the industry. One of the primary contributions of I4.0 technologies, often called digital intelligence, is their ability to enable data-driven traceability, tracking, and decision-making processes. These factors provided the impetus for this research to examine the unique qualities of the network and the product's features, with a particular emphasis on routing between circular coffee shops (CCs) equipped with pre-drying technology and smart bins, to observe the outcomes.

At this point, we would like to formally state the contribution that this research makes to the existing literature:

- The study presents two innovative business models tailored explicitly for SCG management and their variants. These models address the unique challenges and opportunities associated with SCG management, filling a notable literature gap. By employing mathematical modeling for the location vehicle routing problem (LVRP), the research identifies optimal locations for CC, determines their capacity, allocates CS to CC, and optimizes routing among CC to minimize costs and environmental impacts related to waste management.

- Another critical contribution lies in conducting a thorough assessment and comparing the effectiveness and feasibility of the proposed business models for SCG management. Through systematic evaluation under various conditions, the

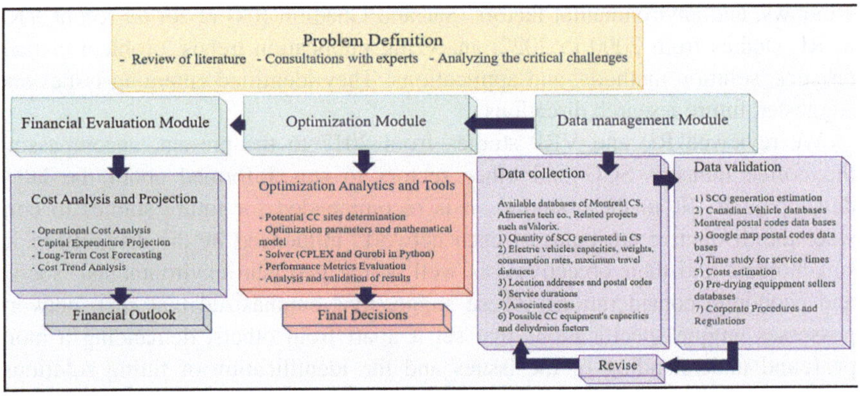

Fig. 2 Decision-making system for SCG RL network

research provides valuable insights into these models' practical applicability and potential impact on waste management practices.

- A third significant contribution involves delving into the impact analysis of uncertainty on SCG management. Focusing on elucidating the system's resilience in confronting unforeseen challenges, the research sheds light on the adaptive capacity and robustness of the proposed business models amidst changing environmental, economic, and operational conditions. This analysis enhances our understanding of the dynamic nature of SCG management and informs strategies for mitigating risks and improving system resilience.

Additionally, the study explores various scenarios to assess the robustness of the proposed model, further enhancing our understanding of its effectiveness under different operational conditions. Furthermore, the research evaluates the project's financial viability over 5 years, considering revenue generation from producing activated carbon using SCG in treatment centers. This comprehensive financial analysis provides insights into the project's long-term sustainability (Fig. 2).

3 Methodology

To tackle our research questions and achieve our objectives, we employ a particular research methodology, illustrated in Fig. 2 , comprising four unique phases. This approach includes a Problem Definition Approach, a Data Management Module, an Optimization Module, and a Financial Evaluation Module.

3.1 Problem Definition Approach

To define this research's problem, we utilized three unique approaches: (A) literature review analysis, (B) engaging in consultations with experts, and (C) analysis of the critical challenges associated with SCG management.

(A) Literature Review Analysis
We evaluated current research on managing RFW alongside technological advancements. We also explored the generation, collection, and treatment processes of SCG and examined the coffee value chain and the integration of I4.0 technologies. Additionally, we reviewed the literature on RL and VRP models in the waste management sector. Our analysis focused on identifying related issues and proposed solutions, how these models are formulated, and the decision variables involved. The goal was to pinpoint gaps in the existing literature.

(B) Engaging in Consultations with Experts
During this phase, we interviewed waste management managers from a Canadian company that handles SCG in Montreal. We aimed to understand the industry's current state and pinpoint stakeholders' most pressing logistics challenges. This collaboration provided a more comprehensive understanding of the challenges and opportunities within the field.

(C) Analysis of the Critical Challenges Associated with SCG Management
By integrating insights from our literature review and expert consultations, we identified several critical challenges in managing SCG. The literature review highlighted theoretical frameworks and technological advancements that offer potential solutions for optimizing SCG management through improved processes. It also revealed gaps in existing models, particularly in VRP applications for waste management. Concurrently, our interviews with experts in the Montreal region provided practical perspectives on these issues, emphasizing challenges such as efficient routing, wet waste storage, and transportation of SCG. Combining these theoretical and practical insights, we identified a comprehensive set of challenges and defined the problem.

3.2 Data Management Module

This module involves collecting and validating data necessary for constructing our case study. We gathered essential data through available databases from Montreal CS, the case company, and related projects. This data collection covered several vital aspects:

1. Information on the amount of SCG produced by CS
2. Detailed specifications regarding electric vehicles (EVs), including their capacity, weight, consumption rates, and maximum travel distance

3. Comprehensive lists of addresses and postal codes for SCG collection points
4. Average times required for SCG collection services at various locations
5. Financial data encompassing the costs of collection, transportation, and processing of SCG
6. Information on the capacities of CC equipment and their efficiency in dehydrating SCG

This data provides a robust foundation for analyzing and optimizing the logistics of SCG management. To ensure the accuracy and reliability of our data, we conducted a thorough validation process. First, we estimated SCG generation by cross-referencing data from local businesses with industry averages. We utilized Canadian vehicle databases to verify the specifications of EVs. Montreal postal codes were validated using official city databases and cross-checked with Google Maps to ensure precise location data. We conducted time studies to measure service durations for SCG collection accurately. Cost estimation involved detailed analysis of financial data, cross-referenced with market rates and operational costs. We consulted databases of pre-drying equipment sellers to verify capacities and dehydration factors. Finally, we reviewed corporate procedures and regulations to ensure compliance with local policies and industry standards, providing a comprehensive and validated dataset for our analysis.

3.3 Optimization Module

This module selects the optimal locations for CC, assigns CS to CC, and generates the optimal routes. Subsequently, the solutions are assessed using specific metrics, and the robustness of the results is analyzed to guide the relevant decision-makers. We identified the problem's assumptions, utilizing insights from the literature review and information gathered about the coffee industry. The mathematical model is formulated to determine the most cost-effective configuration of the RL network.

Through an integrated selection process, we select the potential CC sites from a diverse array of CS. Initially, an electronic survey will be distributed to all CSs to gauge their interest in and willingness to participate in the transformation process. This survey includes questions about current practices, sustainability initiatives, and openness to adopting circular economy principles. Following this, online interviews are conducted with CS managers, where they are evaluated based on criteria such as sustainability practices, business model flexibility, community engagement, and innovation potential. Based on the survey and interview results, CSs are ranked according to their potential for successful transformation, considering factors like infrastructure readiness, managerial commitment, and financial viability. Shortlisted CSs are then visited in person to conduct a detailed assessment of their infrastructure, including waste management systems, energy efficiency, water usage, and supply chain practices. Finally, all gathered data is meticulously analyzed to support the decision and selection of CS for transforming them into CC facilities. This structured approach ensures thorough consideration of all factors and maximizes the effectiveness of the transformation initiative.

Next, we develop the model by defining the objective, constraints, and relevant parameters for which we gathered data. The objective function represents the goal we aim to achieve, while the constraints represent the limitations and requirements the solution must adhere to. We use optimization solvers, specifically CPLEX and Gurobi, to solve the model within the Python programming environment. These tools are chosen for their powerful optimization capabilities, ensuring efficient and accurate solutions that meet the defined constraints and achieve the objective. By leveraging these solvers, we can find the most optimal solution to our model, ensuring that our objective is met within the defined constraints and parameters.

In Business Model 1, focusing on a broad range of CS, employing clustering techniques becomes crucial to efficiently managing computational time challenges. As the number of CS increases, computational time also significantly rises. Therefore, the significance of clustering becomes evident in effectively addressing the mathematical model. The methodology we employed to solve the VRP for many CSs using a hierarchical clustering approach has five steps.

3.4 Financial Evaluation Module

This module focuses on the project's financial outlook, providing decision-makers with analytical insights. We meticulously examined all costs associated with the project. We developed revenue forecasts based on various factors, such as sales projections. Then, we calculated the project's profitability metrics, providing insights into the project's financial performance. We conducted sensitivity analyses to evaluate the project's sensitivity to critical factors such as interest and yield rate changes.

This financial analysis considers a 5-year horizon to provide a balanced view of the project's medium-term viability and performance. This time frame is sufficient to capture the project's initial investment, ramp-up phase, and stabilization of revenue and costs. It allows us to observe the impact of strategic decisions and market conditions over a period long enough to assess trends and make informed predictions while still being short enough to maintain relevance and accuracy in forecasting.

4 Business Model Mathematical Model

4.1 Problem Definition

Our study concentrates on designing various RL network configurations as business models. Within these networks, a set of CSs generates SCG daily, which must be collected and transported to a treatment center to produce activated carbon. The treatment company is responsible for collecting and processing SCG and preparing it for sale in target markets for applications such as water purification and air filtration, as detailed in Business Models 1 and 2. The treatment process involves drying and activating the SCG at elevated temperatures.

The problem involves determining the most efficient strategy for SCG collection. This can be approached from two economic perspectives: minimizing the total cost or maximizing the total profit. Our study tackled the problem from a cost minimization perspective, as our primary interest lies in examining cost efficiency. Revenues and profitability considerations are subsequently addressed in the financial evaluation. To develop a realistic case study, we partnered with the case company located in the Montreal region, which specializes in collecting and treating SCG. Our objective is to explore two configurations of the SCG RL network. These configurations provide decision-makers with a spectrum of potential business model options.

Traditionally, a truck visits all CSs to collect SCG, as shown in Fig. 3. In Business Model 1, the network configuration does not include pre-drying equipment before the treatment center. This means the collected SCG is transported directly from the CS to the treatment center without any intermediate drying process. Consequently, the SCG arrives at the treatment center in its original, wet state, and the drying and activation processes are entirely handled at the treatment facility.

Keeping SCG in CS increases the risk of contamination and socio-environmental effects. In the proposed Business Model 2, SCG is sent out to CC equipped with

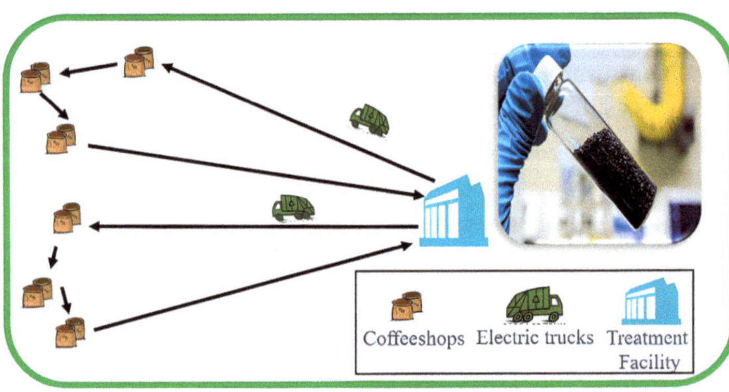

Fig. 3 Schematic view of Business Model 1

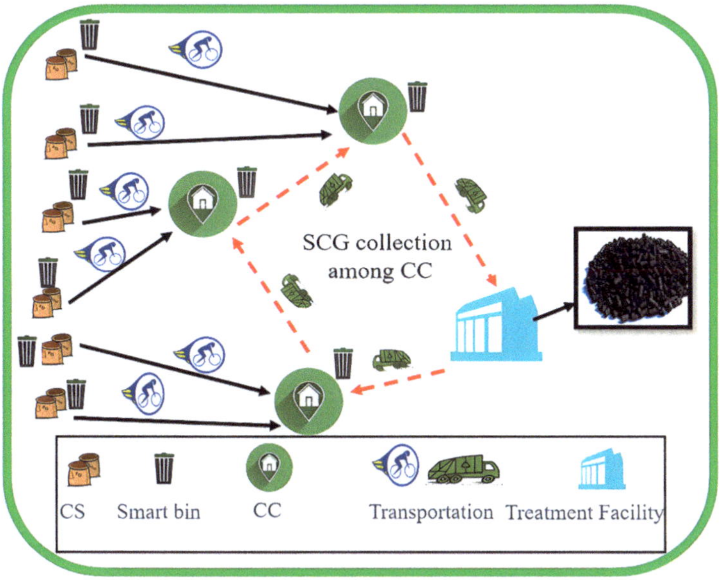

Fig. 4 Schematic view of Business Model 2

pre-drying technology. This plan would involve setting up financial incentives and government subsidies for delivery drivers within the city. In Business Model 2, smart bins in CC are equipped with sensors that help segregate different waste to affect the quality of the final product, data collection on the type and quantity of SCG, and other relevant parameters such as moisture content and temperature. CC can consolidate SCG from multiple CSs into larger volumes that are more economical to transport to a processing facility, sort SCG to remove any nonorganic materials such as plastics or metals that may be mixed with the SCG, and provide temporary storage for SCG until it is transported to a processing facility. Then, dehydrated, compacted SCG is transferred from CC to the treatment facility for processing and producing activated carbon used in different industries.

To better understand the problem, a view of Business Model 2 is depicted in Fig. 4. The decisions derived from this study are expected to lead to finding the optimal location-allocation of CC that reduces transportation costs, risks related to environmental effects, imposed contamination from overload bins on inhabitants, and the final product's quality as much as possible.

- CC: circular coffee shop
- CC h: circular coffee shop h (it means CC in location h)
- CC j: circular coffee shop j (it means CC in location j)

4.2 Assumptions

The development of Business Model 1 is based on the following assumptions:

- The number and location of the CS are known.
- Distances are calculated regarding address and postal codes.
- All CS demands should be fully satisfied.
- The network has only one treatment facility, and its location is known.
- The quantity of SCG in CS is estimated and determined. It is assumed that it is calculated by assessing the CS's size. This assessment considers average coffee sales data, which includes information on the number of coffee cups sold, the volume of those cups, and a conversion factor for the weight of generated SCG.
- The transportation costs of using an EV are estimated and determined.
- The electricity consumption rates of EVs vary and are determined for various operating modes.
- Service time is estimated and determined based on the observations.
- Each EV has a predefined capacity, weight, and maximum allowable length to traverse.
- The velocity of each EV is estimated based on the study's urban area context.
- The fixed cost of each route is estimated and determined.

Additionally, several specific assumptions are considered for Business Model 2:

- A limited amount of SCG regarding capacity issues can be allocated to each CC.
- CC has intelligent bins with notification capabilities.
- Following pre-drying treatment in the CC, the amount of SCG will decrease due to dehydration and the compaction process.
- The inflow into the CC equals the outflow, accounting for the determined dehydration rate.
- The establishment costs of CC equipped with pre-drying technology are estimated and determined based on capacity features (spread over the initial 5 years of use).

- An innovative financial mechanism estimates and determines the transportation costs between CS and CC.
- Candidate locations for setting up a CC are selected from the available CS sites according to a defined procedure.
- The estimated establishment costs of smart bins amount to 30% of the costs of CC equipped with pre-drying technology.

The mathematical model for Business Model 2 includes sets, parameters, decision variables, an objective function, and constraints, as outlined below:

4.2.1 Sets

- i: index of CS as SCG generation point
- j, h: index of potential sites for CC equipped with pre-drying technology
- k: index of routes
- c: index of the capacity of CC

4.2.2 Decision Variables

- y_{hc}, if the CC h with capacity level of c is established in the location of an existing CS
- z_{hi}, if CS i is allocated to CC h
- x_{hjk}, if a route k exists when transporting from CC h to CC j

4.2.3 Parameters

- w_i: amount of SCG generated in CS i (kg)
- Q: capacity of EV (kg)
- L: maximum length that an EV can traverse (km)
- Q_{hc}: capacity of CC h with capacity level of c (kg)
- F_k: fixed cost of route k ($)
- F_{hc}: fixed cost of establishing CC h with capacity level of c ($)

- d_{hj}: distance between CC h and CC j (km)

- t_j: service time in CC j) (min)
- ρ_{idle}: electricity consumption rate per unit time while EV is idle (L/min)
- ρ_{hj}: electricity consumption rate per unit distance while EV is moving (L/km)
- ρ: price of per-unit electricity consumption ($/L)
- α_{ih}: cost of transferring SCG from CS i to CC h ($)
- q_h: the amount of dehydrated SCG which is collected from CC h (kg)
- β: conversion compaction factor

4.2.4 Objective Function

$$\begin{aligned} \text{Min } F = & \sum_{k=1}^{K}\sum_{j=0}^{N}(x_{0jk} * F_k) \\ & + \sum_{h=0}^{N}\sum_{j=0}^{N}(x_{hjk} * d_{hj} * \rho_{hj} * \rho) \\ & + \sum_{j=1}^{N}\sum_{h=0}^{N}(x_{hjk} * t_j * \rho_{\text{idle}} * \rho) \\ & + \sum_{h=0}^{N}\sum_{c=1}^{2}(F_{hc} * y_{hc}) \\ & + \sum_{i=1}^{N}\sum_{h=1}^{N}(\alpha_{ih} * z_{ih}) \end{aligned} \tag{1}$$

4.2.5 Constraints

$$\sum_{k=1}^{K}\sum_{h=0}^{N} x_{hjk} = 1, \quad \forall j = 1, \ldots, N \tag{2}$$

$$\sum_{j=1}^{N} x_{ijk} = 1, \quad \forall h = 1, \ldots, N \tag{3}$$

$$\sum_{h=0}^{N} x_{ijk} = \sum_{j=1}^{N} x_{hjk} = 1, \quad \forall j = 1, \ldots, N;\ k = 1, 2, \ldots, K \tag{4}$$

$$\sum_{j=1}^{N} x_{hjk} * q_h \leq Q, \quad \forall h = 0, 1, \ldots, H \tag{5}$$

$$q_h = (1 - \beta)\left(\sum_{i=1}^{N} z_{ih} * w_i\right) \tag{6}$$

$$\sum_{h=0}^{N}\sum_{j=0}^{N} d_{hj} * x_{hjk} \leq L, \quad \forall k = 1, 2, \ldots, K, \quad h \neq j \tag{7}$$

$$\sum_{h=0}^{N}\sum_{j=0}^{N} x_{hjk} \leq |S| - 1, \quad S \subseteq \{1, 2, \ldots, N\}, \quad \forall k = 1, 2, \ldots, K \tag{8}$$

$$\sum_{h=0}^{N} z_{ih} = 1, \quad \forall i = 1, 2, \ldots, I \tag{9}$$

$$\sum_{c=1}^{2} y_{hc} = 1, \quad \forall h = 1, 2, \ldots, H \tag{10}$$

$$\sum_{i=1}^{N} z_{ih} * w_i \leq Q * y_{hc}, \quad \forall h = 1, 2, \ldots, H, \quad \forall c = 1, 2 \tag{11}$$

$$x_{hjk}, y_{hc}, z_{ih} \in \{0, 1\} \tag{12}$$

The objective function (1) calculates the fixed cost of vehicles, the variable cost of vehicles both while moving among CC and in-service time at CC, the cost of establishing CC, and the cost of transportation of SCG from CS to CC. Constraint (2) shows that each CC must be visited once by a vehicle. Constraints (3) and (4) show that a vehicle begins at the depot and goes back to the treatment facility after the last visit of CC. Equation (5) represents that the amount of SCG for each path cannot be more significant than the vehicle's maximum load. Constraint (6) represents the amount of SCG entering a CC as a percentage of the quantity of dehydrated SCG that exits the CC by collection. Constraint (7) ensures that the total length of each route does not exceed the maximum distance that an EV can traverse. Equation (8) illustrates the subtour elimination. Equation (9) assigns each CS to only one CC. Equation (10) shows that each CC can have one capacity type. Constraint (11) implies that for each established CC, the total quantity of SCG assigned should not be more than the capacity of the CC. Finally, constraint (12) shows the types of variables.

5 Results and Insights

5.1 Case Study Results

The computational experiments were conducted on a workstation with an Intel Core i7-1165 processor running at 2.80 GHz, 16 GB of RAM, and 64-bit Windows 11 Pro employing Python 3.10, CPLEX, and Gurobi. The proposed approach is implemented for a case study in Montreal, Canada. With a population exceeding 1.7 million, Montreal produces considerable SCG daily, demanding significant attention. A treatment facility produces activated carbon from SCG through the activation process under elevated temperatures that can be used in different applications such as water purification, air filtration, and gas separation. The company is looking for an economical and sustainable solution for the SCG collection from CS in Montreal. Each CS typically sells between 4000 and 10,000 cups of coffee weekly, resulting in a daily output of 5 and 7 kg of SCG. For 1000 CS participating in this program, 6271 kg of SCG is generated daily.

In Business Model 1, as depicted in Fig. 5, the collection of 6271 kg of SCG generated daily involves six routes, covering a total distance of 611 km. The collection process takes 65.25 hours, comprising 24 hours for transportation and 41 hours for service time at CS. The transportation cost amounts to $1835.73, which includes $1800 for the routes and $35.73 for the energy consumption of the EV. Consequently, the unit transportation cost is $0.29 per kilogram of SCG. Table 1 provides detailed information for each of the six routes. There is no investment cost since we do not use any intelligent bin or pre-drying equipment.

In Business Model 2, 6271 kg of SCG is transported from CS to CC. However, considering the consolidation rate, only 1500 kg of pre-dried SCG is collected from CC. Figure 6 indicates that 21 of 28 potential sites have been selected to establish CC equipped with medium-sized pre-drying technology, and the existing CSs have been allocated to these selected CCs. Therefore, only one route with a total distance of 89.5 km is enough to collect all pre-dried SCG. Also, the collection process is shorter, 4.9 hours, including 3.5 hours for transportation and 1.4 hours for service time in CC, in Business Model 2 compared with Business Model 1. In Business Model 2, we examine small- and medium-sized pre-drying technology units capable of handling 100 and 300-kg loads. Their annual costs are estimated at $16,000 and $20,000, respectively, spanning a 5-year duration.

In addition, we spend less time on the road, and the risk of EV accidents and contamination does not increase since we can keep the dehydrated SCG safely in CC for a week. Transportation cost is 305.44$, including 300$ for the route and 5.44$ for the energy consumption of the EV. Therefore, the cost of collecting and transporting SCG is 0.29$/kg in Case 1 and 0.04$/kg in Business Model 2. This is considered reasonable and justifiable with the value of the activated carbon and its precious compounds. This makes the overall process viable from a sustainability perspective. Comparing the cost of collection per unit of SCG in both Business models shows that Business Model 2 is better since it has less cost.

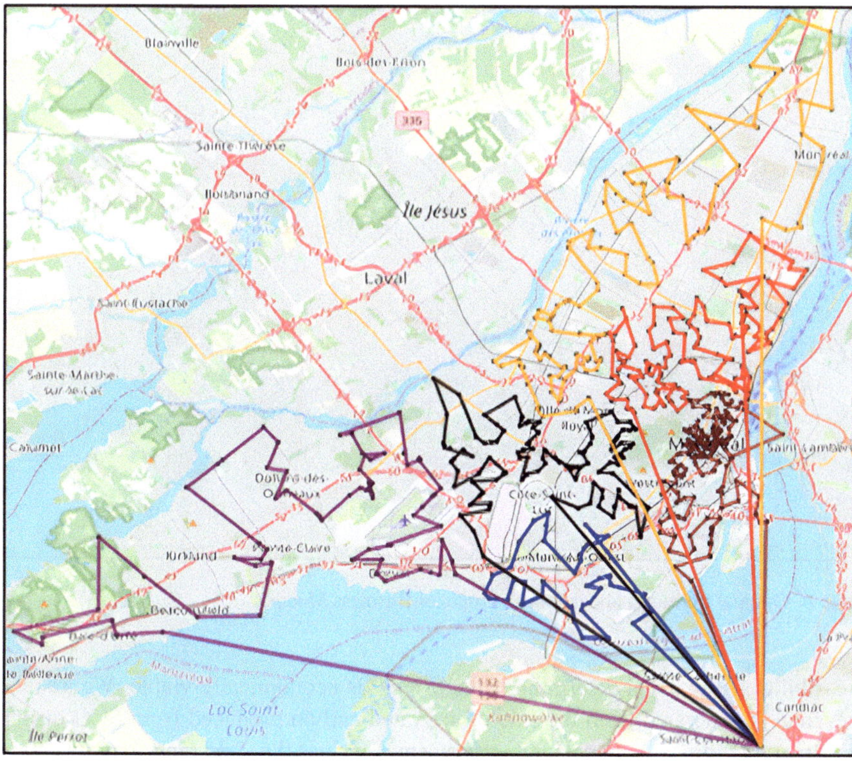

Fig. 5 Optimal routes of Business Model 1

Table 1 Detailed information of each six routes in Business Model 1

Route	SCG amount (kg)	Number of CS	Distance (km)	Time (h)
1	800	120	53.59	4.89
2	1462	230	96.65	10.61
3	447	100	117.70	13.91
4	1288	150	124.00	9.10
5	1258	170	81.23	14.14
6	1016	230	138.11	12.60
Total	6271	1000	611	65.25

5.2 *Cash Flow Projection and Financial Outlook*

The optimization model results provide us with the operational expenses (OPEX). However, capital expenditures (CAPEX) are equally significant. A financial study involves projecting a cash flow, encompassing costs and revenues, over 5 years. This

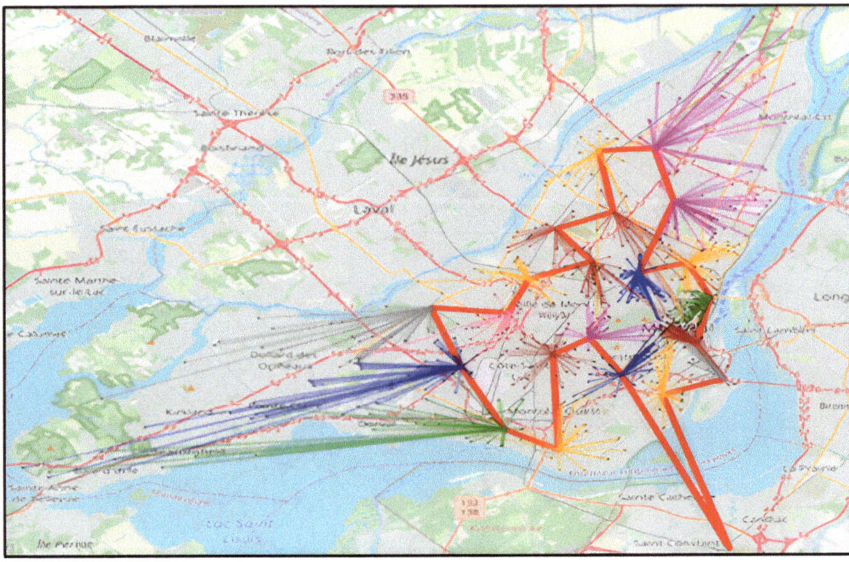

Fig. 6 Optimal locations, allocation, and route of Business Model 2

feasibility study helps determine if the project is economically viable. We assume revenue is generated from the sale of activated carbon produced from SCG during the treatment process. To calculate the revenue, we use the yield range or conversion efficiency, representing the percentage of activated carbon obtained from SCG after the activation process. We assume a rate of 0.22 for Business Model 1 and 0.25 for Business Model 2. The yield rate for Business Model 2 is higher because of pre-drying at CC, which enhances the quality of SCG for the activation process. Pre-drying reduces moisture content, leading to more efficient activation and higher conversion efficiency of SCG into activated carbon.

Operating costs remain constant over 5 years, while capital expenditures are incurred upfront in year 0. Based on the cash flow projection, Business Model 2 demonstrates a better financial outlook over 5 years despite the higher initial investment. The initial investment cost for CC equipment may appear substantial when starting a project. However, it is crucial to recognize that this cost is spread over time and typically allocated to the early phases of the project. Notably, this investment differs significantly from transportation costs and recurring expenses in the system's ongoing operation.

Also, Business Model 1 has less PV than Business Model 2, and Business Model 2 shows an FV of the initial investment. This suggests the potential for growth or return on investment. Considering an interest rate of 5%, Business Model 2 appears to have a more favorable financial outlook regarding present value and potential FV.

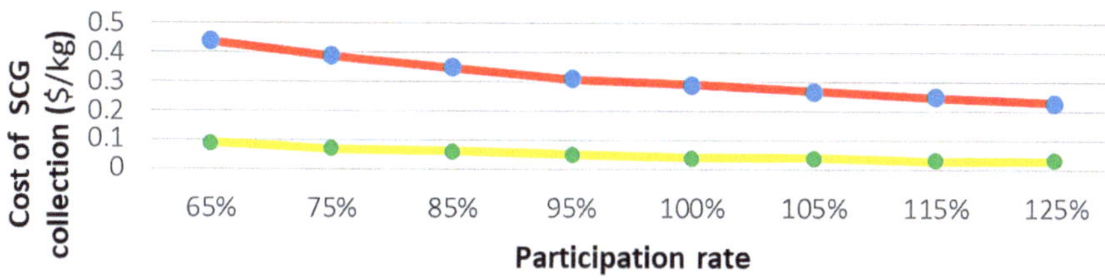

Fig. 7 The behavior of models by changing the participation rate of CS

5.3 Sensitivity Analysis

5.3.1 Change in the Participation Rate of CS in SCG Generation

This analysis examines the program's CS participation rate under optimistic and pessimistic scenarios. Pessimistic scenarios, like accidental SCG disposal, are considered post-network establishment. In these scenarios, the amount of SCG collected is lower than expected. This can lead to underutilization of pre-drying equipment, making investments in new technology potentially unjustifiable. Essentially, the lower participation rates result in higher per-unit collection costs due to the lack of economies of scale. Optimistic scenarios result from enhanced awareness strategies. These scenarios assume higher participation rates. Increased SCG collection leads to lower per-unit collection costs, benefiting from economies of scale. The more SCG collected, the more cost-effective the program becomes. We created scenarios by assigning various participation rates within the realistic bounds of 65–125%, allowing us to analyze a wide range of potential outcomes for the SCG collection program. Therefore, we used a uniform distribution to assign related SCG weights randomly.

Figure 7 illustrates the outcomes of the sensitivity analysis and the performance of Business Models 1 and 2 concerning the collection cost per unit of SCG under different CS participation rate scenarios. Consistently, the red line representing Business Model 1 remains above that of Business Model 2, indicating that Case 1 incurs higher collection costs within the analyzed range of participation. This difference underscores the impact of economies of scale, where increasing the collected amount of SCG in the network leads to decreased collection costs. In pessimistic scenarios where the generated SCG amount is lower than expected, the pre-drying equipment may be underutilized, making the investment in new technology potentially unjustifiable.

The network in Business Model 2 is initially designed with 21 CC. However, in the most optimistic scenario, where successful awareness strategies result in a higher participation rate of CS than anticipated, five additional CCs are required, bringing the total to 26 CCs, as shown in Fig. 8. Therefore, decision-makers must prioritize several strategic actions to address the increased demand for SCG collection effectively. This includes focusing on the city's extension phase of the

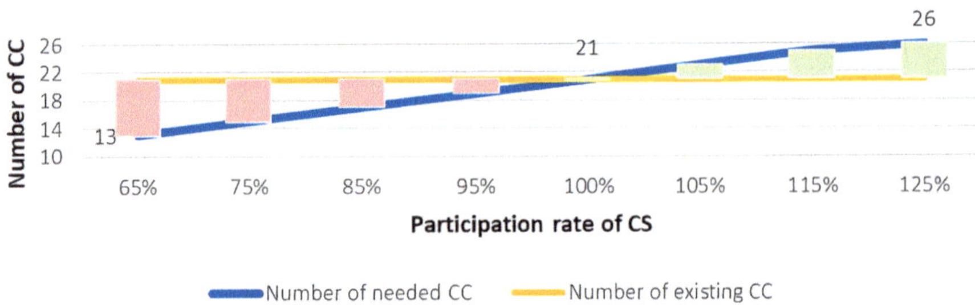

Fig. 8 The effect of changing the participation rate on the number of CCs

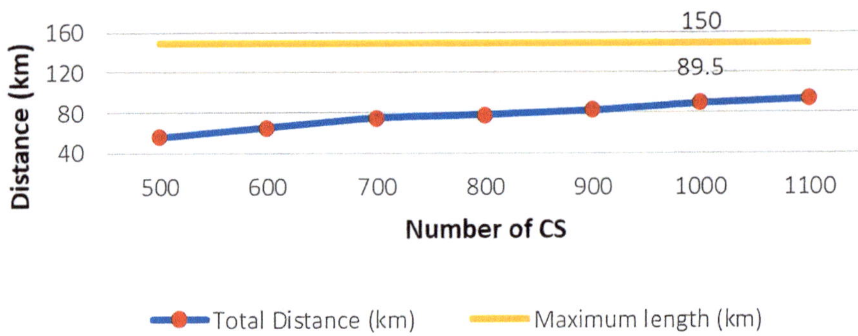

Fig. 9 The effect of change in the number of CSs on distance

project. In addition, continued investment in awareness campaigns is essential to sustain high participation rates among CSs. Conversely, in the most pessimistic scenario, eight CCs remain unused due to the low amount of collected SCG.

5.3.2 Change in the Number of CSs During the Program Contract

Several CSs may exit the program, or new CSs may express interest in joining for different reasons, such as changes in ownership, business priorities, operational challenges, market competition, regulatory requirements, and evolving sustainability initiatives. Figure 9 illustrates the effect of changes in the number of CSs participating in the contract for the SCG collection program on the traversed distance. With the same network configuration in Business Model 2, the total traversed distance decreases when a CS leaves the program.

However, simultaneously, as shown in Fig. 10, the utilization rate of EVs decreases. This observation suggests that while a reduction in CS participation may lead to decreased overall travel distance for collection, it also impacts the utilization rate of EVs. This insight highlights the interconnected nature of operational factors within the SCG collection program and underscores the importance of considering multiple variables when assessing program performance. When CS exits the

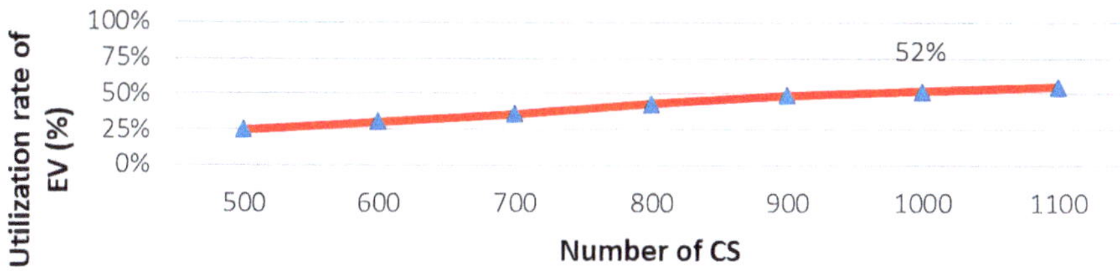

Fig. 10 The effect of changing the number of CS on EV utilization rate

program, the total traversed distance for collection is reduced, indicating potential cost savings. However, the decrease in the utilization rate of EVs suggests the underutilization of resources, which can negatively impact operational efficiency.

Therefore, managers should focus on fostering strong relationships with CS and implementing measures to incentivize their continued participation in the program. This may include providing ongoing support, addressing any concerns or challenges faced by CS, and offering incentives or rewards for participation. Additionally, recruiting new CS to join the program should be balanced with strategies to retain existing participants. Furthermore, managers should continuously monitor CS engagement and satisfaction levels, leveraging data analytics and feedback mechanisms to identify and address issues promptly.

6 Conclusion

This chapter proposes a digital intelligence approach for implementing an SCG recovery system in Montreal. The study suggests optimization models for designing an RL network for collecting SCG. The business model, which is based on reducing the weight of waste and consolidating SCG in CC, is more economically viable in terms of transport.

The research emphasizes the importance of circularity in SCG and stresses the need for intelligent solutions to improve waste management practices. The synthesis of research findings demonstrates the potential of integrating I4.0 technologies throughout the coffee value chain, offering enhanced efficiency and environmental stewardship opportunities. However, determining the economic feasibility and justifying the investment required for implementing innovative and sustainable practices using I4.0 technologies is a crucial challenge.

The research discusses the advantages of incorporating new technologies in SCG collection and addressing uncertainties. CC equipped with smart bins, pre-drying technologies, and EVs for transportation can help create an innovative and sustainable system, mitigating contamination risks and maintaining SCG quality until it is economically collected. A customized model applied to 1,000 CSs in Montreal shows that CC construction enhances SCG quality and efficiency in the collection network, reducing the need for frequent transportation. This RL network addresses strategic and operational decisions on the location of CC, the allocation of CS to CC, and routing among CC, as well as discussing the impacts of disruptive technologies on the parameters and operations.

However, several avenues for future research can further enhance the sustainability and efficiency of this system. Firstly, exploring innovative financial mechanisms for SCG transportation could open doors to new business models. Secondly, considering the production of various products through treatment can help valorize coffee waste.

References

1. Alamsyah A, Widiyanesti S, Wulansari P, Nurhazizah E, Dewi AS, Rahadian D, Tyasamesi P (2023) Blockchain traceability model in the coffee industry. J Open Innov Technol Market Complex 9(1):100008. https://doi.org/10.1016/j.joitmc.2023.100008
2. Almohana AI, Almojil SF, Kamal MA, Alali AF, Kamal M, Alkhatib SE, Algarni M (2022) Theoretical investigation on optimization of biodiesel production using waste cooking oil: machine learning modeling and experimental validation. Energy Rep 8:11938–11951. https://doi.org/10.1016/j.egyr.2022.08.265
3. Aydın Temel F, Cagcag Yolcu O, Turan NG (2023) Artificial intelligence and machine learning approaches in composting process: a review. Bioresour Technol 370:128539. https://doi.org/10.1016/j.biortech.2022.128539
4. Aytaç K, Korçak Ö (2021) IoT based intelligence for proactive waste management in quick service restaurants. J Clean Product 284:125401. https://doi.org/10.1016/j.jclepro.2020.125401
5. Bager SL, Singh C, Persson UM (2022) Blockchain is not a silver bullet for agro-food supply chain sustainability: insights from a coffee case study. Curr Res Environ Sustain 4:100163. https://doi.org/10.1016/j.crsust.2022.100163
6. Batat W (2021) How augmented reality (AR) is transforming the restaurant sector: investigating the impact of "Le Petit Chef" on customers' dining experiences. Technol Forecast Soc Change 172:121013. https://doi.org/10.1016/j.techfore.2021.121013
7. Bedoya MG, Montoya DR, Tabilo-Munizaga G, Pérez-Won M, Lemus-Mondaca R (2022) Promising perspectives on novel protein food sources combining artificial intelligence and 3D food printing for food industry. Trends Food Sci Technol 128:38–52. https://doi.org/10.1016/j.tifs.2022.05.013
8. Bhat JR, AlQahtani SA, Nekovee M (2023) FinTech enablers, use cases, and role of future internet of things. J King Saud Univ Comput Inf Sci 35(1):87–101. https://doi.org/10.1016/j.jksuci.2022.08.033
9. Bordin Yamashita JVY, Leite JPRR (2023) Coffee disease classification at the edge using deep learning. Smart Agric Technol 4:100183. https://doi.org/10.1016/j.atech.2023.100183
10. Bottani E, Tebaldi L, Volpi A (2019) The role of ICT in supporting spent coffee grounds collection and valorization: a quantitative assessment. Sustainability 11(23):6572. https://doi.org/10.3390/su11236572
11. Carassou F (2015) Une récupération spécifique du marc de café aurait-elle une plus-value pour la communauté? Cas de l'île de Montréal (Maîtrise en environnement, université de Sherbrooke) Université de Sherbrooke, Sherbrooke (2015)
12. Chaabane A, Montecinos J, Ouhimmou M, Khabou A (2021) Vehicle routing problem for reverse logistics of end-of-life vehicles (ELVs). Waste Manage 120:209–220. https://doi.org/10.1016/j.wasman.2020.11.008
13. Chai JJK, O'Sullivan C, Gowen AA, Rooney B, Xu JL (2022) Augmented/mixed reality technologies for food: a review. Trends Food Sci Technol 124:182–194. https://doi.org/10.1016/j.tifs.2022.04.021

14. Cheah CG, Chia WY, Lai SF, Chew KW, Chia SR, Show PL (2022) Innovation designs of industry 4.0 based solid waste management: machinery and digital circular economy. Environ Res 213:113619. https://doi.org/10.1016/j.envres.2022.113619
15. Chen CF, Girish VG (2023) Investigating the use experience of restaurant service robots: the cognitive–affective–behavioral framework. Int J Hospital Manage 111:103482. https://doi.org/10.1016/j.ijhm.2023.103482
16. Chen WH, Aniza R, Arpia AA, Lo HJ, Hoang AT, Goodarzi V, Gao J (2022) A comparative analysis of biomass torrefaction severity index prediction from machine learning. Appl Energy 324:119689. https://doi.org/10.1016/j.apenergy.2022.119689
17. Chiang AH, Trimi S, Lo YJ (2022) Emotion and service quality of anthropomorphic robots. Technol Forecast Soc Change 177:121550. https://doi.org/10.1016/j.techfore.2022.121550
18. Clark J, Crandall P, Shabatura J (2018) Wearable technology effects on training outcomes of restaurant food handlers. J Food Protect 81(8):1220–1226. https://doi.org/10.4315/0362-028X.JFP-18-033
19. De Carvalho Alves M, Sanches L, Pozza EA, Pozza AAA, Da Silva FM (2022) The role of machine learning on Arabica coffee crop yield based on remote sensing and mineral nutrition monitoring. Biosyst Eng 221:81–104. https://doi.org/10.1016/j.biosystemseng.2022.06.014
20. Fernandes E, Moro S, Cortez P, Batista F, Ribeiro R (2021) A data-driven approach to measure restaurant performance by combining online reviews with historical sales data. Int J Hospital Manage 94:102830. https://doi.org/10.1016/j.ijhm.2020.102830
21. Flores-Carrasco J, Morillo-Torres D, Escobar JW, Linfati R, Gatica G (2021) The periodic location-routing with selective recycling problem: two solution approaches. IFAC-Papers OnLine 54(1):787–792. https://doi.org/10.1016/j.ifacol.2021.08.092
22. Forcina A, Petrillo A, Travaglioni M, Di Chiara S, De Felice F (2023) A comparative life cycle assessment of different spent coffee ground reuse strategies and a sensitivity analysis for verifying the environmental convenience based on the location of sites. J Clean Product 385:135727. https://doi.org/10.1016/j.jclepro.2022.135727
23. Gabrieli G, Muszynski M, Thomas E, Labbe D, Ruch PW (2022) Accelerated estimation of coffee sensory profiles using an AI-assisted electronic tongue. Innov Food Sci Emerg Technol 82:103205. https://doi.org/10.1016/j.ifset.2022.103205
24. Gebreeyessus GD (2022) Towards the sustainable and circular bioeconomy: insights on spent coffee grounds valorization. Sci Total Environ 833:155113. https://doi.org/10.1016/j.scitotenv.2022.155113
25. George RV, Harsh HO, Ray P, Babu AK (2019) Food quality traceability prototype for restaurants using blockchain and food quality data index. J Clean Product 240:118021. https://doi.org/10.1016/j.jclepro.2019.118021
26. Gruler A, Quintero-Araújo CL, Calvet L (2017) Waste collection under uncertainty: a simheuristic based on variable neighbourhood search. Eur J Ind Eng 11(2):228–255. https://doi.org/10.1504/EJIE.2017.083257
27. Hemmelmayr V, Smilowitz K, De La Torre L (2017) A periodic location routing problem for collaborative recycling. IISE Trans 49(4):414–428. https://doi.org/10.1080/24725854.2016.1267882
28. Hidalgo F, Quiñones-Ruiz XF, Birkenberg A, Daum T, Bosch C, Hirsch P, Birner R (2023) Digitalization, sustainability, and coffee. Opportunities and challenges for agricultural development. Agric Syst 208:103660. https://doi.org/10.1016/j.agsy.2023.103660
29. Hwang J, Choe JY (Jacey), Kim HM, Kim JJ (2021) Human baristas and robot baristas: how does brand experience affect brand satisfaction, brand attitude, brand attachment, and brand loyalty? Int J Hospital Manage 99:103050. https://doi.org/10.1016/j.ijhm.2021.103050
30. Jain A, Bagherwal R (2017) Design and implementation of a smart solid waste monitoring and collection system based on Internet of Things. In: 2017 8th international conference on computing, communication and networking technologies (ICCCNT). IEEE, Delhi, pp 1–5. https://doi.org/10.1109/ICCCNT.2017.8204165
31. Juvan E, Grün B, Dolnicar S (2023) Waste production patterns in hotels and restaurants: an intra-sectoral segmentation approach. Ann Tourism Res Empirical Insights 4(1):100090. https://doi.org/10.1016/j.annale.2023.100090

32. Kang SE, Koo C, Chung N (2023) Creepy vs. cool: switching from human staff to service robots in the hospitality industry. Int J Hospital Manage 111:103479. https://doi.org/10.1016/j.ijhm.2023.103479
33. Kim T, Lee OKD, Kang J (2023) Is it the best for barista robots to serve like humans? A multidimensional anthropomorphism perspective. Int J Hospital Manage 108:103358. https://doi.org/10.1016/j.ijhm.2022.103358
34. Kondamudi N, Mohapatra SK, Misra M (2008) Spent coffee grounds as a versatile source of green energy. J Agric Food Chem 56(24):11757–11760. https://doi.org/10.1021/jf802487s
35. Kouadio L, Deo RC, Byrareddy V, Adamowski JF, Mushtaq S, Phuong Nguyen V (2018) Artificial intelligence approach for the prediction of Robusta coffee yield using soil fertility properties. Comput Electron Agric 155:324–338. https://doi.org/10.1016/j.compag.2018.10.014
36. Leung XY, Chen H, Chang W, Mhlanga L (2022) Is VR game training more effective for hospitality employees? A longitudinal experiment. Tourism Manage Perspect 44:101020. https://doi.org/10.1016/j.tmp.2022.101020
37. Li B, Zhong Y, Zhang T, Hua N (2021) Transcending the COVID-19 crisis: business resilience and innovation of the restaurant industry in China. J Hospital Tourism Manage 49:44–53. https://doi.org/10.1016/j.jhtm.2021.08.024
38. Li M, Cheng WY, Li YC, Wu HM, Wu YC, Lu HW, Lai YC (2021) Deformable, resilient, and mechanically-durable triboelectric nanogenerator based on recycled coffee waste for wearable power and self-powered smart sensors. Nano Energy 79:105405. https://doi.org/10.1016/j.nanoen.2020.105405
39. Li S, Shi C, Sun S, Chan H, Lu H, Nilghaz A, Cao R (2021) From brown to colored: polylactic acid composite with micro/nano-structured white spent coffee grounds for three-dimensional printing. Int J Biol Macromol 174:300–308. https://doi.org/10.1016/j.ijbiomac.2021.01.176
40. Li C, He P, Peng W, Lü F, Du R, Zhang H (2022) Exploring available input variables for machine learning models to predict biogas production in industrial-scale biogas plants treating food waste. J Clean Product 380:135074. https://doi.org/10.1016/j.jclepro.2022.135074
41. Lu X, Pu X, Han X (2020) Sustainable smart waste classification and collection system: a bi-objective modeling and optimization approach. J Clean Product 276:124183. https://doi.org/10.1016/j.jclepro.2020.124183
42. Ma Y, Zhang W, Feng C, Lev B, Li Z (2021) A bi-level multi-objective location-routing model for municipal waste management with obnoxious effects. Waste Manage 135:109–121. https://doi.org/10.1016/j.wasman.2021.08.034
43. Ma E, Yang H, Wang YC, Song H (2022) Building restaurant customers' technology readiness through robot-assisted experiences at multiple product levels. Tourism Manage 93:104610. https://doi.org/10.1016/j.tourman.2022.104610
44. Malefors C, Secondi L, Marchetti S, Eriksson M (2022) Food waste reduction and economic savings in times of crisis: the potential of machine learning methods to plan guest attendance in Swedish public catering during the Covid-19 pandemic. Soc Econ Plan Sci 82:101041. https://doi.org/10.1016/j.seps.2021.101041
45. Manatura K, Chalermsinsuwan B, Kaewtrakulchai N, Kwon EE, Chen WH (2023) Machine learning and statistical analysis for biomass torrefaction: a review. Bioresour Technol 369:128504. https://doi.org/10.1016/j.biortech.2022.128504
46. Mantihal S, Kobun R, Lee BB (2020) 3D food printing of as the new way of preparing food: a review. Int J Gastron Food Sci 22:100260. https://doi.org/10.1016/j.ijgfs.2020.100260
47. Martin-Rios C, Demen-Meier C, Gössling S, Cornuz C (2018) Food waste management innovations in the foodservice industry. Waste Manage 79:196–206. https://doi.org/10.1016/j.wasman.2018.07.033
48. Mayson S, Williams ID (2021) Applying a circular economy approach to valorize spent coffee grounds. Resour Conser Recycl 172:105659. https://doi.org/10.1016/j.resconrec.2021.105659
49. Mohammadi M, Rahmanifar G, Hajiaghaei-Keshteli M, Fusco G, Colombaroni C (2023) Industry 4.0 in waste management: an integrated IoT-based approach for facility location and green vehicle routing. J Ind Inf Integr 36:100535. https://doi.org/10.1016/j.jii.2023.100535

50. Montoya-Munoz AI, Rendon OMC (2020) An approach based on fog computing for providing reliability in IoT data collection: a case study in a colombian coffee smart farm. Appl Sci 10(24):8904. https://doi.org/10.3390/app10248904
51. Montoya-Munoz AI, Silva RACD, Rendon OMC, Fonseca NLSD (2022) Reliability provisioning for fog nodes in smart farming IoT-fog-cloud continuum. Comput Electron Agric 200:107252. https://doi.org/10.1016/j.compag.2022.107252
52. Moreno Cárdenas EL, Zapata-Zapata AD, Kim D (2020) Modeling dark fermentation of coffee mucilage wastes for hydrogen production: artificial neural network model vs. fuzzy logic model. Energies 13(7):1663. https://doi.org/10.3390/en13071663
53. Mutz YS, Do Rosario D, Silva LRG, Galvan D, Stefano JS, Janegitz BC, Conte-Junior CA (2023) Lab-made 3D printed electrochemical sensors coupled with chemometrics for Brazilian coffee authentication. Food Chem 403:134411. https://doi.org/10.1016/j.foodchem.2022.134411
54. Nallan Chakravartula SS, Moscetti R, Bedini G, Nardella M, Massantini R (2022) Use of convolutional neural network (CNN) combined with FT-NIR spectroscopy to predict food adulteration: a case study on coffee. Food Control 135:108816. https://doi.org/10.1016/j.foodcont.2022.108816
55. Ni Z, Bi H, Jiang C, Sun H, Zhou W, Tian J, Lin Q (2022) Investigation of co-combustion of sewage sludge and coffee industry residue by TG-FTIR and machine learning methods. Fuel 309:122082. https://doi.org/10.1016/j.fuel.2021.122082
56. Nozawa C, Togawa T, Velasco C, Motoki K (2022) Consumer responses to the use of artificial intelligence in luxury and non-luxury restaurants. Food Qual. Prefer 96:104436. https://doi.org/10.1016/j.foodqual.2021.104436
57. Offie I, Piadeh F, Behzadian K, Campos LC, Yaman R (2023) Development of an artificial intelligence-based framework for biogas generation from a micro anaerobic digestion plant. Waste Manage 158:66–75. https://doi.org/10.1016/j.wasman.2022.12.034
58. Pardini K, Rodrigues JJPC, Diallo O, Das AK, de Albuquerque VHC, Kozlov SA (2020) A smart waste management solution geared towards citizens. Sensors 20(8):2380. https://doi.org/10.3390/s20082380
59. Park S, Park MK, Heo J, Hwang J, Hwang S, Kim D, Kwak HS (2023) Robot versus human barista: comparison of volatile compounds and consumers' acceptance, sensory profile, and emotional response of brewed coffee. Food Res Int 172:113119. https://doi.org/10.1016/j.foodres.2023.113119
60. Qiao Q, Tao F, Wu H, Yu X, Zhang M (2020) Optimization of a capacitated vehicle routing problem for sustainable municipal solid waste collection management using the PSO-TS algorithm. Int J Environ Res Publ Health 17(6):2163. https://doi.org/10.3390/ijerph17062163
61. Ramos TRP, De Morais CS, Barbosa-Póvoa AP (2018) The smart waste collection routing problem: alternative operational management approaches. Exp Syst Appl 103:146–158. https://doi.org/10.1016/j.eswa.2018.03.001
62. Rodríguez JP, Montoya-Munoz AI, Rodriguez-Pabon C, Hoyos J, Corrales JC (2021) IoT-agro: a smart farming system to Colombian coffee farms. Comput Electron Agric 190:106442. https://doi.org/10.1016/j.compag.2021.106442
63. Sanyal S, Mukhopadhyay S, Xhafa F (2023) Enhanced incentive schemes for volunteers to distribute surplus food in smart cities. Sustain Cities Soc 91:104409. https://doi.org/10.1016/j.scs.2023.104409
64. Sar K, Ghadimi P (2023) A systematic literature review of the vehicle routing problem in reverse logistics operations. Comput Ind Eng 177:109011. https://doi.org/10.1016/j.cie.2023.109011
65. Schmidt CE, Silva ACL, Darvish M, Coelho LC (2019) The time-dependent location-routing problem. Transp Res E Logist Transp Rev 128:293–315. https://doi.org/10.1016/j.tre.2019.06.015
66. Schmidt A, Kabir MWU, Hoque MT (2022) Machine learning based restaurant sales forecasting. Mach Learn Knowl Extract 4(1):105–130. https://doi.org/10.3390/make4010006
67. Strieder MM, Velásquez Piñas JA, Ampese LC, Costa JM, Carneiro TF, Rostagno MA (2023) Coffee biorefinery: the main trends associated with recovering valuable compounds from solid coffee residues. J Clean Product 415:137716. https://doi.org/10.1016/j.jclepro.2023.137716

68. Sun X (2022) A decision-support framework for smart and sustainable reverse logistics network design. The Arctic University of Norway, Norway (2022)
69. Tsai JJ, Chang CC, Huang DY, Lin TS, Chen YC (2023) Analysis and classification of coffee beans using single coffee bean mass spectrometry with machine learning strategy. Food Chem 426:136610. https://doi.org/10.1016/j.foodchem.2023.136610
70. Van Engeland J, Beliën J, De Boeck L, De Jaeger S (2020) Literature review: strategic network optimization models in waste reverse supply chains. Omega 91:102012. https://doi.org/10.1016/j.omega.2018.12.001
71. Van Keulen M, Kirchherr J (2021) The implementation of the circular economy: barriers and enablers in the coffee value chain. J Clean Product 281:125033. https://doi.org/10.1016/j.jclepro.2020.125033
72. Wang X, Wang Y, Tao F, Liu A (2021) New paradigm of data-driven smart customisation through digital twin. J Manufact Syst 58:270–280. https://doi.org/10.1016/j.jmsy.2020.07.023
73. Wen Z, Hu S, De Clercq D, Beck MB, Zhang H, Zhang H, Liu J (2018) Design, implementation, and evaluation of an Internet of Things (IoT) network system for restaurant food waste management. Waste Manage 73:26–38. https://doi.org/10.1016/j.wasman.2017.11.054
74. Workie E, Kumar V, Bhatnagar A, He Y, Dai Y, Wah Tong Y, Fu C (2023) Advancing the bioconversion process of food waste into methane: a systematic review. Waste Manage 156:187–197. https://doi.org/10.1016/j.wasman.2022.11.030
75. Wu H, Tao F, Qiao Q, Zhang M (2020) A chance-constrained vehicle routing problem for wet waste collection and transportation considering carbon emissions. Int J Environ Res Publ Health 17(2):458. https://doi.org/10.3390/ijerph17020458
76. Wu H, Tao F, Yang B (2020) Optimization of vehicle routing for waste collection and transportation. Int J Environ Res Publ Health 17(14):4963. https://doi.org/10.3390/ijerph17144963
77. Zhang L, Xing X, Liu Y, Shi W, Wang M (2023) Directional methanolysis of kitchen waste for the co-production of methyl levulinate and fatty acid methyl esters: catalytic strategy and machine learning modeling. Bioresour Technol 367:128274. https://doi.org/10.1016/j.biortech.2022.128274
78. Zohourfazeli H, Chaabane A, Jabbarzadeh A (2023) Restaurant waste management using Industry 4.0 technologies. United Arab Emirates
79. Zohourfazeli H, Chaabane A, Jabbarzadeh A (2024) In search of the optimal facility location and vehicle routing for spent coffee grounds waste management. Morocco
80. Zohourfazeli H, Sabaghpourfard A, Chaabane A, Jabbarzadeh A (2024) Circular coffee shops' smart approach in Montreal. IEEE, Malta
81. Zohourfazeli H, Sabaghpourfard A, Chaabane A, Jabbarzadeh A (2024) Coffee waste smart transformation: innovating sustainable business models, Canada

Optimizing Circular Economy: Simulation and Optimization in Stochastic Inventory Control of Perishable Dairy Products

Çağrı Sel and Tuğrul Bayraktar

Abstract Efficient management of pasteurized fluid milk inventory is a need in the dairy industry to maximize profit. Beyond this, decreasing waste is a sustainability goal in the dairy market. Fresh milk is highly perishable and should be sold before it reaches its shelf life. The existing literature has long focused on deterministic models; however, supply chain uncertainties, demand fluctuations, and shelf-life variability have attracted attention recently. Optimization techniques and simulation models have complementary strengths that make stochastic inventory control easier to manage. Our study presents a simulation-optimization approach that combines a simple heuristic and simulations to develop robust inventory management strategies for managing multiple perishable commodities with fixed shelf life. The proposed approach aims to aid decision-makers in the dairy industry as a tool to manage perishable goods and efficiently promote sustainability in dairy marketing. We discuss a case study to reveal managerial insights from which the dairy and other perishable product industries can benefit.

Keywords Perishable product · Waste management · Inventory control · Simulation

1 Introduction

Perishable products have a limited shelf life and require proper storage and handling to prevent spoilage. These products can be essential for many industries and consumer needs as they offer high quality, variety, and nutritional value. However, managing perishable products can be challenging due to inventory management, shipping, packaging, waste reduction, food safety, and environmental impact. To effectively manage perishable inventories, it is essential to understand their

Ç. Sel · T. Bayraktar (✉)
Karabuk University, Department of Industrial Engineering, Karabuk, Turkey
e-mail: cagrisel@karabuk.edu.tr; tugrulbayraktar@karabuk.edu.tr

B. Jarboui et al. (eds.), *Data-Driven Waste Revolution*, Computational Intelligence Methods and Applications, https://doi.org/10.1007/978-981-96-5890-9_4

characteristics, benefits, and drawbacks and implement best practices and strategies for proper handling and storage.

The dairy industry produces and processes dairy products such as milk, cheese, yoghurt, butter, and ice cream. One of the main challenges in the dairy industry is to control spoilage caused by microbial growth or chemical reactions. Pasteurization involves heating milk to a specific temperature to kill harmful bacteria and slow spoilage. However, it does not prevent spoilage entirely; even pasteurized milk must be refrigerated and consumed before its expiration date, usually 12–14 days. Factors such as exposure to light, high temperature, microbial contamination, and chemical reactions can cause spoilage that can negatively impact milk's taste, odor, texture, and nutritional value, reducing dairy products' quality and safety [7]. This can cause economic losses, health risks, and consumer dissatisfaction. If pasteurized fluid milk or its derivatives expire, they should be disposed of to avoid health risks and dissatisfaction or recycled to minimize economic loss.

It is essential to regularly monitor the stock replenishment and production levels of dairy products to minimize economic losses caused by waste. According to the Food and Agriculture Organization (FAO), approximately 20% of global milk production is lost or wasted throughout the supply chain [8]. This results in economic losses for producers and consumers while causing significant environmental impact through greenhouse gas emissions, water consumption, and land use. Over the past few decades, due to supply chain challenges and waste management issues, the dairy industry has had to adopt more effective techniques to optimize inventory parameters, such as order quantity and safety stock levels. The dairy industry widely uses mathematical modeling to determine the optimal order quantities, safety stock levels, replenishment policies, and inventory locations for dairy products. Mathematical modeling can also assist in assessing the trade-offs between inventory expenses and service levels and evaluating the impact of inventory decisions on other performance indicators, such as customer satisfaction, production efficiency, and environmental sustainability.

However, inventory policies can vary regarding reorder point, reorder quantity, safety stock, and lead time, depending on the level of demand uncertainty. For instance, high demand uncertainty may necessitate a higher safety stock to prevent stock-outs but may also raise storage costs. As a result, inventory policies should be tailored to match the demand uncertainty and optimize the total inventory cost. Demand uncertainties can have a significant impact on the performance and profitability of a supply chain. This can lead to wastage of resources, inventory obsolescence, stock-outs, lost sales, and customer dissatisfaction. The mathematical modeling approaches can help balance responsiveness and efficiency, aligning the supply chain with market conditions despite demand uncertainty. The mathematical modeling approaches can help balance responsiveness and efficiency and enable the supply chain to adapt to market conditions despite the uncertainty of demand. Deterministic modeling assumes that demand is known and constant, which allows for calculating the optimal inventory levels and order quantities based on cost parameters. On the other hand, stochastic modeling recognizes that demand is unpredictable and varies. Choosing appropriate modeling approaches for a supply chain depends on factors such as lead time, demand variability, cost structure, and customer expectations.

The dairy industry has significantly benefited from developing new inventory management concepts for dairy products. These gains are exemplified by numerous studies in the literature as follows:

1. Shelf-life integrated models can optimize production planning using setup times, minimum batch sizes, process capacity [11], product type, working time, and changeovers [46] to minimize the production cost and wastage while meeting demand.
2. The price of raw milk may impact the demand for direct sales and input costs within the sector. Analyzing how prices are transmitted vertically can help determine accurate prices by identifying long-term relationships between stakeholders [28]. Governmental regulations can also impact the milk price by placing either a minimum or maximum price, implementing taxes or subsidies, or enforcing quality standards [25]. Sharing information within an integrated dairy supply chain can reduce operational costs and increase efficiency when dealing with uncertainties [44].
3. Forging a well-designed supply chain network that assembles all parties involved can effectively reduce operational costs. The purpose of this supply chain should be overall economic benefits that can be achieved by optimizing the distance traveled [41], selecting the best locations for facilities [1], and adjusting the product quality parameters that impact customer satisfaction [2].
4. Managing return flows for dairy products can be achieved through a closed-loop supply chain (CLSC), which involves remanufacturing [59] or reusing wasted products [24] within the industry. Additionally, integrating buyback strategies into CLSC can also help to convert these wasted products into input material for other industries [33].
5. Inventory management in the dairy industry incorporates logistics issues such as environmental impact and uncertain environment, leading to reduced uncertainty and improved sustainability [31, 42, 47, 55, 58].
6. Sensitivity analysis is a helpful tool to evaluate the impact of changes in key parameters such as traffic conditions [51], demand variance, product holding cost, production rate, intermediate product holding cost, and shortage cost [48]. It helps to assess the robustness of the model, identify areas of uncertainty, and determine where additional data or research is needed.

The quantity of product that goes to waste is determined by the movements of stock caused by both the manufacturer and the customers. It is challenging for the dairy industry to manage dairy products with a short shelf life due to the uncertainty of customer purchase decisions, which can result in wastage. Decision-makers must carefully work operational actions related to pasteurized fluid milk in the dairy supply chain within a short time frame to minimize product loss. On the other hand, since there is a high possibility of pasteurized milk wastage, consumers should choose the product package size wisely to avoid leftovers. Manufacturers of pasteurized milk have the power to control the market by determining the levels of production and pricing. Manufacturers can increase customer satisfaction and profitability by producing milk in various sizes according to customer demand. This approach allows manufacturers to cater to their customer's specific needs and preferences instead of offering a standard size that may not be suitable for everyone. By doing so, manufacturers can reduce waste, lower costs, improve quality, and enhance their market reputation and competitiveness.

This research aims to focus on inventory management in the dairy supply chain, which has unique characteristics that distinguish it from other food supply chains: (1) Unlike many other food product inputs, pasteurized fluid milk is highly perishable. (2) The production ratio of pasteurized fluid milk with different packaging sizes affects the number of expired products that lead to milk waste. In terms of managing highly perishable inventory, it is necessary to optimize the production level for dairy products with a short shelf life and uncertain daily consumption.

This chapter presents a comprehensive analysis of the literature on inventory management strategies for pasteurized fluid milk and its derivatives, focusing on waste management. Additionally, it discusses a real-life case study of a typical inventory management problem for perishable products in the dairy industry. The goal is to determine the most efficient production ratio for pasteurized fluid milk in different packaging sizes, focusing on reducing inventory handling costs and waste. The problem is formulated and solved using a recent simulation-based optimization approach.

This chapter is organized as follows: Firstly, a brief taxonomy of the literature is presented. Then, the subsequent sections offer a state-of-the-art assessment of the inventory management methods for perishable dairy products to identify research directions toward minimizing wastage. Following this, a real-life stochastic inventory management problem of perishable dairy products with a fixed shelf life is presented as a case study. Lastly, the overall results are discussed, and conclusions are presented.

2 A Brief Taxonomy of the Literature on Inventory Management Practices in the Dairy Industry

The papers that have been reviewed were selected by conducting a Web of Science search with the keyword "dairy" along with the following keywords: perishable product, inventory management, shelf life, waste management, pricing, and buyback. A total of 54 research papers written in English were accessed, and a taxonomic analysis of these studies is briefly discussed from the perspective of waste management. Note that none of the simulation-based studies on inventory management for dairy products incorporates waste management strategies. In fact, none of the simulation-based studies accounts for any inventory management strategies in the dairy industry except for two. Mykrantz and Bozic [36] examine

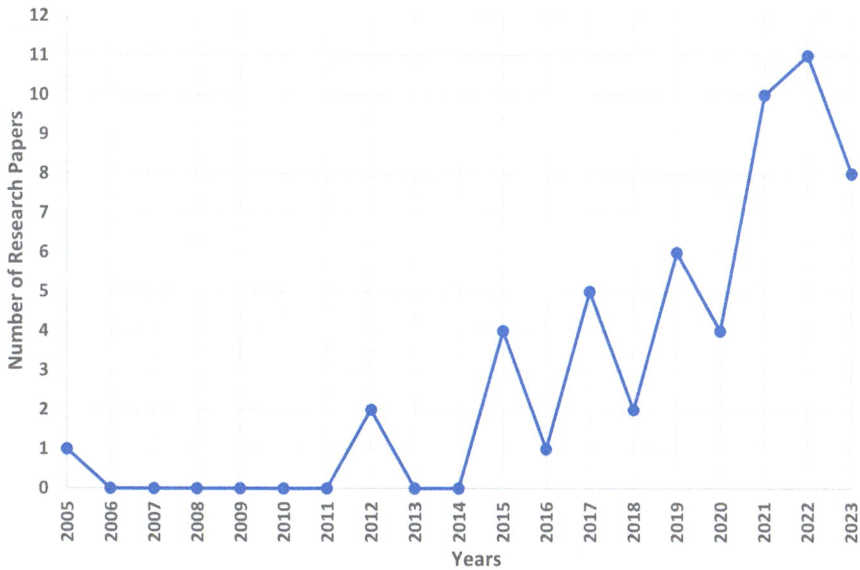

Fig. 1 Number of research papers per year (Source: Author)

a novel milk pricing method based on the concept of opportunity cost for milk used in cheese and whey to potentially improve the functioning of Federal Milk Marketing Orders (FMMOs) and the US dairy industry. Khalafi et al. [25] discuss a game-theoretic approach for analyzing financial intervention in supply chain management in the dairy industry. However, the shelf life of dairy products and waste management perspective are not taken into account in these studies. To provide a comprehensive understanding, the literature is analyzed and discussed from a broader viewpoint.

Figure 1 represents the number of research papers that are published each year. Table 1 presents the outlets that have shown interest in the topic. According to the number of published research papers, the Journal of Cleaner Production ranks first on the list with seven papers. Other journals, such as "Computers & Industrial Engineering" and "International Journal of Production Economics," are also pioneers in the field. These journals are commonly categorized in the following titles: "Computer/Management Science and Operations Research," "Industrial and Environmental Engineering," and "Green Sustainable Science Technology." The researchers in the fields of "Engineering," "Operations research management science," "Environmental science," and "Computer science" have shown relatively more interest on the topic (see Table 2).

As of February 2024, Table 3 displays the total number of citations received by the top 25 research papers and their impact on the related literature. The research papers are sorted in a descending order based on the number of citations they have received, i.e., Wang et al. [54], Soysal et al. [49, 50], Entrup et al. [11], and so forth.

Table 1 Journals publishing the research papers

Journals	Records	Research papers
Journal of Cleaner Production	7	Yavari and Zaker [59]; Yavari and Geraeli [58]; Tat et al. [52]; Sha et al. [47]; Jouzdani and Govindan [21]; Kayikci et al. [23]; Momeni et al. [33]
Computers and Industrial Engineering	3	Singh and Kuthambalayan (2022) [48]; Ahmadi-Javid et al. [2]; Herbon [19]
International Journal of Production Economics	3	Soysal et al. [49]; Pauls-Worm et al. [37]; Wang et al. [55];
Computers and Operations Research	2	Soysal et al. [50]; Mousavi and Mahdi [34]
Environmental Science and Pollution Research	2	Talouki et al. [51]; Abbasian et al. [1]
European Journal of Operational Research	2	Herbon [20]; Ganguly and Robb [15]
Heliyon	2	Yaghoubi and Akrami [57]; Roa et al. [42]
Omega	2	Wang and Li [54]; Santos et al. [45]
Rairo—Operations Research	2	Lu et al. [30]; Krommyda et al. [27]
Scientia Iranica	2	Marandi and Zegordi [32]; Gitinavard et al. (2019) [16]
Advanced Material Research	1	Fard and Yegane [12]
Agricultural Economics	1	Lajdova and Bielik [28]
Annals of Operations Research	1	Li, Teng and Chang [29]
Applied Mathematical Modelling	1	Grillo et al. [17]
Computers and Chemical Engineering	1	Sel et al. [46]
Environment Development and Sustainability	1	Khalafi et al. [25]
Expert Systems with Applications	1	Salas-Navarro et al. [44]
International Journal of Food Engineering	1	Rinaldi et al. [41]
International Journal of Logistics Management	1	Chaudhuri et al. [9]
International Journal of Management Science and Engineering Management	1	Khalafi et al. [24]
International Journal of Production Research	1	Entrup et al. [11]
Journal of Agribusiness in Developing and Emerging Economies	1	Routroy and Behera [43]
Journal of Agricultural and Applied Economies	1	Mykrantz and Bozic [36]

(continued)

Table 1 (continued)

Journals	Records	Research papers
Journal of Industrial and Management Optimization	1	Kaya and Bayer [22]
Journal of Industrial Information Integration	1	Foroozesh et al. [14]
Journal of Manufacturing Systems	1	Polotski et al. [40]
Journal of Modelling in Management	1	Fattahi and Tanhatalab [13]
Journal of the Operational Research Society	1	Chen and Chen [10]
Kybernetes	1	Khazaeli et al. [26]
Management Systems in Production Engineering	1	Alrawabdeh [5]
Mathematics	1	Ali et al. [4]
Procedia Manufacturing	1	Transchel and Hansen [53]
Sustainable Production and Consumption	1	Aka and Akyuz [3]
The International Journal of Advanced Manufacturing Technology	1	Polotski et al. [39]
Transportation Journal	1	Majidi et al. [31]
Transportation Research Part C—Emerging Technologies	1	Wu et al. [56]
Waste Management	1	Muriana [35]

Source: Author

The list of keywords used in the research papers is presented in Table 4. Performing such an analysis enables us to identify the most important keywords related to the topic. The study of Wang et al. [54], which is the most cited research in this field, addresses food waste and improving profit margins of grocery retailers by offering an optimization model for pricing and ordering decisions. This research focuses on setting prices for perishable products by considering quality losses that occur during the supply chain process. The aim is to reduce prices to stimulate demand while considering the trade-off between sales revenue and associated costs. Sensitivity analysis was conducted to examine the effect of product quality value and timing on model performance.

Various methods are proposed in the literature to enhance inventory management for perishable products. These methods include considering service levels and customer satisfaction for highly perishable goods, taking inventory and waste costs into account (e.g., Soysal et al. [49, 50]), creating a sustainable environment for dairy supply (Yavari and Geraeli [58]; Jouzdani and Govindan [21]) and providing managerial insights for dairy supply chain operations (Sel et al. [46]).

Table 2 Research areas in the field

Research areas	Records
Operations Research Management Science	18
Industrial Engineering	12
Environmental Sciences	11
Green Sustainable Science Technology	9
Environmental Engineering	8
Management	8
Computer Science Interdisciplinary Applications	7
Manufacturing Engineering	7
Multidisciplinary Engineering	4
Agricultural Economics Policy	3
Computer Science Artificial Intelligence	2
Economics	2
Electrical Electronic Engineering	2
Mathematics Interdisciplinary Applications	2
Multidisciplinary Sciences	2
Automation Control Systems	1
Computer Science Cybernetics	1
Computer Science Information Systems	1
Chemical Engineering	1
Environmental Studies	1
Food Science Technology	1
Multidisciplinary Materials Science	1
Mathematics	1
Mechanics	1
Nanoscience Nanotechnology	1
Transportation	1
Transportation Science Technology	1

Source: Author

The increasing academic interest in recent years highlights the importance of sustainability and waste management in the dairy industry, particularly in relation to perishable products. From our keyword analyses, it can be observed that a few of the papers in the field address a buyback strategy, which demonstrates a possible gap in the literature.

3 A State-of-the-Art Assessment of the Literature

The scarcity of natural resources has made it imperative for organizations to prioritize waste management in the supply chain, especially during the production phase. Managing the supply chain can be difficult due to factors such as maintaining service levels, keeping customers satisfied, and dealing with demand uncertainties.

Table 3 Top 25 most cited research papers

No.	Authors	Total citations
1	Wang and Li [54]	246
2	Soysal et al. [50]	176
3	Soysal et al. [49]	153
4	Entrup et al. [11]	128
5	Jouzdani and Govindan [21]	126
6	Routroy and Behera [43]	94
7	Herbon and Khimelnitsky [20]	91
8	Yavari and Geraeli [58]	85
9	Chaudhuri et al. [9]	80
10	Yavari and Zaker [59]	68
11	Sel et al. [46]	58
12	Pauls-Worm et al. [37]	32
13	Wu et al. [56]	30
14	Grillo et al. [17]	29
15	Ali et al. [4]	27
16	Tat et al. [52]	20
17	Marandi and Zegordi [32]	19
18	Sha et al. [47]	17
19	Muriana [35]	16
20	Abbasian et al.[1]	14
21	Kayikci et al. [23]	13
22	Khalafi et al. [24]	12
23	Wang et al. [55]	12
24	Yaghoubi and Akrami [57]	11
25	Mousavi et al. [34]	10

Source: Author

However, it can be even more challenging to determine the appropriate supply level for perishable products, as the shelf life of the products plays a vital role in shaping the structure of the entire supply chain. For example, in the dairy industry, ultra-high temperature (UHT) milk has a shelf life of up to 6–9 months, while pasteurized milk expires in 12–14 days. Variations in product shelf life affect supply chain decision variables such as reorder point, order quantity, and lead time. As shelf life shortens, demand uncertainty increases, making it difficult for decision-makers to balance supply and demand by determining accurate variable values.

Numerous research papers prioritize lot-sizing and scheduling decisions without considering perishability. However, Entrup et al. [11] can be regarded as one of the first studies considering perishability. To address the shelf-life integrated planning and scheduling problems in yogurt production, they have developed several mixed-integer linear programming (MILP) models. Since the product starts deteriorating right after being produced, placing it on shelves as soon as possible is essential. The models aim to find the optimal batch size for the fermentation process, reducing the setup times and increasing the availability of fresh products on the shelf. These

Table 4 The keywords used in the research papers

Research paper	Keywords
Abbasian et al. [1]	Food supply chain; location inventory routing (LIR); sustainability; resiliency; perishable products
Ahmadi-Javid et al. [2]	Distribution; perishable products; food quality degradation; temperature control; energy management; vehicle routing problem with time windows (VRPTW); general variable neighborhood search (GVNS)
Aka and Akyuz [3]	Inventory and production model; food supply chains; food waste; fuzzy parameters; perishability
Ali et al. [4]	Supply chain network; inventory; product's decay and deterioration; fuzzy set; nonlinear multi-objective optimization
Alrawabdeh [5]	Probabilistic inventory; demand; age discrimination; newsvendor; supply chain
Chaudhuri et al. [9]	Asia; Data analytics; literature review; logistics; cold chain; ICT infrastructure; content analysis based
Chen and Chen [10]	Markdown; perishable food; quality loss function; shelf life; supply chain inventory management
Entrup et al. [11]	Fresh food production; mixed-linear integer programming; perishables; shelf life; scheduling; yoghurt
Fard and Yegane [12]	Smart packaging; deteriorating items; supply chain
Fattahi and Tanhatalab [13]	Optimization; inventory management; mathematical programming; inventory routing problem; IRP; Lagrangian relaxation; transshipment; chance-constrained programming
Foroozesh et al. [14]	Sustainable-resilient supply chain network; design; circular economy; interval-valued fuzzy sets; compromise decision-making method; interval-valued fuzzy-robust programming; AUGMECON2 method
Ganguly and Robb [15]	Inventory; food waste; households; modeling
Gitinavard et al. [16]	Multi-echelon supply chain; Pareto-optimal solution; perishable products; group decision analysis; possibilistic approach
Grillo et al. [17]	Order-promising process; available to promise; lack of homogeneity in the product; fruit supply chain; perishability
Herbon [20]	Inventory; dynamic pricing; differential pricing policy; additive demand; perishable items
Herbon [19]	Expiring products; shelf-life extension; perishable inventory; ordering and pricing policy
Jouzdani and Govindan [21]	Supply chain network design; sustainability; perishable food; goal programming; chance-constrained programming; dairy product
Kaya and Bayer [22]	Dynamic pricing; inventory control; perishable products; age-dependent demand; dynamic programming
Kayikci [23]	Dynamic pricing; data analytics; food supply chain; waste management; perishable products; IoT sensor
Khalafi et al. [24]	Green supply chain; perishable goods industries; revenue-sharing contract; remanufacturing; fuzzy mathematical programming
Khalafi et al. [25]	Game theory; perishable product; fresh green supply chain; policymaker financial intervention; freshness-keeping effort
Khazaeli et al. [26]	Quality perception; perishable agricultural products; Arrhenius equation; integration; cold supply chain; case study

(continued)

Table 4 (continued)

Research paper	Keywords
Krommyda et al. [27]	Inventory; EOQ; expiration date; food donation
Lajdová and Bielik [28]	Asymmetry; co-integration; milk; price transmission; VECM
Li et al. [29]	Supply chain; price-and-age-dependent demand; advance-cash-credit payments; discounted cash-flow analysis
Lu et al. [30]	Agricultural products; time-varying deterioration rate; dynamic pricing; preservation technology investment; periodic ordering
Majidi et al. [31]	Sustainable supply chain; production routing; workforce planning; pricing; perishable products
Marandi and Zegordi [32]	Production and distribution; permutation flow shop scheduling; vehicle routing problem; integration; mixed-integer programming; particle swarm optimization
Momeni et al. [33]	Circular supply chain management; regenerated expired products; coordination mechanism; buyback contract; cooperative investment
Mousavi et al. [34]	Production routing problem; pandemic; perishable product; freshness; matheuristic algorithm
Muriana [35]	Food losses recovery; waste management; supply chain optimization; food donation; stochastic shelf life
Mykrantz and Bozic [36]	Butter; cheese; dry whey; milk; nonfat dry milk; opportunity cost; price; protein
Pauls-Worm et al. [37]	Production-inventory planning; long lead time; perishable product; nonstationary stochastic demand; fill rate
Polotski et al. [39]	Manufacturing systems; perishable products; limited shelf life; policy optimization; stochastic processes; simulations
Polotski et al. [40]	Manufacturing systems; perishable products; periodic demand; production policy adjustment; hedging curve
Rinaldi et al. [41]	Dairy transportation; food; nearest neighbor algorithm; practical tool; vehicle routing problem
Roa et al. [42]	Sustainability; decision support system (DSS); supply chain; FePIA
Routroy and Behera [43]	Agriculture; agricultural products; supply chain management
Salas-Navarro et al. [44]	Vendor managed inventory; supply chain; deteriorating item; probabilistic demand
Santos et al. [45]	Minimum life on receipt; two-echelon; food waste; freshness
Sel et al. [46]	Integrated planning and scheduling; mixed-integer linear programming; constraint programming; hybrid approach; perishability; dairy supply chain
Sha et al. [47]	Sustainable supply chain; dairy products; robust optimization; heuristic algorithm
Singh and Kuthambalayan [48]	Marketing/operations interface; perishables; shelf life; stochastic 0–1 integer nonlinear programming; Benders' decomposition
Soysal et al. [50]	Inventory routing; horizontal collaboration; energy consumption; perishability; demand uncertainty
Talouki et al. [51]	Robust optimization; dynamic green vehicle routing problem (DGVRP); augmented ϵ—constraint; perishable products; green traffic

(continued)

Table 4 (continued)

Research paper	Keywords
Tat et al. [52]	Pharmaceutical supply chain; coordination; pharmaceutical waste; buyback and shortage risk-sharing contract; governmental penalty
Transchel and Hansen [53]	Perishable product inventory control; random lead-time; service-level constraints
Wang et al. [54]	Product tracing and tracking; quality deterioration; pricing; perishable food; shelf life; waste reduction
Wang et al. [55]	Sustainability; food quality; graphic evaluation and review technique; Bayesian approach; multi-objective optimization
Wu et al. [56]	Catering service; high-speed rail; public railway plan; location routing model; Cross-entropy algorithm; time-oriented neighbor heuristic
Yaghoubi and Akrami [57]	Systems engineering; industrial engineering; mathematical modeling; computational intelligence; process modeling; industry management; supply chain management; ant colony optimization algorithm; particles swarm optimization algorithm; perishable raw material
Yavari and Geraeli [58]	Green closed-loop supply chain; heuristic algorithm; robust optimization; perishable goods
Yavari and Zaker [59]	Interdependent two layer; resilient green-closed-loop supply chain; location inventory; perishable products; disruption

Source: Author

models consider various constraints and criteria, such as inventory limits, capacity constraints, demand fulfillment, and minimum fermentation batch sizes. Krommyda et al. [27] proposed a generic model for all perishable products instead of focusing on a particular one, as Ertrup et al. did previously. They expanded the model by using dynamic decision stages to include optimal order quantity and reorder point considering demand, production capacity, and expiration date constraints. Dynamic programming enables them to derive optimal decisions at each stage for obtaining the optimum values to maximize profit. Santos et al. [45] expand the standard perishable product-integrated lot-sizing and scheduling problem by using a flexible multiparty logistics network design strategy with minimum life of receipt (MLOR).

The perishability of products has a significant impact on the supply chain. It can limit the lead time, order quantity, reorder point, and production level. Decision-makers must balance the inventory and stock-out costs by determining the correct amount of perishable products under given limitations. Nevertheless, controlling inventory flawlessly in the real world can be challenging due to demand uncertainties, especially for perishable products. Decision-makers cannot manage stock-out costs after all products have been sold. However, they can use various strategies to attract customers' attention and avoid struggling to sell perishable products before expiration. While strategies such as inventory level-dependent pricing and buyback programs can be implemented on the market side, on the production side, it is essential to forecast market demand to determine the production level of each product to avoid overstocking.

Many researchers have emphasized the importance of pricing strategies to reduce waste for perishable goods. Although Entrup et al. [11] highlighted the importance of prizing strategy, Wang et al. [54] provide a more in-depth analysis of this topic by using a deterministic model to reduce food waste while improving profit margins of grocery retailers. The proposed solution approach suggests pricing perishable products based on the quality loss that occurs during supply chain processes. This strategy aims to reduce the price and stimulate the demand to generate more revenue from the release while considering the trade-off between sales revenue and the costs involved. Lu et al. [30] investigate the problem of determining preservation technology investment and the periodic ordering strategy besides the optimal dynamic pricing policy for agricultural products with limited shelf life with a time-varying deterioration rate. Chen and Chen [10] focused on determining the optimal time for providing promotional discounts on perishable food products. They have proposed a mathematical model that helps identify the most suitable markdown policies within supply chain settings for such products. The model considers several factors, including the expiration date of the product, the markdown rate, and the costs related to processing unsold products.

Recent studies indicate a growing interest in dairy products due to their perishable nature. Ali et al. [4] highlight the issue of inventory losses and higher costs related to the supply chain complexity of dairy products. The research focuses on a three-period scenario where the demand is known, and the authors propose a deterministic fuzzy multi-objective model to minimize the organization's cost and the total cost of deterioration. The model considers the procurement and distribution of all dairy products during the planning horizon to reduce costs. Khalafi et al. [24] conducted a study to explore the revaluation of expired perishable dairy products. They also used a deterministic fuzzy multi-objective model to consider several factors, including product return rates, expiration dates, collection and recycling costs, and product value after recycling. This study is one of the few that focuses specifically on recycling expired perishable dairy products. The model can be used by managers and decision-makers to effectively reuse perishable dairy products near expiration dates. By doing so, they can reduce disregarded costs and promote sustainable closed-loop supply chains (SCLSC). Momeni et al. [33] have introduced a novel coordination mechanism for buyback contracts between manufacturers and retailers. This mechanism enables retailers to return expired products to the manufacturer for revaluation at a predetermined buyback price. The researchers analyzed the impact of various supply chain parameters, such as holding and disposal costs, to provide insights into the operational and financial performance of the supply chain. They formulated an optimization model based on game theory to determine the best pricing and ordering policies for each supply chain member under different scenarios.

Jouzdani and Govindan [21] and Abbsian et al. [1] focus on the structure of the dairy supply chain in terms of economic, social, and environmental concerns rather than solely product revaluation. Jouzdani and Govindan have proposed a model that employs linearization techniques to prepare it for the multi-choice goal programming (MCGP) method. The MCGP model involves multiplying binary variables, making it a nonlinear model that can be expressed linearly. Experts' opinions determine the aspiration level of each objective, including economic, social, and environmental concerns. The practical advantage of this method is that it provides a single preferable solution rather than a set of Pareto-optimal solutions that may confuse decision-makers. The goal programming (GP) model includes system and goal constraints. The latter is added to minimize deviations from the aspiration levels to obtain the solution. The MCGP approach allows decision-makers to consider multiple target values for each goal. Abbasian et al. have proposed a dynamic multi-objective model for designing dairy supply chain networks. The approach involves selecting optimal locations for facilities, assigning retailers to these facilities, and developing a plan for vehicle distribution over multiple periods. The primary objective of this approach is to minimize the supply chain's total cost and environmental impact.

When a product has a fixed shelf life and a known demand, inventory management problems can be solved efficiently using deterministic methods. However, in situations where the shelf life of the same product varies, such as in the grocery market, deterministic models may not be able to deal with the complexity of the problem. Lu et al. [30], Chen and Chen [10], and Jouzdani and Govindan [21] dealt with perishable products with varying deterioration rates besides the one with a fixed shelf life. Lu et al. [30] and Chen and Chen [10] have proposed a problem-solving approach that involves using iterative algorithms to optimize input parameters to handle the complexity of the problem. On the other hand, Jouzdani and Govindan [21] used the chance-constrained concept, along with a linearization approach, to determine the varying shelf life and optimize the problem deterministically.

Aka and Akyuz [3] reduced the uncertainty of parameters such as demand, costs, and deterioration rate by fuzzification to simplify the problem for deterministic solutions. They use the concepts of "expected value" and "expected interval" of a fuzzy number for operations of ranking fuzzy numbers, and Pishvaee and Torabi's [38] approach was used for the defuzzification process. Additionally, the study considers some parameters as fuzzy in the second stage and reveals different solution alternatives in this way. Ahmadi-Javid et al. [2] and Khazaeli et al. [26] offer insights for optimizing quality in complex food supply chains. Ahmadi-Javid et al. [2] proposed a hybrid heuristic algorithm that combines a guided variable neighborhood search (GVNS) and a dynamic tabu search (DTS) aiming for cost minimization for perishable food products, including milk and its derivatives. On the other hand, Khazaeli et al. [26] proposed a mixed-integer nonlinear programming (MNLP) that uses the Arrhenius equation to calculate the approximate freshness level of perishable products. Using the Arrhenius equation, they simplify the problem with a deterministic model that aims to maximize profits for perishable products with a short shelf life.

The problem of managing inventory becomes more complex when demand is uncertain and decision-making for pricing strategies is increasingly complicated. Deterministic models may not be the best approach to optimize decision variables when dealing with a complex supply chain with uncertain demand for perishable products with short shelf life. In such cases, stochastic models and simulation models can be used to solve inventory management problems effectively. Muriana [35] conducted a study to determine the most effective way to recover food and reduce waste in the supply chain while maximizing profit. The study used an exact mathematical formulation and simulation-based optimization to identify the optimal values for factors significantly impacting food recovery. A sensitivity analysis was also conducted to test the optimal solution's robustness under different demand scenarios. Herbon [19] also developed an exact mathematical formula using a nested formulation to optimize subproblems by trying different input parameter values such as replenishment quantity, price, extension duration ratio, and cycle length. The formula finds the optimum inventory policy and pricing strategy for a perishable product with demand that depends on both time and price. The product is subject to a perishability constraint and a time-to-sell constraint. The primary objective is to maximize expected profit per unit of time by determining the optimal selling price and the optimal time to begin selling the product. The formula considers inventory holding cost, salvage value, cost of price change, and the demand that depends on the price.

Stochastic modeling is a technique that can help determine the likelihood of uncertain variables, such as demand in a supply chain, occurring. It provides a way to calculate the probability of different outcomes, even when a degree of randomness or unpredictability is involved. Li et al. [29] and Kaya and Bayer [22] used stochastic dynamic programming to predict the most appropriate timing for a change in the price and quantity of the orders. Li et al. [29] developed a two-stage dynamic programming model to optimize the supply chain of perishable products. In the first stage, the model computes the optimal order quantity at each stage of the supply chain. In the second stage, it determines the optimal pricing decision that maximizes the total expected profit. The model is based on a stochastic demand function that depends on the product's price. It considers various factors such as inventory holding costs, stock-out costs, salvage value, perishability, and demand variability and uses them to determine optimal pricing and inventory policies. Additionally, the model can be extended to incorporate other decision variables like transportation, allocation, and coordination, to analyze their impact on the optimal pricing and inventory policy. Kaya and Bayer have developed a model to manage the inventory replenishment in perishable products by considering stochastic age- and price-dependent demand function in a continuously reviewed inventory system. This approach focuses on balancing the decreased demand rate over time and the cost of wasting the current stock with the cost of ordering a new batch. This trade-off was not present in models where demand is independent of the age of the products. Stochastic dynamic programming with an approximate value iteration determines the optimal price and inventory replenishment policies.

By using stochastic models, researchers and analysts can gain insights into the potential risks and opportunities associated with different scenarios and make more informed decisions based on this information. Robust optimization may be preferred over stochastic models due to its easier applicability in uncertain environments. It is interesting to note that robust optimization is widely used in studies related to perishable dairy products. Gitinavard et al. [16] used robust optimization to minimize the total transportation costs and total time-window costs for dairy product customers and distribution centers. Yavari and Geraeli [58], Sha et al. [47], Talouki et al. [51], and Roa et al. [42] also employed this technique to balance the trade-off between economic, social, and environmental impacts of selected dairy supply chain variables. Meanwhile, Foroozesh et al. [14] used robust optimization with fuzzy set interval values to enhance the efficiency and effectiveness of decision-making processes in various sectors. It is worth noting that certain studies, namely Gitinavard et al. [16], Yavari and Geraeli [58], Sha et al. [47], and Talouki et al. [51], have assumed that the shelf life of dairy products is fixed. However, Foroozesh et al. [14] and Roa et al. [42] have not mentioned any expiration date in their solution approach. Furthermore, Roa et al. [42] have only taken the pricing policy into account as a decision variable for a sustainable dairy supply chain. Future research may explore robust optimization integrated dairy inventory management, taking into account varying shelf-life and pricing policies in uncertain environments.

4 Solving a Real Problem with Quantitative Techniques: A Case Study

Sustainable milk production requires a balance between economic benefit, public health, and environmental impact. The primary goal of dairy milk producers and sellers is to minimize inventory costs to maximize profits. However, daily milk should be consumed within approximately two weeks since it is a highly perishable product. This emphasizes the necessity of determining the most appropriate inventory level within a short time frame. On the other hand, products that have reached the end of their shelf life and cannot be sold are not recyclable and are disposed of, as stated in the case study. This results in the wastage of the scarce natural resources used in production.

The practical application of theoretical knowledge is just as important as the theoretical developments themselves. Therefore, we provide an example of how a common real-world issue faced by company decision-makers can be addressed using methods given in the literature. For this purpose, we have selected a stochastic inventory control problem that can be encountered in the dairy industry operating in urban areas. Facilities that produce daily milk distribute their products according to orders received from demand points. In the case study we are examining, we assume that a facility produces and locally distributes to a population of approximately 70,000. The facility makes weekly deliveries to groceries, markets, and supermarkets.

The main goal is to minimize the wastage of pasteurized milk while minimizing the stock-out cost. We converted annual sales into weekly demand by considering seasonal effects. The weekly inventory cost is calculated by considering the estimated weekly consumption and the actual daily consumption. The demand level of three parts of the dairy market differs with the customer base.

4.1 Problem Description

The inventory manager might use the stochastic dynamic program (SDP) to solve the problem defined above optimally. However, it is well-documented that adopting SDP as a solution approach is not practical in most cases due to its relatively complex structure [6]. Therefore, our aim here is to introduce an easy-to-implement solution approach based on the simulation of the inventory system. To this end, we propose an inventory simulation under which the objective function value is computed by running different input stochastic variables. Algorithm 1 aims to determine the optimal production quantities of pasteurized milk using the Monte Carlo simulation technique, taking into account the daily consumption habits of pasteurized milk customers. The simulation study considers different production rates as system parameters, where the costs of pasteurized milk, while on the shelf and the sales profit, dependent on the sales price, are crucial for the sustainability of milk production. Producing pasteurized milk in the correct quantities can contribute to sustainability by avoiding unnecessary production. The study applies a simulation-based optimization approach to determine the best production rate for two volumes (1 and 2 liters) of pasteurized milk products.

The sale of pasteurized milk is prioritized based on the FIFO (first in, first out) rule, ensuring freshness and quality. The amount of stock at the end of period t depends on the stock from the previous period, the actual demand D_t, and the delivery of the ordered quantity Q_t. The notation $(\cdot)^+ = max(\cdot, 0)$ shortens the expressions. When given the state $\mathbf{I}_{t-1}$ and the action Q_t, the system's state transition is described by the following expressions. At the end of period t, waste, denoting items aged A, is defined as

$$W_t = \left(I_{t-1}^{A-1} - D_t\right), \qquad t = 1, \ldots, T. \tag{1}$$

Inventory of age $a \in \{2, \ldots, A-1\}$ that can still be used in the next period:

$$I_t^a = \left(I_{t-1}^{A-1} - \left(D_t - \sum_{j=a}^{A-1} I_{t-1}^j\right)\right)^+, \qquad t = 1, \ldots, T; a = 2, \ldots, A-1. \tag{2}$$

And the inventory of fresh products:

$$I_t^1 = Q_t - \left(D_t - \sum_{a=1}^{A-1} I_{t-1}^a\right)^+, \qquad t = 1, \ldots, T. \tag{3}$$

This study utilizes Monte Carlo simulation to estimate $E(\mathbf{I}_t)$ in a replenishment cycle. With this sample-based estimation, we can calculate the anticipated total cost of replenishment cycles and create a simulation-based expansion of Silver's heuristic. The heuristic obtained is not dependent on the specific demand distribution being examined. Specifically, for a cycle (t, r), the method selects N independent demand samples for each period $k = t, ..., r$. It then computes sample average estimates $\widehat{E}(\mathbf{I}_k)$ of $E(\mathbf{I}_k)$, for $k = t, \ldots, r$, based on the order quantity Q and the initial inventory $\mathbf{I}_{t-1}$, using the inventory dynamics outlined in Eqs. 1–3. These estimates $E(\mathbf{I}_k)$ can be directly converted into a Monte Carlo estimation $\widehat{C}^N_{t,r}(\mathbf{I}_{t-1}, Q)$ of the expected total cost for the cycle (t, r) [18].

$$\widehat{C}^N_{t,r}(\mathbf{I}_{t-1}, Q) = \sum_{k=t}^{r}\left(h\widehat{E}\left(I_k^1\right)^+ + h\sum_{a=2}^{A-1}\widehat{E}\left(I_k^a\right) + p\widehat{E}\left(-I_k^1\right)^+ + w\widehat{E}\left(I_k^A\right)\right). \tag{4}$$

This cost estimate is used in a simulation-based inventory optimization algorithm (Algorithm 1) to compute the optimal production quantities Q^* for 1- and 2-liter pasteurized milk.

Algorithm 1 The simulation-based inventory optimization

Require: Problem parameters, $\mathbf{t}$, $\mathbf{I_{t-1}}$
Ensure: $Q^* \geq 0$

$r \leftarrow t$;
$c^* \leftarrow \infty$;
while $c \leq c^*$ *and* $r < t + A$ **do**
    $Q \leftarrow argmin_{Q \geq 0}\left[\widehat{C}^N_{t,r}(\mathbf{I}_{t-1}, Q)\right]$
    $c \leftarrow \left(\widehat{C}^N_{t,r}(\mathbf{I}_{t-1}, Q)\right)/(r - t - 1)$
    if $c \leq c^*$ **then**
        $c^* \leftarrow c$
        $Q^* \leftarrow Q$
    end if
end while
$r \leftarrow r + 1$

4.2 *Case Study*

The planning horizon consists of 52 periods, each corresponding to one week. The business we are considering produces and supplies milk, sourcing raw milk and producing pasteurized milk. The numerical analysis considers supply and production time negligible and zero. The business is capable of producing sufficient quantities. The shelf life of the pasteurized milk is two weeks. Based on the literature, the daily milk consumption per household is assumed to follow a normal distribution with an average of 0.011 liters and a standard deviation of 0.007 liters. The price of a one-liter carton of pasteurized milk is set at \$1.06, while the price of a two-liter carton is \$2.08. It is assumed that the space occupied by a two-liter carton of pasteurized milk is twice that of a one-liter carton, and the holding cost will vary

Table 5 Profit and amount of wasted milk in one-liter and two-liter carton pasteurized milk production rates

Production rates (1–2 L)	Total profit	Amount of wasted milk
0.90–0.10	115283.16 ± 2035.94	5739.73 ± 1184.38
0.85–0.15	**115879.39 ± 2482.74**	**5593.00 ± 1463.56**
0.80–0.20	114832.86 ± 2557.47	5969.23 ± 1355.40
0.75–0.25	114777.12 ± 3158.47	5998.33 ± 1754.97
0.70–0.30	114963.40 ± 2961.99	5870.07 ± 1572.89
0.65–0.35	114451.02 ± 2411.87	5981.97 ± 1479.26
0.60–0.40	114342.79 ± 2453.26	6248.67 ± 1288.66
0.55–0.45	114591.12 ± 2344.00	5902.47 ± 1290.55
0.50–0.50	113459.72 ± 2711.20	6477.87 ± 1458.09
0.45–0.55	113666.45 ± 2606.28	6431.43 ± 1341.05
0.40–0.60	114261.50 ± 2417.32	6102.33 ± 1440.37
0.35–0.65	113621.86 ± 2590.78	6359.93 ± 1433.50
0.30–0.70	113123.62 ± 2623.55	6692.47 ± 1502.75
0.25–0.75	113338.37 ± 2595.48	6488.60 ± 1444.65
0.20–0.80	112382.44 ± 2375.80	6937.33 ± 1371.75
0.15–0.85	112775.08 ± 3669.34	6670.67 ± 2153.99
0.10–0.90	113109.03 ± 2247.73	6555.40 ± 1280.77

accordingly, with the holding cost per one-liter carton assumed to be an average of $0.006.

Using this data, a simulation study of different production rate scenarios was conducted based on the weekly demand of a settlement with a population of 70,000 (approximately 17,500 households) and the projected annual demand. The results of this simulation are presented in Table 5. Pasteurized milk is assumed to be distributed weekly due to its short shelf life. Given these costs, an easy-to-use solution is proposed for determining businesses' supply or production plans.

In Table 5, the total profit is presented in current US dollars, while the amount of unsold milk indicates the total volume of one-liter and two-liter cartons of pasteurized milk that expire before being sold. The table shows that when a production ratio of 85% one-liter and 15% two-liter cartons is selected, the highest profit is achieved ($p < .05$), and the least number of milk cartons reach their expiration and become waste.

In the second phase, when production ratios of one-liter cartons are assigned values within the 80–90% range, the results shown in Table 6 are obtained. When examining Table 6, it is observed that the highest profit and the lowest amount of wasted milk are achieved when 87% of production is allocated to one-liter cartons and 13% to two-liter cartons of pasteurized milk. Based on the results of the simulation study, it can be concluded that this 87–13% ratio represents the most sustainable pasteurized milk production scenario ($p < .05$).

Table 6 Profit and amount of wasted milk for 80–90% production rate range of one-liter pasteurized milk

Production rates (1–2 Lt.)	Total profit	Amount of wasted milk
0.89–0.11	114795.37 ± 2633.76	5856.53 ± 1468.24
0.88–0.12	115073.25 ± 2156.26	5866.20 ± 1255.32
0.87–0.13	**115627.76 ± 2348.59**	**5720.07 ± 1409.80**
0.86–0.14	114538.80 ± 2724.58	6137.63 ± 1496.63
0.85–0.15	114574.42 ± 3106.76	6264.80 ± 1652.84
0.84–0.16	114733.50 ± 2683.38	5891.83 ± 1433.64
0.83–0.17	114418.17 ± 3290.18	6385.43 ± 1692.39
0.82–0.18	113903.29 ± 2166.94	6170.83 ± 1111.03
0.81–0.19	115053.80 ± 2605.31	5834.73 ± 1421.68

5 Conclusion

Managing perishable products, especially in the dairy industry, is important for maintaining a balance between profitability and sustainability. Academic literature suggests that deterministic and stochastic models are fundamental for optimizing inventory and pricing decisions in uncertain environments. Deterministic approaches are beneficial when demand and shelf life are predictable, while stochastic models, dynamic programming, and robust optimization are more suitable for handling complex supply chains with variable demand and perishability. Recent studies indicate a growing interest in integrating sustainability considerations, such as waste reduction and the promotion of closed-loop supply chains, into inventory management practices. This research highlights the importance of optimizing inventory management within the dairy supply chain, particularly in the context of pasteurized milk. Finding a middle ground between minimizing waste and maximizing profit is important for ensuring sustainable operations.

This study shows that adjusting production ratios for one-liter and two-liter milk cartons can significantly impact profitability and waste reduction. The results suggest that a production ratio of 87% for one-liter cartons and 13% for two-liter cartons results in the highest profit and the least waste, making it the most sustainable choice for milk producers. This ratio aligns with the dairy industry's economic efficiency and sustainability objectives. The study's use of stochastic inventory control methods demonstrates how businesses can consider demand uncertainties and other factors affecting production and distribution.

Given the growing global demand for sustainable practices, particularly in food supply chains, these optimization techniques will become even more essential. The research offers valuable insights into tailoring inventory policies to specific demand scenarios, enabling decision-makers in the dairy industry to make well-informed choices that support economic and environmental goals.

"Enhancing these models in future research could involve the incorporation of more complex variables, such as consumer behavior, seasonal fluctuations, and government policies. Additionally, broadening the scope to encompass other perishable products could enhance the applicability of these findings across different sectors of the food industry. Subsequent studies could benefit from further exploration of robust optimization techniques that incorporate different shelf-life and pricing policies, thus ensuring a more adaptable and efficient approach to inventory management in uncertain environments."

Ultimately, sustainable milk production necessitates continual adjustments to production and inventory strategies to maintain the delicate balance between supply, demand, and perishability. This study establishes a basis for businesses aiming to optimize their operations and decrease the environmental impact of wasted dairy products.

References

1. Abbasian M, Sazvar Z, Mohammadisiahroudi M (2023) A hybrid optimization method to design a sustainable resilient supply chain in a perishable food industry. Environ Sci Pollut Res 30:6080–6103
2. Ahmadi-Javid A, Mansourfar M, Lee C, Liu L (2023) Optimal distribution of perishable foods with storage temperature control and quality requirements: an integrated vehicle routing problem. Comput Ind Eng 182:109215
3. Aka S, Akyüz G (2021) An inventory and production model with fuzzy parameters for the food sector. Sustain Product Consum 26:627–637
4. Ali S, Barman H, Kaur R, Tomaskova H, Roy S (2021) Multi-product multi echelon measurements of perishable supply chain: fuzzy non-linear programming approach. Mathematics 9:2093
5. Alrawabdeh W (2021) Multi-period age-discriminated perishable inventory. Manage Syst Product Eng 29:97–105
6. Bertsimas D, Thiele A (2006) A robust optimization approach to inventory theory. Oper Res 54:150–168
7. Bylund G (2015) Tetra pak processing systems AB dairy processing handbook (Tetra Pak Processing Systems AB Lund, 2015), Section: 482 s. : illustrations
8. Campbell C, Feldpausch G (2022) Invited review: the consumer and dairy food waste: an individual plus policy, systems, and environmental perspective. J Dairy Sci 105:3736–3745
9. Chaudhuri A, Dukovska-Popovska I, Subramanian N, Chan H, Bai R (2018) Decision-making in cold chain logistics using data analytics: a literature review. Int J Logist Manage 29:839–861
10. Chen S, Chen C (2021) Dynamic markdown decisions based on a quality loss function in on-site direct-sale supply chains for perishable food. J. Oper Res Soc 72:822–836
11. Entrup M, Günther H, Van Beek P, Grunow M, Seiler T (2005) Mixed-integer linear programming approaches to shelf-life-integrated planning and scheduling in yoghurt production. Int J Product Res 43:5071–5100
12. Fard A, Yegane B (2012) Investigation of the impact of smart packaging on performance in deteriorating items supply chain. Adv Mater Res 403:3344–3349
13. Fattahi P, Tanhatalab M (2022) Stochastic inventory-routing problem with lateral transshipment for perishable product. J Model Manage 17:539–568
14. Foroozesh N, Karimi B, Mousavi S, Mojtahedi M (2023) A hybrid decision-making method using robust programming and interval-valued fuzzy sets for sustainable-resilient supply chain network design considering circular economy and technology levels. J Ind Inf Integr 33:100440
15. Ganguly S, Robb D (2022) An analytical model to characterize consumption and wastage of fresh fruit and in households. Eur J Oper Res 300:151–163

16. Gitinavard H, Ghodsypour S, Shirazi M (2019) A bi-objective multi-echelon supply chain model with Pareto optimal points evaluation for perishable products under uncertainty. Sci Iran 26:2952–2970
17. Grillo H, Alemany M, Ortiz A, Fuertes-Miquel V (2017) Mathematical modelling of the order-promising process for fruit supply chains considering the perishability and subtypes of products. Appl. Math. Model 49:255–278
18. Gutierrez-Alcoba A, Rossi R, Martin-Barragan B, Hendrix E (2017) A simple heuristic for perishable item inventory control under non-stationary stochastic demand. Int J Product Res 55:1885–1897
19. Herbon A (2023) Shelf-life extension under implementation costs. Comput Ind Eng 180:109241
20. Herbon A, Khmelnitsky E (2017) Optimal dynamic pricing and ordering of a perishable product under additive effects of price and time on demand. Eur J Oper Res 260:546–556
21. Jouzdani J, Govindan K (2021) On the sustainable perishable food supply chain network design: A dairy products case to achieve sustainable development goals. J Clean Product 278:123060
22. Kaya O, Bayer H (2021) Pricing and lot-sizing decisions for perishable products when demand changes by freshness. J Ind Manage Optim 17:3113–3129
23. Kayikci Y, Demir S, Mangla S, Subramanian N, Koc B (2022) Data-driven optimal dynamic pricing strategy for reducing perishable food waste at retailers. J Clean Product 344:131068
24. Khalafi S, Hafezalkotob A, Mohammaditabar D, Sayadi M (2020) Multi-objective fuzzy programming of remanufactured green perishable products using supply contracts. Int J Manage Sci Eng Manage 15:274–287
25. Khalafi S, Hafezalkotob A, Mohammaditabar D et al (2024) A game theoretic approach for analyzing financial intervention models in dairy green supply chains with freshness-keeping effort. Environ Dev Sustain 26:17131–17156. https://doi.org/10.1007/s10668-023-03330-w
26. Khazaeli S, Jabalameli MS, Sahebi H (2024) Bi-objective model for multi-level supply chain by focusing on quality of agricultural products: a case study. Kybernetes 53(10):3484–3521. https://doi.org/10.1108/K-05-2022-0745
27. Krommyda I, Tatsis V, Skouri K (2020) Optimal ordering and disposal decisions for products with a fixed shelf life. RAIRO-Oper Res 54:1453–1465
28. Lajdova Z, Bielik P (2015) The evidence of asymmetric price adjustments. Agric Econ Zemedelska Ekon 61:105–115
29. Li R, Teng J, Chang C (2021) Lot-sizing and pricing decisions for perishable products under three-echelon supply chains when demand depends on price and stock-age. Ann Oper Res 307:303–328
30. Lu J, Zhang J, Jia X, Zhu G (2019) Optimal dynamic pricing, preservation technology investment and periodic ordering policies for agricultural products. RAIRO-Oper Res 53:731–747
31. Majidi A, Farghadani-Chaharsooghi P, Al-E-Hashem S (2022) Sustainable pricing-production-workforce-routing problem for perishable products by considering demand uncertainty; a case study from the dairy industry. Transp J 61:60–102
32. Marandi F, Zegordi S (2017) Integrated production and distribution scheduling for perishable products. Sci Iran 24:2105–2118
33. Momeni M, Jain V, Govindan K, Mostofi A, Fazel S (2022) A novel buy-back contract coordination mechanism for a manufacturer-retailer circular supply chain regenerating expired products. J Clean Product 375:133319
34. Mousavi R, Bashiri M, Nikzad E (2022) Stochastic production routing problem for perishable products: modeling and a solution algorithm. Comput Oper Res 142:105725
35. Muriana C (2015) Effectiveness of the food recovery at the retailing stage under shelf life uncertainty: an application to Italian food chains. Waste Manage 41:159–168
36. Mykrantz J, Bozic M (2022) Opportunity cost: an economic concept that may improve the functioning of federal milk marketing orders and the U.S. dairy industry. J Agric Appl Econ 54:606–620

37. Pauls-Worm K, Hendrix E, Alcoba A, Haijema R (2016) Order quantities for perishable inventory control with non-stationary demand and a fill rate constraint. Int J Product Econ 181:238–246
38. Pishvaee M, Torabi S (2010) A possibilistic programming approach for closed-loop supply chain network design under uncertainty. Fuzzy Sets Syst 161:2668–2683
39. Polotski V, Gharbi A, Kenne J (2021) Production control of unreliable manufacturing systems with perishable inventory. Int J Adv Manufact Technol 116:2473–2496
40. Polotski V, Gharbi A, Kenne J (2022) Production control in manufacturing systems with perishable products under periodic demand. J Manufact Syst 63:288–303
41. Rinaldi M, Bottani E, Solari F, Montanari R (2022) The vehicle routing problem in the dairy sector: a case study. Int J Food Eng 18:239–252
42. Roa A, Escobar J, Montoya M (2023) Robust design of a logistics system using FePIA procedure and analysis of trade-offs between CO2 emissions and net present value. Heliyon 9:e18444
43. Routroy S, Behera A (2017) Agriculture supply chain a systematic review of literature and implications for future research. J Agri Develop Emerg Econ 7:275–302
44. Salas-Navarro K, Romero-Montes J, Acevedo-Chedid J, Ospina-Mateus H, Florez W, Cardenas-Barron L (2023) Vendor managed inventory system considering deteriorating items and probabilistic demand for a three-layer supply chain. Exp Syst Appl 218:119608
45. Santos M, Martins S, Amorim P, Almada-Lobo B (2022) On the impact of adjusting the minimum life on receipt (MLOR) criterion in food supply chains. Omega. 112:102691
46. Sel C, Bilgen B, Bloemhof-Ruwaard J, Vorst J (2015) Multi-bucket optimization for integrated planning and scheduling in the perishable dairy supply chain. Comput Chem Eng 77:59–73
47. Sha F, Kazemi A, Chaghooshi A, Sazvar Z, Mahdiraji H (2021) A robust multi-objective optimization model for inventory and production management with environmental and social consideration: a real case of dairy industry. J Clean Product 294:126230
48. Singh N, Kuthambalayan T (2022) Integrating operations and marketing decisions to manage perishability risks with target minimum remaining shelf-life available to consumers. Comput Ind Eng 163:107812
49. Soysal M, Bloemhof-Ruwaard J, Haijema R, Van Der Vorst J (2015) Modeling an inventory routing problem for perishable products with environmental considerations and demand uncertainty. Int J Product Econ 164:118–133
50. Soysal M, Bloemhof-Ruwaard J, Haijema R, Vorst J (2018) Modeling a green inventory routing problem for perishable products with horizontal collaboration. Comput Oper Res 89:168–182
51. Talouki R, Javadian N, Movahedi M (2021) Optimization and incorporating of green traffic for dynamic vehicle routing problem with perishable products. Environ Sci Pollut Res 28:36415–36433
52. Tat R, Heydari J, Rabbani M (2020) A mathematical model for pharmaceutical supply chain coordination: reselling medicines in an alternative market. J Clean Product 268:121897
53. Transchel S, Hansen O (2019) Supply planning and inventory control of perishable products under lead-time uncertainty and service level constraints. Proc Manufact 39:1666–1672
54. Wang X, Li D (2012) A dynamic product quality evaluation based pricing model for perishable food supply chains. Omega 40:906–917
55. Wang H, Zhan S, Ng C, Cheng T (2020) Coordinating quality, time, and carbon emissions in perishable food production: a new technology integrating GERT and the Bayesian approach. Int J Product Econ 225:107570
56. Wu X, Nie L, Xu M (2017) Designing an integrated distribution system for catering services for high-speed railways: a three-echelon location routing model with tight time windows and time deadlines. Transp Res C Emerg Technol 74:212–244
57. Yaghoubi A, Akrami F (2019) Proposing a new model for location - routing problem of perishable raw material suppliers with using meta-heuristic algorithms. Heliyon 5:e03020
58. Yavari M, Geraeli M (2019) Heuristic method for robust optimization model for green closed-loop supply chain network design of perishable goods. J Clean Product 226:282–305
59. Yavari M, Zaker H (2019) An integrated two-layer network model for designing a resilient green-closed loop supply chain of perishable products under disruption. J Clean Product 230:198–218

Sustainability with 4R Principle on Multi-objective Waste Management Under Fermatean Fuzzy Environment

Sankar Kumar Roy, Shyamali Ghosh, and Josef Jablonsky

Abstract The purpose of this chapter is to establish a strong correlation between two corresponding sections as waste management and sustainability. A multi-objective waste management (MOWM) plan from ground level can find zero waste by applying 4R (reduce, recycle, reuse, and refuse) principle. From collection center(s) to treatment plant(s), the waste items are fluctuated by the generators' activity and the management capability of treatment plants. These fluctuations outcome with an uncertainty which is tackled here by selecting Fermatean fuzzy environment on source and demand parameters. Time, distance, and risk factor are sensitive issues for waste management. To curb these issues we analyze several sustainable factors in the proposed model such as circular economy, carbon emission with CAP policy, and time management with time window restriction. The Pareto-optimal solutions of the MOWM problem are attained by applying fuzzy programming, Fermatean fuzzy programming, and global criterion method. A comparison analysis is done at the end of result discussion. Final conclusions with pros and cons are depicted with future directions.

Keywords Multi-objective waste management · Sustainability · Carbon CAP policy · Circular economy · Fermatean fuzzy set and Fermatean fuzzy programming

1 Introduction

Multi-Objective Waste Management (MOWM) Starting ground to ending task of waste management (WM) is completed by various stages and procedures

S. K. Roy (✉) · S. Ghosh
Department of Applied Mathematics, Midnapore, West Bengal, India

J. Jablonsky
Department of Econometrics, Faculty of Informatics and Statistics, Prague University of Economics and Business, Prague, Czech Republic
e-mail: jablon@vse.cz

B. Jarboui et al. (eds.), *Data-Driven Waste Revolution*, Computational Intelligence Methods and Applications, https://doi.org/10.1007/978-981-96-5890-9_5

without disturbing natural or human activities. The focus of WM is to follow the social, economic, and environmental aspects of sustainability. Different factors of economic sustainability may be profit, WM cost, shifting cost, operational charge, penalty charge of time window, penalty for non-collected waste, revenue, etc. Social sustainability is to be finding optimum workload deviation, minimum health risk, or minimum elapsed time. The environmental aspect of sustainability is to reduce carbon emission from transportation, treatment plants, or use of electric vehicle in replace of fossil fuel consumed vehicle. Therefore, multiple criteria are introduced in a WM problem when hazardous waste are disposed of one side and some secondary recycled products generated by developing a green environment in the selected sites of WM. A single objective function is not enough to tackle such a situation. An MOWM problem, extension of single-objective WM problem is able to optimize all the objective functions together by fulfilling several criteria selected as constraints. An MOWM problem through a collection networks was discussed by Asl et al. [2]. A sustainable MOWM problem with sustainability was evaluated by Ghosh et al. [10]. Ghosh et al. [11] introduced an MOWM problem to develop sustainability. Rabbani et al. [20] evaluated a location-routing MOWM problem. Tirkolaee et al. [24] discussed an urban MOWM problem for waste collection by arc routing problem.

MOWM with 4R Principle Several types of waste items are to be managed by proper analysis and application such that most of the negative effects can be reduced in an intellectual way. An MOWM problem can be completed in a sustainable way when collection to treatment process of generated waste items follows 4R principle. In reduction technique, MOWM always tries to keep no residue of generated waste and also tries to collect and then supplies the waste into their treatment plants, as soon as possible. A transfer station (TS) [13] collects the haphazard waste from several source centers, sorts out them in their category, and transfers the material into their safety treatment plants [8]. The collected wastes are reduced in treatment plants by recycling [16], landfilling, and disposing hazardous waste with generating a small amount of solid residue that again reuses in landfilling. When the generators of waste items reuse the secondary recycled product, then they can avoid or reject the optional new product and, at the same time, maximize the economic growth that finds the circular economy of the supporting system. Again if the public avoids to use petroleum by using alternative resource as renewable bioenergy generated from waste item [19] and use compost by refusing chemical fertilizer, etc., then 4R principle follows global development. Therefore an MOWM problem is able to complete a circular economy by providing a green and smart city [17] with some resources/energies/products to the waste generated areas that have initially dumped the waste items with a risk.

Sustainability on MOWM Without sustainability any MOWM cannot be a supporting factor of global development. In most of the cases MOWM only focuses on social, economic, and environmental sustainability from several factors of sustainability. For economic sustainability, the proposed MOWM always tries to find maximum profit with minimum investment cost (collection, separation, operation, and transportation charge) and find a circular economy. For social sustainability, MOWM can find maximum job opportunity, minimum health risk, minimum time investment, and minimum workload deviation. For environmental sustainability, WM tries to reduce total pollution of air, water, and land caused by hazardous waste items. Air is mostly effected by the CO_2 emission from transportation [12], treatment plant, and recycle plant of any WM. Mondal and Roy [18] defined a sustainable multi-objective opened- and closed-loop supply chain. With regard to sustainability, Dong et al. [6] implemented CO_2 reduction plan in the global ICT industry. Ghosh et al. [14] presented a carbon emission reduction strategy on a transportation problem with multi-objective optimization. Zhang et al. [29] analyzed a sustainable freight transportation to find zero carbon measure. Using carbon cap and trade policy [11] on CO_2 gas reduction, an MOWM system improved environmental and economic sustainability. Therefore the proposed approach finds a correlation between sustainability and MOWM.

Time Window on MOWM The loop of MOWM is completed by a long duration because the operating time is larger than the shifting time of WM. To collect waste materials as soon as possible and to decrease the negative impact of these items on the surrounding area, a time window (TW)[14] restriction is needed. In this restriction, traveling cost and time are reduced by a penalty charge, which is imposed to the processing system when the transferring time does not belong to the initial time interval; In another case, when shifting time belongs to the provided time interval, then profit of the system does not effect by the penalty. TW restriction is applied on a multi-objective optimization problem by Ghosh et al. [9]. Ghannadpour and Zandieh [7] analyzed TWs through a multi-objective game-theoretical approach. To maximize the service quality to customer, Srivastava and Singh [23] imposed a TW restriction with the variant of vehicle routing problem.

MOWM with Fermatean Fuzzy The total amount of generated waste items is not fixed due to the several activities of generators in several seasons. Again, for automated natural disposal of some organic wastes, these are not fully treated. Therefore source of waste and demand of treatment plants are fluctuated and uncertainty appeared on the related data. To control both truthfulness and falsity of the imprecise data, intuitionistic fuzzy set (IFS), proposed by Atanassov [3], is applied, where the sum of membership and nonmembership grades is limited by 1. Yager and Abbasov [28] extended IFS [15] by presenting Pythagorean fuzzy set (PFS), when the sum of membership and nonmembership grades is greater than 1, but the sum of square of membership and nonmembership grades is less than or equal to one. Again, Yager [27] proposed a classical set of these uncertainties named as q-rung orthopair fuzzy set, where the sum of qth power of supporting factor to "yes" and supporting factor to "no" criteria is less than or equal to 1. When the value of q increases, then the space for expressing the uncertainty with membership and nonmembership grades becomes larger. Fermatean fuzzy set (FFS) is a q-rung orthopair fuzzy set by considering the value of $q = 3$. Das et al. [4] formulated and then solved multi-objective green solid transportation location problem with carbon policy under type-2 fuzzy logic. A sustainable supply chain management with Fermatean fuzzy-based objectives was proposed by Deliktas et al. [5]. To control more inconsistency of data, this chapter introduces FFS on the source and demand parameter.

Aims of Study The motivations for this chapter are outlined in the following after studying the related previous works:

- Extensive residency, urban population, and industrialization create a giant amount of waste items which create a burden in social, environment, and economic sites. To inhibit the harmful effects of waste items on the adjacent area and to control as much garbage as possible, a proper WM plan is therefore needed.
- When WM completes a process, it faces with several criteria that are defined as objectives. Therefore, it is very tough to manage all of the criteria in a sustainable way by compromising a single-objective WM problem.
- Total waste items are not fixed in several seasons and individual activities of public works. As a result, information of waste materials is inconsistent, incomplete, and insufficient. Moreover, in some situations, the uncertainty cannot be controlled by fuzzy set, IFS, or PFS. An example is discussed to better illustrate for motivating Fermatean fuzzy MOWM problem.
  Example: Selecting a certain type of waste source point that generates a type of waste item. From average knowledge and general information, it is found that the average amount of generated waste items is 30 ton per month. DM(s) judges about the membership values that the statement is true as 0.9 and the statement is false as 0.6. Therefore, the membership grade and nonmembership grade do not follow the criteria of fuzzy set, IFS, and PFS, due to sum, and the square sum of both membership and nonmembership grades is greater than 1. But $(0.9)^3 + (0, 6)^3 < 1$, and this information can be only expressed by FFS as $\langle 30; (0.9), (0.6) \rangle$. Thus, FFS is capable to deal with this complex uncertain situation of an MOWM problem.

Contribution The main contributions of this chapter are summarized as follows:

- To protect social, economic, and environmental site from the worst effect of hazardous waste item, this chapter introduces an MOWM plan and shows a correlation between WM and sustainability.
- To tackle a multi-criteria decision-making on WM, 4R principle is applied in this chapter.
- To mitigate CO_2 emission, this chapter applies carbon cap with tax policy, which is more effectable rather than carbon tax.
- To tackle the hazardous wastes within a small time duration, time minimization with TW restriction is implemented in this chapter.
- To overcome the limitation of Pythagorean fuzzy set, this chapter includes Fermatean fuzzy environment on the selected data.
- Finally we select the final Pareto-optimal solution by comparing the results of two fuzzy methods, namely fuzzy programming (FP) and Fermatean fuzzy

programming (FFP), and a non-fuzzy method, namely global criterion method (GCM).

The remainder part of this chapter is designed as follows: Sect. 2 displays the basic elementary concept based on Fermatean fuzzy environment. Notations with assumptions and model formulation are introduced in Sect. 3. Section 4 briefly describes the approaches, namely, FP, FFP, and GCM. In Sect. 5, numerical results are determined with discussion analysis. Section 6 presents the conclusions with the advantages and limitations and the future research scope.

2 Preliminaries

To establish the proposed model, this section describes some fundamental definitions and operations based on Fermatean fuzzy set (FFS) and Fermatean fuzzy number (FFN). General operations of crisp set are insufficient to numerical requirements of FFN. For algebraic calculation, i.e., to combine, separate or compare the Fermatean fuzzy parameters of MOWM, it is needed to find deterministic value. The score function, accuracy function, or any ranking function provides deterministic value by calculating the FFN in terms of membership and nonmembership grades. These calculations are implemented by addition, multiplication, or other operations which are not similar to the basic operations of crisp data. For this cause, fundamental definitions and operations based on FFS and FFN are defined here.

Definition 1 ([22]) An FFS $\tilde{A}^f$ on universal set U is stated as $\tilde{A}^f = \{\langle u, \mu_{\tilde{A}^f}(u), \gamma_{\tilde{A}^f}(u)\rangle : u \in U\}$, where $\mu_{\tilde{A}^f}(u), \gamma_{\tilde{A}^f}(u) : U \rightarrow [0, 1]$ are the membership and nonmembership functions, respectively, which satisfy $0 \leq \mu^3_{\tilde{A}^f}(u) + \gamma^3_{\tilde{A}^f}(u) \leq 1$ for all $u \in U$. The degree of indeterminacy of an element u in the set $\tilde{A}^p$ is defined as the function $\pi_{\tilde{A}^f}(u) = \sqrt[3]{1 - \mu^3_{\tilde{A}^f}(u) - \gamma^3_{\tilde{A}^f}(u)}$.
If U has one element, then $\tilde{A}^f$ is called as Fermatean fuzzy number (FFN). For simplicity, an FFN is denoted as $\tilde{A}^f = (\mu_{\tilde{A}^f}, \gamma_{\tilde{A}^f})$.

Definition 2 ([22]) Consider $\tilde{A}^f = (\mu_{\tilde{A}^f}\gamma_{\tilde{A}^f})$ and $\tilde{B}^f = (\mu_{\tilde{B}^f}, \gamma_{\tilde{B}^f})$ are two FFNs in the universal set U, and let $r \geq 0$ be a parameter. The following arithmetic operations are defined for FFNs:

1. Sum: $\tilde{A}^f \oplus \tilde{B}^f = \left(\sqrt[3]{\mu^3_{\tilde{A}^f} + \mu^3_{\tilde{B}^f} - \mu^3_{\tilde{A}^f}\mu^3_{\tilde{B}^f}}, \gamma_{\tilde{A}^f}.\gamma_{\tilde{B}^f}\right)$
2. Inner product: $\tilde{A}^f \odot \tilde{B}^f = \left(\mu_{\tilde{A}^f}.\mu_{\tilde{B}^f}, \sqrt[3]{\gamma^3_{\tilde{A}^f} + \gamma^3_{\tilde{B}^f} - \gamma^3_{\tilde{A}^f}\gamma^3_{\tilde{B}^f}}\right)$
3. Scalar product: $r.\tilde{A}^f = \left(\sqrt[3]{1 - (1 - \mu^3_{\tilde{A}^f})^r}, (\gamma_{\tilde{A}^f})^r\right)$
4. Power: $(\tilde{A}^f)^r = \left((\mu_{\tilde{A}^f})^r, \sqrt[3]{1 - (1 - \gamma^3_{\tilde{A}^f})^r}\right)$

Definition 3 ([22]) Let $\tilde{A}^f = (\mu_{\tilde{A}^f}, \gamma_{\tilde{A}^f})$ be an FFN. Then the score function $S(\tilde{A}^f)$ and the accuracy function $M(\tilde{A}^f)$ of $\tilde{A}^f$ are defined as
$S(\tilde{A}^f) = (\mu^3_{\tilde{A}^f} - \gamma^3_{\tilde{A}^f}),\ M(\tilde{A}^f) = (\mu^3_{\tilde{A}^f} + \gamma^3_{\tilde{A}^f})$.
It is remarked that $S(\tilde{A}^f) \in [-1, 1]$ and $M(\tilde{A}^f) \in [0, 1]$.

Definition 4 ([1]) A triangular Fermatean fuzzy number (TFFN) on universal set U is defined as $\hat{A}^f = \langle(a, b, c); \mu_{\hat{A}^f}, \gamma_{\hat{A}^f}\rangle$, where

$$\mu_{\hat{A}^f} = \begin{cases} \omega_1 \frac{x-a}{b-a}, & \text{if } a < x < b, \\ \omega_1, & \text{if } x = b, \\ \omega_1 \frac{c-x}{c-b}, & \text{if } b < x < c, \end{cases} \quad \gamma_{\hat{A}^f} = \begin{cases} 1 - (1-\omega_2)\frac{x-a}{b-a}, & \text{if } a < x < b, \\ \omega_2, & \text{if } x = b, \\ \omega_2 + (1-\omega_2)\frac{x-b}{c-b}, & \text{if } b < x < c. \end{cases}$$

Hence, ${\omega_1}^3 + {\omega_2}^3 \in [0, 1]$.

The score function and the accuracy function are potential types of defuzzified values of FFNs. Now, a ranking function using the score function [21] is explained that finds all positive values in terms of TFFN.

Definition 5 Let $\hat{A}^f = \langle(a, b, c); \mu_{\hat{A}^f}, \gamma_{\hat{A}^f}\rangle$ be a TFFN. Then the ranking index is defined as

$$\Re(\hat{A}^f) = \frac{a+b+c}{3}\left(\frac{1 + \mu^3_{\hat{A}^f} - \gamma^3_{\hat{A}^f}}{2}\right).$$

Let $\hat{A}^f$ and $\hat{B}^f$ be TFFNs; then the following comparisons are followed by ranking operation:

Case (i) $\Re(\hat{A}^f) > \Re(\hat{B}^f) \implies \hat{A}^f >_{\Re} \hat{B}^f$.
Case (ii) $\Re(\hat{A}^f) < \Re(\hat{B}^f) \implies \hat{A}^f <_{\Re} \hat{B}^f$.
Case (iii) $\Re(\hat{A}^f) = \Re(\hat{B}^f)$ and $M(\hat{A}^f) = M(\hat{B}^f) \implies \hat{A}^f =_{\Re} \hat{B}^f$.

3 Problem Background

Here, we formulate a mathematical model of MOWM problem which uses 4R principle to develop sustainability and to find a circular economy. The WM process is controlled by a system (as a third party) that has got the responsibility from municipality/industry/commercial area. The system tries to find maximum profit from revenue of waste management (RWM) that deduced by investment cost, called as waste management charge (WMC) and penalty charge of TW (PCTW) restriction. The revenue is obtained from secondary recycled products and from landfilling. The investment charge (WMC) is considered for waste collection, transportation, loading, unloading, separation, and operational charge of treatment

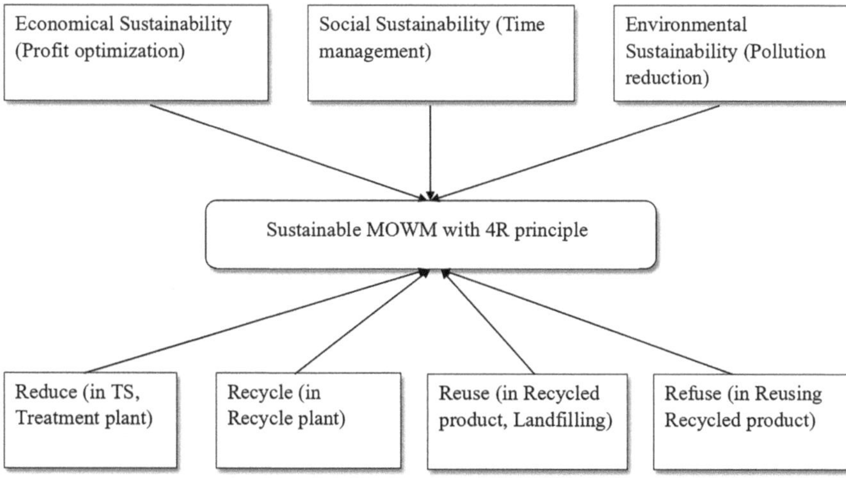

Fig. 1 Design of a sustainable MOWM with 4R principle (proposed by the authors)

plant. TS collects the haphazard waste items and then transfers into their treatment plants after separating them according to their nature. In incineration plant, some hazardous waste items are disposed, and then a solid residue with a huge amount of CO_2 gas is produced. Generated solid residue is also reused in landfilling for getting revenue. The running vehicles avoid violating TW restriction for paying the PCTW. Recycle plant provides the secondary recycled products to the users who are able to refuse some items by reusing alternate secondary products. Therefore circular economy supports to the economic sustainability of WM, and therefore, the system considers profit as the first objective function in the proposed model.

Time is the vital issue of any WM. Therefore to reduce all types of negative effect of waste item, minimum elapsed time is always required. Minimum elapsed time satisfies the social sustainability which is selected as the second objective function.

From collection to treatment plant(s), also at the end point of public users, most of the WM process emits CO_2 gas. Mainly, transportation and treatment plants are a major part to pollute air caused by high emission of carbon. To find environmental sustainability, and as well as to find economic opportunity, carbon emission is reduced by "carbon cap with tax policy" which is selected as the third objective function. The performing stages of the formulated model are graphically described in Fig. 1.

Notations and Assumptions of the Proposed Study

Here we provide a list of notations with their description in Table 1, and the required assumptions are presented to formulate the proposed model of MOWM problem.

Table 1 Useful notations with descriptions

Index	Description
I	Index for source of waste items ($i = 1, 2, \ldots, I$)
J	Index for transfer station ($j = 1, 2, \ldots, J$)
K	Index for recycle plant ($k = 1, 2, \ldots, K$)
L	Index for landfill area ($l = 1, 2, \ldots, L$)
M	Index for incineration plant ($m = 1, 2, \ldots, M$)
N	Index for recycled product selling market center ($n = 1, 2, \ldots, N$)
Decision variable	
x_{ij}	Amount of waste items that to be transported from ith place to jth TS
$y^1_{jk}, y^2_{jl}, y^3_{jm}$	Amount recyclable, landfilling, and disposable waste items that to be shifted from jth TS to kth recycle plant, lth landfill area, and mth incineration plant, respectively
z^1_{ml}	Amount of solid residue is transferred from mth incineration plant to lth landfill are
z^2_{kn}	Amount of recycled product is transferred from kth recycle plant to nth market center
T	Total elapsed time.
η^{11}_{ij}	Binary variable taking the value "1" if $x^1_{ij} > 0$ and "0" otherwise
η^{21}_{jk}	Binary variable taking the value "1" if $y^1_{jk} > 0$ and "0" otherwise
η^{22}_{jl}	Binary variable taking the value "1" if $y^2_{jl} > 0$ and "0" otherwise
η^{23}_{jm}	Binary variable taking the value "1" if $y^3_{jm} > 0$ and "0" otherwise
η^{31}_{ml}	Binary variable taking the value "1" if $z^1_{ml} > 0$ and "0" otherwise
η^{32}_{kn}	Binary variable taking the value "1" if $z^2_{kn} > 0$ and "0" otherwise
Parameter	
$\hat{a}^f_i$	Available amount of waste items at ith source area
$\hat{b}^{f1}_j, \hat{b}^{f21}_k, \hat{b}^{f22}_l, \hat{b}^{f23}_m$	Demand capability of jth TS, kth recycle plant, lth landfill area, and mth incineration plant, respectively
$\hat{b}^{3f}_n$	Demand capability of nth recycled product selling market center
$\phi \in [0, 1]$	Conversion factor waste to residue
I^L, I^R	Revenue from landfill area and recycled product market, respectively
$\theta_1, \theta_2, \theta_3$	Tax for carbon emission to transportation, to recycle plant, and to incineration plant
$e^1_{CO_2}, e^2_{CO_2}, e^3_{CO_2}$	Rate of carbon emission of transportation (per unit item and per unit distance), of recycle plant, and of incineration plant (per unit waste item), respectively
Z_s	Objective function, ($s = 1, 2, 3$)
α	Extra permit charge for excess carbon emission per unit item
CAP	Carbon cap (maximum allowable carbon emission)
P_c	Penalty charge per unit extra carbon to cap
$[ET_w, LT_w]$	Time interval with ET_w as the early time service and LT_w as the late time service for hard TW constraint
P^e_w, P^l_w	PCTW for early time service and for late time service, respectively

(continued)

Table 1 (continued)

Index	Description
$c_{ij}^{11}, c_{jk}^{21}, c_{jl}^{22}, c_{jm}^{23}$	WMC per unit waste item that transported from ith source to jth TS and from jth TS to kth recycle plant, lth landfill area, and mth incineration plant, respectively
c_{ml}^{31}, c_{kn}^{32}	WMC per unit solid residue and recycled product that to be transported from mth incineration plant to lth landfill area and from kth recycle plant to nth recycled market, respectively
d_{ij}^{11}, t_{ij}^{11}	Distance and transportation time from ith source to jth TS
d_{jk}^{21}, t_{jk}^{21}	Distance and transshipment time from jth TS to kth recycle plant
d_{jl}^{22}, t_{jl}^{22}	Distance and shipment time from jth TS to mth incineration plant
d_{jm}^{23}, t_{jm}^{23}	Distance and transportation time from jth TS to mth landfill area
d_{ml}^{31}, t_{ml}^{31}	Distance and transporting time from mth incineration plant to lth landfill area
d_{kn}^{32}, t_{kn}^{32}	Distance and shifting time from kth recycle plant to nth market center

Assumptions

- TW restriction is chosen only when transporting waste items from source to TS and from TS to treatment plant. The operating times of all treatment plants are ignored here.
- Carbon tax with cap policy is applied on CO_2 emission reduction from transportation, and tax per unit rate of carbon emission is imposed on other cases.
- WMC includes waste collection charge, transportation cost, loading-unloading cost, waste separation charge, and all operations charge of treatment plants.
- PCTW is described as

$$P_w = \begin{cases} P_w^e, \text{ if } & T_w < ET_w, \\ 0, \text{ if } & ET_w \leq T_w \leq LT_w, \\ P_w^l, \text{ if } & T_w > LT_w. \end{cases}$$

- Carbon cap is defined as

$$[CAP - C_1]^+ = \begin{cases} CAP - C_1, & \text{if } CAP > C_1, \\ 0, & \text{if } CAP \leq C_1. \end{cases}$$

C_1 is the total carbon emission during transportation.

3.1 Implementation of MOWM Model

This subsection implements the proposed MOWM model under Fermatean fuzzy environment. The source and demand parameters are selected as TFFNs. To formulate the mathematical model (i.e., Model 1), we present each elementary part separately. Main elementary properties of this model are to maximize profit of the total system through the first objective function, minimize elapsed time by the second objective function, and minimize the total carbon emission of the system through the third objective function. Other criteria are presented through the constraints.

In Fig. 1, it is shown how 4R principle is incorporated in the formulated model. First R: Reduce is incorporated in the first objective function, when WMC includes an investment cost for reducing waste items through disposing, incineration, landfilling, and recycling. Also, the pollution factor of WM as carbon emission is reduced in the third objective function. Second R: Recycle is introduced as a treatment of waste items by getting revenue and investing WMC in the first objective function. Third R: Reuse is occurred when recycled items are used by providing revenue RWM in the first objective function. Fourth R: Refuse is incorporated by using secondary recycled items in place of new item or using CNG, electricity by refusing fossil fuel, and revenue RWM obtained in the first objective function. 4R principle is directly or indirectly incorporated in the first objective function.
Model 1

$$\text{maximize } \; Z_1 = \Big(RWM - WMC - PCTW \Big). \tag{1}$$

The first objective function, i.e., the total profit, is taken by Eq. (1), where
$RWM = \sum_L (\sum_J y^2_{jl} + \sum_M z^1_{ml}) I^L + \sum_K \sum_N z^2_{kn} I^R$.
$WMC = \sum_I \sum_J x_{ij} c^{11}_{ij} + \sum_J \big(\sum_K y^1_{jk} c^{21}_{jk} + \sum_L y^2_{jl} c^{22}_{jl} + \sum_M y^3_{jm} c^{23}_{jm}\big) + \sum_L \sum_M z^1_{ml} c^{31}_{ml} +$
$\sum_K \sum_N z^2_{kn} c^{32}_{kn}$.
$PCTW = \sum_{w=1}^{5} P_w \big[(ET_w - T_w)^+ + (T_w - LT_w)^+\big]$.

$$\text{minimize } \; Z_2 = T. \tag{2}$$

The second objective function, i.e., the total elapsed time which is to be minimized, is expressed by Eq. (2). The total elapsed time of the MOWM is $T = T_1 + T_2 + T_3 + T_4 + T_5 + T_6$.
$T_1 = \sum_I \sum_J t^{11}_{ij} \eta^{11}_{ij}$, $T_2 = \sum_J \sum_K t^{21}_{jk} \eta^{21}_{jk}$, $T_3 = \sum_J \sum_L t^{22}_{jl} \eta^{22}_{jl}$, $T_4 = \sum_J \sum_M t^{23}_{jm} \eta^{23}_{jm}$,
$T_5 = \sum_L \sum_M t^{31}_{ml} \eta^{31}_{ml}$, $T_6 = \sum_K \sum_N t^{32}_{kn} \eta^{32}_{kn}$.

$$\text{minimize } \; Z_3 = \theta_1 C_1 + \theta_2 C_2 + \theta_3 C_3 + \alpha P_c \big[C_1 - CAP\big]^+. \tag{3}$$

The third objective function is indicated by Eq. (3), the total carbon emission charge which is to be minimized. In Eq. (3), the first part is the total tax of carbon emission during all transportations from source to destination, and the second and third parts are the total tax of carbon emission from recycle and incineration treatment plants.

The fourth part is the total penalty charge with extra permit charge for excess carbon than CAP value. Total carbon emission of MOWM: $C = C_1 + C_2 + C_3$;
$C_1 = e^1_{CO_2}[\sum_J (\sum_I x_{ij} d^{11}_{ij} + \sum_K y^1_{jk} d^{21}_{jk} + \sum_L y^2_{jl} d^{22}_{jl} + \sum_M y^3_{jm} d^{23}_{jm}) + \sum_M \sum_L z^1_{ml} d^{31}_{ml}$
$+ \quad \sum_K \sum_N z^2_{kn} d^{32}_{kn}]$.
$C_2 = \sum_J \sum_K y^1_{jk} e^2_{CO_2}$.
$C_3 = \sum_J \sum_M y^3_{jm} e^3_{CO_2}$.

$$\text{subject to } \ T_w \in [ET_w, LT_w], \ w = 1, 2, 3, 4, 5. \tag{4}$$

Constraints (4) are the restriction of the soft TW.

$$\sum_J x^1_{ij} \leq \hat{a}^f_i \ \forall \, i. \tag{5}$$

The total source availability condition of ith source point is described by constraint (5).

$$\sum_I x^1_{ij} \geq \hat{b}^f_j \ \forall \, j. \tag{6}$$

Demand capability of jth TS is defined by constraints (6).

$$\sum_K y^1_{jk} + \sum_L y^2_{jl} + \sum_M y^3_{jm} \leq \hat{b}^{f1}_j \ \forall \, j. \tag{7}$$

Total source availability of jth TS is ensured by constraints (7).

$$\sum_I x_{ij} \geq \sum_K y^1_{jk} + \sum_L y^2_{jl} + \sum_M y^3_{jm} \ \forall \, j. \tag{8}$$

Constraints (8) confirm that the total collected waste by TS is always greater than or equal to the total distributed waste from TS.

$$\sum_J y^1_{jk} \geq \hat{b}^{f21}_k \ \forall \, k. \tag{9}$$

$$\sum_J y^2_{jl} + \sum_M z^1_{ml} \geq \hat{b}^{f22}_l \ \forall \, l. \tag{10}$$

$$\sum_J y^3_{jm} \geq \hat{b}^{f23}_m \ \forall \, m. \tag{11}$$

Demand capability of treatment plants is shown by the constraints (9), (10), and (11), respectively.

$$\sum_{N} z_{kn}^{2} \leq \hat{b}_{k}^{f21} \; \forall \, k. \tag{12}$$

$$\sum_{K} z_{kn}^{2} \geq \hat{b}_{n}^{f3} \; \forall \, n. \tag{13}$$

Constraints (12) and (13) ensure the source availability of kth recycled product center and demand capability of nth recycle market, respectively.

$$\sum_{L} z_{ml}^{1} = \phi \sum_{J} y_{jm}^{3} \; \forall \, m. \tag{14}$$

Constraints (14) present the total amount of solid residue generated in mth incineration plant.

$$x_{ij} \geq 0, \, y_{jk}^{1} \geq 0, \, y_{jl}^{2} \geq 0, \, y_{jm}^{3} \geq 0, \, z_{ml}^{1} \geq 0, \, z_{kn}^{2} \geq 0, \forall \, i, \; j, \; k, \; l, \; m, \; n. \tag{15}$$

Constraints (15) define the nonnegativity restriction of the variables. The feasibility conditions of this MOWM problem are set as $\sum_{I} \hat{a}_{i}^{f} \geq \sum_{J} \hat{b}_{j}^{f1}$, $\sum_{J} \hat{b}_{j}^{f1} \geq \sum_{K} \hat{b}_{k}^{f21} + \sum_{L} \hat{b}_{l}^{f22} + \sum_{M} \hat{b}_{m}^{f23}$, $\sum_{K} \hat{b}_{k}^{21} \geq \sum_{N} \hat{b}_{n}^{f3}$, and $\sum_{M} \hat{b}_{m}^{f23} \geq \sum_{L} \hat{b}_{l}^{f22}$.

Deterministic MOWM Model

Model 1 of the proposed MOWM problem cannot be evaluated directly due to existing uncertainty on source and demand parameters. Using the ranking function from Definition 5, Model 1 is converted into deterministic form, i.e., Model 2 is shown as follows:

Model 2

$$\begin{aligned}
\text{maximize} \;\; & Z_1 \\
\text{minimize} \;\; & Z_2 \\
\text{minimize} \;\; & Z_3 \\
\text{subject to} \;\; & \text{constraints (4), (8), (14)-(15)},
\end{aligned}$$

$$\sum_{J} x_{ij} \leq \Re(\hat{a}_{i}^{f}) \; \forall \, i, \tag{16}$$

$$\sum_{I} x_{ij} \geq \Re(\hat{b}_{j}^{f1}) \; \forall \, j, \tag{17}$$

$$\sum_{K} y_{jk}^{1} + \sum_{L} y_{jl}^{2} + \sum_{M} y_{jm}^{3} \leq \Re(\hat{b}_{j}^{f1}) \; \forall \, j, \tag{18}$$

$$\sum_{J} y_{jk}^{1} \geq \Re(\hat{b}_{k}^{f21}) \; \forall \, k, \tag{19}$$

$$\sum_{L} y_{jl}^2 + \sum_{M} z_{ml}^1 \geq \Re(\hat{b}_l^{f22}) \; \forall \, l, \tag{20}$$

$$\sum_{J} y_{jm}^3 \geq \Re(\hat{b}_m^{f23}) \; \forall \, m, \tag{21}$$

$$\sum_{N} z_{kn}^2 \leq \Re(\hat{b}_k^{f21}) \; \forall \, k. \tag{22}$$

$$\sum_{K} z_{kn}^2 \geq \Re(\hat{b}_n^{f3}) \; \forall \, n. \tag{23}$$

The feasibility conditions of this MOWM problem are set as $\sum_I \Re(\hat{a}_i^f) \geq \sum_J \Re(\hat{b}_j^{f1})$, $\sum_J \Re(\hat{b}_j^{f1}) \geq \sum_K \Re(\hat{b}_k^{f21}) + \sum_L \Re(\hat{b}_l^{f22}) + \sum_M \Re(\hat{b}_m^{f23})$, $\sum_K \Re(\hat{b}_k^{21}) \geq \sum_N \Re(\hat{b}_n^{f3})$, and $\sum_M \Re(\hat{b}_m^{f23}) \geq \sum_L \Re(\hat{b}_l^{f22})$.

4 Methodology for Fermatean Fuzzy Programming

To find the Pareto-optimal solution of the proposed MOWM, various fuzzy and non-fuzzy techniques are available from the literature survey. Some non-fuzzy techniques are goal programming [15], weighted sum approach, global criterion method [13], ϵ-constraint method [11], etc. Also, several fuzzy techniques are fuzzy programming (FP) [14], intuitionistic fuzzy programming [15], neutrosophic linear programming [11], TOPSIS approach [12], neutrosophic hesitant fuzzy programming [8], Pythagorean hesitant fuzzy programming [10], Fermatean fuzzy programming (FFP), etc. Here two fuzzy techniques are chosen as FP and FFP, and non-fuzzy technique is GCM. FFP is an extended fuzzy method based on FFS. This programming maximizes Fermatean fuzzy membership grade and minimizes Fermatean fuzzy nonmembership grade, and, then decision-maker (DM) finds the efficient solution in a technical way. The solution procedures of FP approach and GCM are elaborated in Ref. [14]. Using the FFP solution criteria, we propose the following stepwise procedure:

- **Step 1:** Determine the crisp model using ranking function on FF uncertain model, and evaluate each objective function individually with subject to all constraints.
- **Step 2:** Determine positive ideal solution (PIS) and negative ideal solution (NIS), and design the membership functions corresponding to each objective function in FF environment as follows:
  For $s = 2, 3$

$$T_s(Z_s) = \begin{cases} 1, & \text{if} \quad Z_s \leq L_s, \\ \frac{U_s - Z_s}{U_s - L_s}, & \text{if} \quad L_s \leq Z_s \leq U_s, \\ 0, & \text{if} \quad Z_s \geq U_s, \end{cases}$$

$$F_s(Z_s) = \begin{cases} 0, & \text{if} \quad Z_s \le L_s, \\ \frac{Z_s - L_s}{U_s - L_s}, & \text{if} \quad L_s \le Z_s \le U_s, \\ 1, & \text{if} \quad Z_s \ge U_s. \end{cases}$$

For $s = 1$

$$T_s(Z_s) = \begin{cases} 0, & \text{if } Z_s \le L_s, \\ \frac{Z_s - L_s}{U_s - L_s}, & \text{if } \quad L_s \le Z_s \le U_s, \\ 1, & \text{if } Z_s \ge U_s, \end{cases}$$

$$F_s(Z_s) = \begin{cases} 1, & \text{if } Z_s \le L_s, \\ \frac{U_s - Z_s}{U_s - L_s}, & \text{if } L_s \le Z_s \le U_s, \\ 0, & \text{if } Z_s \ge U_s. \end{cases}$$

For $s = 2, 3$, U_s = NIS for Z_s, and L_s = PIS for Z_s. For $s = 1$, U_s = PIS, and L_s = NIS;

- **Step 3:** To get the maximum degree for satisfaction and minimum degree for rejection, the FFP model for solving multi-objective decision making problem is expressed as follows:
  Model 3

$$\begin{aligned} \text{maximize} \quad & \frac{\eta_1^3 + \eta_2^3 + \eta_3^3}{3} - \frac{\zeta_1^3 + \zeta_2^3 + \zeta_3^3}{3} \\ \text{subject to} \quad & [T_s(Z_s)]^3 \ge \eta_s^3 \ (s = 1, 2, 3), \\ & [F_u(Z_s)]^3 \le \zeta_s^3 \ (s = 1, 2, 3), \\ & \eta_s^3 \ge \zeta_s^3, \\ & \eta_s^3 + \zeta_s^3 \in [0, 1], \eta_s^3 \in [0, 1], \zeta_s^3 \in [0, 1], \\ & \text{constraints} \quad (4), (8), (14)\text{–}(15), (16)\text{–}(23). \end{aligned}$$

  Here η_s and ζ_s are the membership grades and the nonmembership grades corresponding to each objective function.
- **Step 4:** To find the Pareto-optimal solution(s) of Model 3, an iterative procedure with parameters η_s and ζ_s is applied. LINGO 19.0 iterative scheme is used for modeling and solving the proposed model(s).

Theorem 1 *If $(x^*, y^*, z^*, \eta_s^*, \zeta_s^*)$ is an optimal solution of Model 3, then it is also a Pareto-optimal (an efficient) solution of Model 2 and Model 1.*

Proof The proof of this theorem is apparent from reference [11].

5 Numerical Illustration

Two numerical examples are analyzed in this chapter to show the effectiveness of the proposed model.

Example 1 In this example we choose $I = 3$ as the number source center of waste items and $J = 2$ as the number of TS. Centers of treatment plants are selected as $K = 2$, $L = 2$, $M = 2$. $N = 2$ is selected as the number of recycled product selling market. All the required parameters are presented in Tables 2, 3, 4, 5, and 6. The source and demand parameters are considered as TFFN, and defuzzified value of the parameters is also defined in Table 2. Here we select the carbon cap $C = 2500$ gm and all the parameters unit, e.g., cost in Rupees (INR), time in hour (h), and distance in kilometer (km).

Example 2 In this example, we take the carbon cap $C = 3500$ gm, and all other parameters are remained the same as in Example 1.

Results and Discussion

The results of two numerical examples with three approaches, namely FP, FFP, and GCM, and discussion analysis are presented in this subsection.

To find Pareto-optimal solution, each crisp objective function is solved individually and to determine the general solutions. From this general solution, each objective function value is found, and a pay-off matrix is formulated in Table 7. From pay-off matrix, PIS (i.e., maximum value of the first objective function

Table 2 Supply and demand of sources, TSs, treatment plants, recycled markets, and their ranking values

$\hat{a}_1^f = \langle(200, 220, 240); 0.75, 0.70\rangle;$	$\Re(\hat{a}_1^f) = 118.668$
$\hat{a}_2^f = \langle(210, 230, 250); 0.8, 0.7\rangle;$	$\Re(\hat{a}_2^f) = 134.435$
$\hat{a}_3^f = \langle(240, 245, 250); 0.8, 0.65\rangle;$	$\Re(\hat{a}_3^f) = 151.5815$
$\hat{b}_1^f = \langle(200, 230, 260); 0.85, 0.65\rangle;$	$\Re(\hat{b}_1^f) = 154.054$
$\hat{b}_2^f = \langle(190, 220, 250); 0.75, 0.75\rangle;$	$\Re(\hat{b}_2^f) = 110$
$\hat{b}_1^{f21} = \langle(20, 40, 60); 0.9, 0.6\rangle;$	$\Re(\hat{b}_1^{f21}) = 30.26$
$\hat{b}_2^{f21} = \langle(35, 50, 65); 0.85, 0.55\rangle;$	$\Re(\hat{b}_2^{f21}) = 36.195$
$\hat{b}_1^{f22} = \langle(30, 40, 50); 0.96, 0.45\rangle;$	$\Re(\hat{b}_1^{f22}) = 35.872$
$\hat{b}_2^{f22} = \langle(25, 30, 35); 0.89, 0.66\rangle;$	$\Re(\hat{b}_2^{f22}) = 21.261$
$\hat{b}_1^{f23} = \langle(80, 90, 100); 0.78, 0.63\rangle;$	$\Re(\hat{b}_1^{f23}) = 55.098$
$\hat{b}_2^{f23} = \langle(65, 70, 75); 0.92, 0.54\rangle;$	$\Re(\hat{b}_2^{f23}) = 56.742$
$\hat{b}_1^{f3} = \langle(50, 65, 80); 0.75, 0.69\rangle;$	$\Re(\hat{b}_1^{f3}) = 35.5355$
$\hat{b}_2^{f3} = \langle(45, 55, 65); 0.9, 0.55\rangle;$	$\Re(\hat{b}_2^{f3}) = 42.9715$

Table 3 WMC c_{ij}^{11}, c_{jk}^{21}, c_{jl}^{22}, c_{jm}^{23}, c_{ml}^{31}, c_{kn}^{32}

Source-Destination	c_{ij}^{11}	c_{jk}^{21}	c_{jl}^{22}	c_{jm}^{23}	c_{ml}^{31}	c_{kn}^{32}
1 − 1	12	25	18	19	9	10
2 − 1	13	20	19	22	11	13
3 − 1	9	–	–	–	–	–
1 − 2	11	20	10	18	10	15
2 − 2	10	22	11	21	14	12
3 − 2	14	–	–	–	–	–

Table 4 Elapsed time t_{ij}^{11}, t_{jk}^{21}, t_{jl}^{22}, t_{jm}^{23}, t_{ml}^{31}, t_{kn}^{32}

Source-Destination	t_{ij}^{11}	t_{jk}^{21}	t_{jl}^{22}	t_{jm}^{23}	t_{ml}^{31}	t_{kn}^{32}
1 − 1	0.90	1.0	1.14	0.68	0.50	0.40
2 − 1	1.30	0.80	1.6	0.90	0.68	0.66
3 − 1	1.10	–	–	–	–	–
1 − 2	0.80	0.66	1.02	0.56	0.84	1.46
2 − 2	0.50	1.12	1.14	0.44	1.14	1.0
3 − 2	1.0	–	–	–	–	–

Table 5 Distance d_{ij}^{11}, d_{jk}^{21}, d_{jl}^{22}, d_{jm}^{23}, d_{ml}^{31}, d_{kn}^{32}

Source-Destination	d_{ij}^{11}	d_{jk}^{21}	d_{jl}^{22}	d_{jm}^{23}	d_{ml}^{31}	d_{kn}^{32}
1 − 1	9	15	20	12	9	6
2 − 1	13	12	28	16	12	10
3 − 1	11	–	–	–	–	–
1 − 2	12	10	18	10	15	22
2 − 2	8	17	20	8	20	15
3 − 2	10	–	–	–	–	–

Table 6 Carbon emission rate, tax, extra permit charge for excess carbon, penalty charge, revenue, rate of ash production, and time interval

$e_{CO_2}^1 = 0.5$ gm/km, $e_{CO_2}^2 = 0.6$ gm/kg, $e_{CO_2}^3 = 0.8$ gm/kg, $\theta_1 = 0.60$, $\theta_2 = 0.50$, $\theta_3 = 0.75$, $\alpha = 0.6$, $P_c = 0.5$, $P_w^e = 4$, $P_w^l = 6$, $I^L = 80$, $I^R = 70$, $\phi = 25\%$, $[ET_w, LT_w] = [1, 2]$, $w = 1, 2, 3, 4, 5$

Table 7 Pay-off matrix

Approach	Z_1	Z_2	Z_3
Z_1	5840.571*	252.3861′	2268.7371′
Z_2	322.245	198.1537*	1760.969
Z_3	322.245′	198.1537	1760.969*

and minimum value of second and third objective functions, noted in $*$ symbol) and NIS (minimum value of the first objective function and maximum value of second and third objective functions, noted in $'$ symbol) are obtained. For FP, only membership function, and for FFP, membership and nonmembership functions, both are formulated by PIS and NIS. In GCM, the Euclidean distance between ideal solution and PIS is minimized. Pareto-optimal solutions are selected when the highest membership grade for FP and FFP and the lowest nonmembership

Table 8 Objective values of two examples

Approach	Example 1 (Cap C=2500)	Example 2 (Cap C=3500)
FP	(3592.44,10.48,2157.656)	(3592.44,10.48,1967.451)
FFP	(3077.448,10.48,2098.0462)	(3077.448,10.48,1927.711)
GCM	**(318.286, 9.82, 1847.9333)**	**(318.286, 9.82, 1760.969)**

grade for FFP are obtained among a set of membership and nonmembership grades corresponding to each objective function. The optimal results are obtained by LINGO 19.0 iterative scheme. The objective values in both examples are summarized in Table 8.

From Table 7, it is seen that the second objective value is equal in FP and FFP. It is also evident that the first objective value is always preferable in FP, whereas the second and third objective values are preferable in GCM for both examples. Therefore, among the set of Pareto-optimal solutions provided in three approaches, it is clear that GCM (highlighted in bold face) is a selected approach by the priority of second and third objective values, and the obtained result is declared as the final Pareto-optimal solution of both examples. By observing the objective values in two cases of the considered approaches in Table 7, the main differences between the solutions and the applicability of carbon cap policy are determined. For the effect of this policy, it is clear that there exists a trade-off between CAP value and the total carbon emission charge. For lower CAP value carbon emission charge increases, whereas for greater CAP value carbon emission charge decreases. But, the profit and elapsed time are the same in both the examples. This conflicting situation occurs due to the economic opportunity of carbon cap policy. For upper CAP value from the total carbon emission, the users (who emit carbon) find a relaxation from penalty of extra carbon than CAP value; only pay the tax per unit rate of emission for regular carbon emission. Hence, zero penalty with carbon tax decreases the total emission charge and economic sustainability found by the user. For lower CAP value, total emission easily exceeds to CAP, and therefore the system pays penalty charge for extra emission. In this issue, the positive penalty charge increases the total emission charge.

Each treatment plant of this MOWM also emits carbon, and a regular tax is paid per unit carbon. This tax is decreased only when the rate of carbon emission reduces from the treatment plants. Therefore carbon tax with cap policy supports to develop environmental sustainability and economic sustainability by reducing CO_2 emission and providing the limit index of carbon emission. Also the pollution reduction from carbon emission is a part of social sustainability. Here, other two objective functions as maximum profit and minimum elapsed time provide economic sustainability and social sustainability, respectively. Maximum profit supported by4R principle finds a circular economic situation when reduce and refuse proceed by reusing recycled product made from waste items. Therefore, the proposed MOWM finds a strong correlation with three types of sustainability.

6 Conclusions with Pros and Cons and Future Scopes

Conclusions This study has been incorporated an MOWM problem under 4R principle. Fermatean fuzzy environment has been used to tackle the uncertainty of the data in the developed model. Three selected contradictory objective functions as profit, elapsed time, and carbon emission have been optimized together by finding a strong correlation between WM and sustainability. Here the transporting time of waste item has been managed by TW restriction, and 4R principle, i.e., reduce, recycle, reuse, refuse, has followed the circular economy on the whole WM system. The applicability of the proposed model has been analyzed by finding Pareto-optimal solution of two real-life decision-making problems in the presence of carbon cap with tax policy, and the final Pareto-optimal solution has been improved by comparing the results of FP, FFP, and GCM. The conclusion of proposed MOWM has highlighted as follows:

(i) A strong correlation between MOWM and sustainability has been found by converging three sustainable factors from the direction of social, economic, and environmental view.
(ii) Optimum profit from 4R principle has been found a circular economy.
(iii) Time management of TW restriction has been utilized to drop the negative affect of waste items on social sites.
(iv) Carbon cap with tax policy rather than only tax or only cap policy has supported both economic and environmental sustainability.
(v) Fermatean fuzzy rather than fuzzy or intuitionistic fuzzy or Pythagorean fuzzy has tackled more space of uncertainty on the depicted MOWM problem.
(vi) The final Pareto-optimal solution has been marked for comparing the Pareto-optimal solutions provided by two fuzzy techniques and one non-fuzzy technique, without limiting in a single technique.
Pros and Cons: (i) This study is not only limited to waste collection and treatment but also follows several objectives together by establishing a model of MOWM problem.
(ii) Not only an MOWM problem but also the proposed discussion finds a relation between sustainability and WM by optimizing several sustainable factors.
(iii) From 4R principle, the proposed study finds a circular economy.
(iv) The presented topic of this chapter is not only tried to reduce CO_2 emission but also provides an economic opportunity by the application of carbon cap with tax policy.
(v) Not only selecting a specific space of uncertainty as fuzzy, IF or Pythagorean fuzzy, this study tackles more space of uncertainty by considering Fermatean fuzzy environment on the source and demand parameters.
(vi) One limitation of this study is to select only carbon cap with tax policy without analyzing the other policy of carbon emission. Also, zero carbon emission system cannot be implemented anywhere.

Future Scopes Extending the limitations of this chapter, researchers can find several opportunities for future research scope on arc routing problem or location routing problem for the shortest distance of a robust green WM [25]. A new environment such as Fermatean hesitant fuzzy or Fermatean probabilistic hesitant fuzzy may be introduced with comparing results on TOPSIS approach.

Acknowledgments The author Josef Jablonský acknowledges the financial support of the Internal Grant Agency of the Faculty of Informatics and Statistics, Prague University of Economics, project no. 10/2024.

References

1. Akram M, Shah SMU, Al-Shamiri MMA, Edalatpanah SA (2022) Fractional transportation problem under interval-valued Fermatean fuzzy sets. AIMS Mat 7(9):17327–17348
2. Asl HF, Makui A, Jabbarzadeh A, Barzinpour F (2018) Solving a multi-objective sustainable waste collection problem considering a new collection network. Oper Res 20:1977–2015
3. Atanassov K (1986) Intuitionistic fuzzy sets. Fuzzy Sets Syst 20(1):87–96
4. Das SK, Roy SK, Weber GW (2020) Application of type-2 fuzzy logic to a multi-objective green solid transportation-location problem with dwell time under carbon tax, cap and offset policy: Fuzzy vs. Non-fuzzy Techniques. IEEE Trans Fuzzy Syst 28(11):2711–2725
5. Deliktas D, Karagoz S, Simic V, Aydin N (2023) A stochastic Fermatean fuzzy-based multi-choice conic goal programming approach for sustainable supply chain management in end-of-life buildings. J Clean Prod 382:135305.
6. Dong K, Wang J, Taghizadeh-Hesary F (2022) Assessing the embodied CO_2 emissions of ICT industry and its mitigation pathways under sustainable development: a global case. Appl Soft Comput 131:109760
7. Ghannadpour SF, Zandieh F (2020) A new game-theoretical multi-objective evolutionary approach for cash-in-transit vehicle routing problem with time windows (A real-life case). Appl Soft Comput 93:106378
8. Ghosh S, Roy SK (2023) Closed-loop multi-objective waste management through vehicle routing problem in neutrosophic hesitant fuzzy environment. Appl Soft Comput 148:110854
9. Ghosh S, Küfer KH, Roy SK, Weber GW (2022) Type-2 zigzag uncertain multi-objective fixed-charge solid transportation problem: time window vs. preservation technology. Cent Eur J Oper Res 31(4):362
10. Ghosh S, Küfer KH, Roy SK, Weber GW (2022) Carbon mechanism on sustainable multi-objective solid transportation problem for waste management in Pythagorean hesitant fuzzy environment. Complex Intell Syst 8(4):4115–4143
11. Ghosh S, Roy SK, Weber GW (2023) Interactive strategy of carbon cap-and-trade policy on sustainable multi-objective solid transportation problem with twofold uncertain waste management. Ann Oper Res 326:157–197
12. Ghosh S, Roy SK, Fügenschuh A (2022) The multi-objective solid transportation problem with preservation technology using Pythagorean fuzzy sets. Int J Fuzzy Syst 24(6):2687–2704
13. Ghosh S, Roy SK (2021) Fuzzy-rough multi-objective product blending fixed-charge transportation problem with truck load constraints through transfer station. RAIRO Oper Res 55:S2923–S2952
14. Ghosh S, Roy SK, Verdegay JL (2022) Fixed-charge solid transportation problem with budget constraints based on carbon emission in neutrosophic environment. Soft Comput 26(21):11611–11625

15. Ghosh S, Roy SK, Ebrahimnejad A, Verdegay JL (2021) Multi-objective fully intuitionistic fuzzy fixed-charge solid transportation problem. Complex Intell Syst 7(2):1009–1023
16. Haseli G, Torkayesh AE, Hajiaghaei-Keshteli M, Venghaus S (2023) Sustainable resilient recycling partner selection for urban waste management: Consolidating perspectives of decision-makers and experts. Appl Soft Comput 137:110120
17. Mingaleva Z, Vukovic N, Volkova I, Salimova T (2020) Waste management in green and smart cities: a case study of Russia. Sustainability 12:94
18. Mondal A, Roy SK (2021) Multi-objective sustainable opened-and-closed-loop supply chain under mixed uncertainty during COVID-19 pandemic situation. Comput Ind Eng 159:107453
19. Pan XH, Wang YM, He SF., Labella A, Martinez L (2023) An interval type-2 fuzzy ORESTE method for waste-to-energy plant site selection: A case study in China. Appl Soft Comput 136:110092
20. Rabbani M, Heidari R, Farrokhi AH, Rahimi N (2018) Using metaheuristic algorithms to solve a multi-objective industrial hazardous waste location-routing problem considering incompatible waste types. J Clean Prod 170:227–241
21. Sahoo L (2021) Some score functions on Fermatean fuzzy sets and its application to bride selection based on TOPSIS method. Int J Fuzzy Syst Appl 10(3):18–29
22. Senapati T, Yager RR (2020) Fermatean fuzzy sets. J Ambient Intell Humaniz Comput 11:663–674
23. Srivastava G, Singh A (2023) Two evolutionary approaches with objective-specific variation operators for vehicle routing problem with time windows and quality of service objectives. Appl Soft Comput 134:109964
24. Tirkolaee EB, Goli A, Pahlevan M, Kordestanizadeh RM (2019) A robust bi-objective multi-trip periodic capacitated arc routing problem for urban waste collection using a multi-objective invasive weed optimization. Waste Manag Res 37(11):1089–1101
25. Tirkolaee EB, Mahdavi I, Esfahani MMS, Weber GW (2020) A robust green location-allocation inventory problem to design an urban waste management system under uncertainty. Waste Manag 102:340–350
26. Wu H, Gao Y (2023) An ant colony optimization based on local search for the vehicle routing problem with simultaneous pickup delivery and time window. Appl Soft Comput 139:110203
27. Yager RR (2017) Generalized orthopair fuzzy sets. IEEE Trans Fuzzy Syst 25(5):1222–1230
28. Yager RR, Abbasov AM (2013) Pythagorean membership grades, complex numbers, and decision making. Int J Intell Syst 28(5):436452
29. Zhang Z, Zhang H, Zhou L (2023) Zero-carbon measure prioritization for sustainable freight transport using interval 2 tuple linguistic decision approaches. Appl Soft Comput 132:109864

Optimizing the Municipal Waste Management to Support Circular Economy Program at Regency Level in Indonesia: Case Study in Banyumas Regency

Elisabeth Rianawati, Yosef Barita Sar Manik, Subkhi Abdul Aziz, Didit Okta Pribadi, Anthony Halog, Evita Mahar Dewi, Baihaqi Muhammad, Muhammad Asa, and Atik Munggiarti

Abstract Global waste production is projected to reach 3.4 billion metric tons by 2050, emphasizing the need for improved waste management. The circular economy (CE) model, which focuses on extending the lifespan of raw materials through recycling and bioconversion technologies, presents a sustainable framework for waste management. In countries such as Indonesia, challenges in adopting CE principles hinder economic diversification and value creation. This chapter focuses on Banyumas Regency, Indonesia, where innovative initiatives and waste management mobile applications like Salinmas and JEKNYONG have successfully implemented the CE model. The chapter describes the practical implementation of waste management in Banyumas and uses network and linear programming analysis

---

E. Rianawati (✉)
Resilience Development Initiative, Bandung, Indonesia

Y. B. Sar Manik
Environmental Engineering Program, President University, Bekasi, Indonesia
e-mail: yosef.manik@president.ac.id

S. A. Aziz · D. Okta Pribadi
Research Organization for Governance, Economy, and Community Welfare, National Research and Innovation Agency, Jakarta, Indonesia
e-mail: subkhi.aziz@outlook.co.id

A. Halog
School of Earth and Environmental Sciences, University of Queensland, St Lucia, QLD, Australia
e-mail: a.halog@uq.edu.au

E. Mahar Dewi · B. Muhammad · M. Asa
Resilience Development Initiative, Bandung, Indonesia
e-mail: evita.dewi@rdi.or.id; baihaqi.muhammad@rdi.or.id; mhdasa@rdi.or.id

A. Munggiarti
Banyumas Regency Regional Development Planning, Research and Development Agency, Banyumas, Indonesia

B. Jarboui et al. (eds.), *Data-Driven Waste Revolution*, Computational Intelligence Methods and Applications, https://doi.org/10.1007/978-981-96-5890-9_6

to optimize garbage transportation routes. Through a combination of qualitative and quantitative approaches, including desk reviews, key informant interviews, and field surveys, the study provides an in-depth understanding of the unique characteristics of waste management in Banyumas. Furthermore, the chapter emphasizes the sustainability impact of initiatives on economic, social, and environmental aspects, drawing lessons from global leaders in circular economy implementation. Lastly, the chapter provides valuable insights into optimizing circular economy applications for sustainable waste management.

Keywords Banyumas Regency · JEKNYONG · Network and linear programming analysis · Salinmas · Integrated waste management facilities

1 Introduction

1.1 Global Perspective

Effective solutions are urgently needed to address climate change. According to Assessment Report 6 (AR6) by the United Nations Intergovernmental Panel on Climate Change (IPCC), researchers emphasize the necessity for stronger and more directed efforts to combat the climate crisis [10]. One critical area where immediate action can yield significant benefits is waste management. The Synthesis Report (SYR) of the AR6 underlines that waste management is intrinsically linked to climate change due to the greenhouse gas (GHG) emissions from waste treatment and disposal processes. Landfills, for example, are significant sources of methane, a potent greenhouse gas contributing to global warming. Reducing methane emissions from waste and wastewater is highlighted as a key mitigation option, costing 100 USD per tCO2-eq or less. Implementing this measure could cut global emissions by at least half the 2019 level by 2030.

Furthermore, the IPCC underscores that water and waste management infrastructure can critically contribute to achieving deep emission reductions and advancing climate-resilient development in urban systems [9]. Effective waste management strategies, such as reducing, reusing, and recycling materials, not only mitigate greenhouse gas emissions but also enhance carbon sequestration, reduce energy consumption, and lessen the demand for virgin materials. Therefore, integrating waste management into broader climate action plans can accelerate progress toward achieving the targets set out in the Paris Agreement and other international climate commitments. By adopting comprehensive waste management strategies, countries can address both the environmental and socioeconomic aspects of climate change, paving the way for sustainable and resilient urban development.

A focal component of waste management efforts is the adoption of a circular economy approach, which aims to minimize waste and maximize the use of existing resources. Implementing a circular economy allows waste to be transformed into valuable inputs for new products, reduces the extraction of raw materials, and creates sustainable economic opportunities. This shift not only supports environmental goals but also drives innovation and economic resilience, making it a crucial climate strategy. In addition, the current decade is expected to see significant technological advancements that will revolutionize the waste industry. Integrating Internet of Things (IoT) applications in waste collection, such as iPads for drivers and the Radio Frequency Identification (RFID) tags on containers, is becoming increasingly common. These technologies are also being adopted in transfer stations and the deployment of electric vehicles. The widespread implementation of Remote Monitoring and Control (RMC) systems for various facilities will enhance operational efficiency and environmental management. These innovations will play a pivotal role in advancing both the circular economy and sustainable waste management practices, helping to create a more resilient and adaptive waste management system for the future.

1.2 Circular Economy Initiatives in Indonesia

Indonesia faces a significant waste management challenge, with the nation producing 23.1 million tons of waste annually as of 2023, as reported by SIPSN. Over half of this waste (50.3%) ends up in landfills, while only 16.5% is recycled or reused to prevent environmental strain. Predominantly composed of 41.54% food waste and 11.78% wood and branches, Indonesia's waste profile highlights organic materials as the most common types. To address these issues, Indonesia implemented Law No. 18 of 2008 on Waste Management, establishing a legal framework aimed at improving waste management nationwide. Under this law, responsibilities are delineated among national, provincial, and municipal governments: The national government facilitates infrastructure development and coordinates national initiatives, while municipal and provincial governments oversee local waste management operations. Municipal waste begins its journey at temporary garbage disposal sites (Tempat Penampungan Sementara/TPS), which are categorized into 3R facilities which are promoting reduction, reuse, and recycling, multi-compartmental TPS, and those operating on a waste type-based schedule. This procedure is critical in ensuring that waste is appropriately managed before final disposal at designated landfill sites (Tempat Pemrosesan Akhir/TPA) [7].

In the context of the circular economy, Indonesia, through the Low Carbon Development Initiative (LCDI) platform led by the National Development Planning Agency (Badan Perencanaan Pembangunan Nasional/Bappenas), is spearheading the implementation of circular economy principles in national policy. Since the early 2020s, LCDI has initiated several studies, including "Food Loss and Waste in Indonesia to Support the Implementation of Circular Economy and Low Carbon Development" and "The Future is Circular—Concrete Steps towards Circular Economy Initiatives in Indonesia." Recently, in 2024, Bappenas issued the "Roadmap and National Action Plan for Indonesia's Circular Economy 2025–2045" to outline national initiatives aimed at promoting circular economy practices across five priority sectors: food, retail (with a focus on plastic packaging), electronics, construction, and textiles. This document aims to establish measurable and achievable national targets. Several of these priority sectors also contribute to addressing Indonesia's waste management challenges, such as food and agricultural waste, e-waste, and plastic waste.

In order to effectively tackle waste management challenges, it is crucial to make use of digital solutions. By employing digital applications, significant savings in time and resources can be achieved, improving the efficiency of waste management systems and contributing to sustainability and smarter operations [1]. These technologies allow for real-time monitoring, data-driven decision-making, and improved operational transparency, leading to more effective environmental management. Indonesia is actively adopting various digital solutions to enhance the national waste management system. Both the government and private sectors are utilizing digital applications for different purposes. These digital solutions can be categorized into two approaches: web- and application-based. While both approaches aim to use data for better decision-making or provide improved services, they serve different functions. The following section will delve into the specific types of applications used in Indonesia.

Government-Led Initiatives

The National Waste Management Information System (Sistem Informasi Pengelolaan Sampah Nasional/SIPSN), launched by the Ministry of Environment and Forestry (MoEF) in 2021, provides a comprehensive view of the waste management facilities in Indonesia. The platform provides both distribution maps showing the locations of waste banks (Bank Sampah), recycling centers (Pusat Daur Ulang/PDU), and other facilities across the country, as well as tabular data detailing waste generation, sources, composition, and the amount of waste managed in each area.

SIPSN promotes collaboration between national and city or district governments. MoEF acts as the data aggregator, while city and district governments contribute by reporting their waste management profiles to the national database. This transparency allows users to understand how these management facilities are distributed and their effectiveness across all cities and districts in Indonesia. This data enables the Government of Indonesia (GoI) to prioritize policy interventions and allocate budgets effectively.

Other than SIPSN, MoEF also released the Waste Bank Management Information System (Sistem Informasi Manajemen Bank Sampah/Simba). While SIPSN contains information on Indonesia's overall waste management landscape, Simba provides detailed information on one specific type of waste management facility: a waste bank that can be defined as a community.

Other Digital Initiatives

Differing from government-led digital initiatives focused on information portals, the private sector also contributes to waste management by providing user-oriented digital solutions. Waste management applications in Indonesia, developed by the

Table 1 Comparison of digital solutions for waste management in Indonesia

Feature	Waste collection and recycling	Refill and reuse
Examples	eRecycle, Octopus, Rapel, Kepul, Rekosistem	Siklus
Primary function	Connects users with waste collectors for recyclable pickup	Provides refillable household products, reducing single-use plastic
Waste focus	Recyclable materials (plastics, paper, metals, etc.)	Specific refillable products (detergents, cooking oil, cleaning supplies)
Scheduling	Schedule pickup for recyclable materials	Not applicable—refills are typically ordered on-demand
Material categorization	Guides users on proper sorting and categorization of recyclables	Not applicable—users order predefined refillable products
Reward systems	May offer points or rewards for recycling pickup	Not applicable—focuses on reducing waste, not generating rewards
Waste tracking	May offer tracking of collected recyclables	Not applicable—refill products are delivered, not collected
Payment integration	May allow payment for collection services or reward redemption	May allow payment for refillable products and deposits
Subscription management	Not applicable—users schedule pickup as needed	May offer subscriptions for regular deliveries of refillable products
Deposit management	Not applicable	May offer deposit systems for reusable containers (pay deposit, refunded upon return)
Delivery tracking	Not applicable	May offer tracking of refillable product deliveries
Educational content	May offer limited information on proper recycling practices	May offer educational content on the benefits of refilling and reducing waste

Source: Author Analysis, 2024

private sector, encompass two key functionalities: on-demand waste collection with a material recovery focus and subscription-based refillable product delivery services. Table 1 explains the differences between these functionalities and provides examples of each.

On-demand waste collection with a material recovery focus is a core service offered by applications like eRecycle or Octopus. These applications connect users with collectors for recyclable pickup, often leveraging dynamic resource allocation and service scheduling to optimize collection routes. Additionally, waste stream segregation and classification features guide users in properly sorting their recyclables. Gamification, through points or rewards, may incentivize user participation. Real-time waste management monitoring and tracking, if available, allows users to follow the movement of collected recyclables.

On the other hand, subscription-based refillable product delivery services tackle waste generation at its source. Applications like Siklus focus on refill and reuse by offering subscriptions for refillable household products. Automated replenishment ensures users receive refills based on their needs, while customer relationship management (CRM) features handle subscriptions and communication. Deposits on reusable containers encourage their return through a dedicated deposit and return system. Real-time delivery management and tracking allow users to monitor refill deliveries.

1.3 Case Study: Circular Economy Initiatives in Banyumas Regency

Banyumas Regency, a region known for its vibrant community and rich cultural heritage, spans 1,327.59 square kilometers and is home to approximately 1.81 million residents (Central Bureau of Statistics, 2022). The regency, located in Central Java, is characterized by its diverse landscapes, ranging from urban areas to agricultural lands and natural reserves. In recent years, Banyumas has faced significant challenges in managing the increasing volumes of waste generated by its growing population and economic activities. Based on SIPSN data from 2023, the region generated an estimated 197,758.42 tons of waste per year, assuming an average individual waste production of 0.32 kg per day. This considerable amount of waste poses serious environmental and health risks, necessitating urgent and innovative solutions.

Prior to 2019, waste management in Banyumas primarily followed a "collect-transport-dump" model. Unsegregated waste from households, markets, shops, shopping centers, and public facilities was collected and transported to TPSs scattered throughout the region, including numerous illegal TPSs. The accumulated waste was then transported to the TPAs or landfills. Additionally, in 2014, residents protested the closure of the Gunungtugel TPA due to pollution concerns, and in 2018, the Kaliori TPA was also closed following similar demands. These closures were necessitated by pollution and threats to the health and well-being of surrounding communities.

Recognizing the need for change, Banyumas' government endeavored to realize sustainable waste management in mid-2018. This involved a shift away from the traditional "collect-transport-dump" model. Temporary storage facilities were replaced with Integrated Waste Management Facilities (Tempat Pengolahan Sampah Terpadu/TPST) equipped for sorting, composting, and recycling. Educational campaigns and incentives were launched to encourage residents to reduce waste generation and embrace reuse practices. Waste banks, already present in some communities, were expanded and strengthened to facilitate community participation in waste segregation, recycling, and composting.

The waste management efforts in Banyumas resulted in 99% of waste being properly managed and only 1% left unmanaged. This achievement is attributed to activities such as recovery, recycling, waste to energy, and other initiatives, with some waste ultimately being sent to landfills. The use of digital solutions such as JEKNYONG and Salinmas supports the improvement of the waste supply chain, making it more integrated between users as waste generators and waste collectors such as waste banks or industries. Moreover, Banyumas also produces various upcycled products such as maggots, compost, and refuse-derived fuel (RDF), which helps eliminate waste while generating revenue. Specifically, for RDF, Banyumas has secured agreements with cement industries such as PT Solusi Bangun Indonesia and PT Semen Bima to deliver the RDF product to their cement kilns. For all these initiatives, Banyumas was chosen as the host of the Smart Green ASEAN Cities (SGAC) program, earning international recognition.

2 Lessons Learned

2.1 Circular Economy: KOLEKT Application

KOLEKT is a mobile application developed by the Netherlands-based Circular Action and launched in 2017. It is considered a successful waste management tool, as it bridges the gap between waste producers and collectors, leading to a more transparent and efficient waste management system. The app uses a user-friendly interface and is available in multiple languages, such as English, Portuguese, Indonesian, Spanish, Arabic, and Vietnamese. This makes it accessible to a diverse range of users globally. KOLEKT can be used by waste generators, collectors, processors/recyclers, and brand owners/PROs. Each user has a role in the waste collection value chain. All users can buy and sell waste materials, except waste generators (sell/offer only) and recyclers (buy only) [4].

KOLEKT has several strong points, including:

1. The user-friendly interface and availability in multiple languages, such as English, Portuguese, Indonesian, Spanish, Arabic, and Vietnamese. This makes the apps accessible to a diverse range of users globally. This could potentially open a bigger market for waste.
2. Multi-material management. Materials to be sold or bought in the apps include plastics, paper, cartons, metals, rubber, e-waste, glass, foam, medical waste, construction waste, biomass, and many sub-materials. This waste can appear as shredded, sorted, clean, baled, bagged, mixed, or contaminated conditions. The variety of materials sold in the apps will increase and add to the amount of recycled waste.
3. Offer/sell and request/buy. The apps offer more innovative trade mechanisms. The app allows users to sell materials with an expiry date on the offer or "sell-now" face-to-face. Interested buyers can place bids and negotiate prices when

sellers still have active offers. Users looking for materials can post a request with the expiry date or "buy now" face-to-face. Interested sellers can bid to deliver and negotiate the prices for the waste they will provide as requested. These bid and negotiation features will allow sellers and buyers to get better economic benefits accordingly. All the offered and requested materials should include weight, price per kg, GPS location, and photo condition of waste. There are also options to deliver or pick up the waste by the buyers or requesters.
4. Map and search features. Users can find the exact location of available materials for sale.
5. The apps can be accessed with or without an Internet connection. Users can post offers, request materials, and view items that they viewed before going offline, which will be re-synchronized when they go online. This feature allows the apps to be used in locations with unstable or scattered Internet connections.
6. People without smartphones can make transactions. If someone without a smartphone wants to sell or buy waste, they can have an account and log in to the apps using KOLEKT ID and photo verification. Therefore, one smartphone can be used for unlimited accounts.

2.2 Circular Economy: Cambodia

Cambodia has two ongoing initiatives in utilizing digital tool waste management: Green Cambodia and Samraam.

1. Green Cambodia
Green Cambodia was launched in 2020. This project was funded by USAID-Finances Tech for Green Cities (T4GC) and implemented in four provinces, including Battambang. In December 2021, this application reached around 14,900 downloads, 15,500 sign-ups, and 400 citizen reports [2]. The application has key features:

- Report
  Users can report waste disposal or management problems in real time. They can attach images and add links to the location (geotagging). The report is then sent to the related service provider to be resolved.
- Find
  The app provides residents easy access to waste collection schedules and service providers.
- Learn
  The Green Cambodia app is a reporting tool and a valuable learning resource. It features a dedicated learning section, the Green Cambodia learning page, which hosts a wealth of educational resources. Users can access waste disposal guides and educational videos contributed by the local community or field experts, making it a comprehensive learning platform.
- Marketplace

Users can make transactions to buy, sell, or donate recyclable and reusable material.

With these features, the app promotes collaborative efforts in waste management, empowering citizens to be active and educated while encouraging responsible waste management practices and promoting a circular economy.

2. Samraam
This is an application designed to enhance solid waste management. The app provides several valuable features for citizens, including real-time tracking of waste collection trucks via GPS. Real-time tracking feature enables users to plan and avoid overflowing bins. Samraam also offers transparent waste collection schedules with notifications about the truck's approaching arrival. Furthermore, the app allows users to report issues such as clogged bins, illegal dumping, or waste buildup. Additionally, Samraam integrates with popular payment platforms like Pay Way and Wing Pay, allowing users to conveniently view and pay their waste collection bills within the app, thus eliminating the need for physical invoices. Moreover, the app includes a service request feature for various types of waste, such as household waste, garden trimmings, and construction debris. Overall, Samraam's citizen-centric approach promotes transparency and accountability in waste collection services, enabling residents to actively maintain a clean and healthy environment.

2.3 What to Learn

This section will visualize the points of excellence from the best practices selected in the previous section. The points will be considered in capturing aspects recommended for improving the features of waste management applications in Banyumas Regency.

3 Operation Research Application for JEKNYONG and Salinmas, Circular Economy Practice in Banyumas Regency

3.1 Upstream to Downstream of Banyumas Regency Waste Management

Waste management in Banyumas Regency is currently divided into upstream, middle, and final processes. Upstream waste management is carried out by each household. Each household plays a crucial part by separating organic and inorganic waste and managing it separately. For organic waste, the community's efforts are

Table 2 Multi-criteria analysis for the features of selected best practices of waste management applications

Criteria	KOLEKT	Green Cambodia	Samraam
Easy to use	Yes	Yes	Yes
Offline feature	Yes	No	No
Map feature	Yes, for waste or materials location	Yes, photo geotagging for waste management problems	Yes, tracking the waste truck
Live tracking	No	Yes	Yes
Users must have a smartphone to make an account	No, people without smartphone can make their account	No, website available	Yes, the website is available but cannot be accessed
Educational Resources	No	Yes	No
Digital payment	Yes, coins	Yes, online payment (unidentified)	Yes, various choices of digital payment methods

Source: Author Analysis, 2024

further showcased by utilizing Salinmas. Meanwhile, inorganic waste utilizes the JEKNYONG application and waste banks.

Then the waste management process in the center is carried out by the Community Self-Help Group (Kelompok Swadaya Masyarakat/KSM). KSMs are local government partners and technical operators of waste management in settlements. KSMs process household waste in waste banks, TPST, PDU, or Reduce, Reuse, and Recycle-Waste Management Unit (Tempat Pengolahan Sampah Reduce, Reuse, dan Recycle/TPS3R) on a manual, mechanical, thermal, and biological basis. Manual waste processing is carried out by sorting high-value waste that can be sold to the recycling industry. Mechanical-based processing is done after manual sorting to pick up low-value or mixed waste. Then, the sorted results are processed with a shredding machine. The results of the shredding machine can be used as waste pulp, raw materials for refuse-derived fuel, and materials for other upcycling products (paving blocks, plastic tiles, and pots). Then, there is biological processing, namely, with the help of biological agents, to convert organic waste into compost and maggot feed ingredients. Final waste management is the process of destroying residual waste that can no longer be processed through the pyrolysis or combustion process. This final waste processing can be carried out at the Environment-and-Education-Based Landfill (Tempat Pembuangan Akhir Berbasis Lingkungan dan Edukasi/TPA BLE) unit and several TPSTs in Banyumas Regency.

The efforts to manage waste from upstream to downstream in Banyumas Regency have succeeded in increasing participation from various parties to realize zero waste and have shown significant results in reducing waste disposed of in landfills. In 2018, the waste disposed of in landfill was 300 tons per day. In 2022, it was reduced to 35 tons per day, and those would still be processed at the TPA BLE.

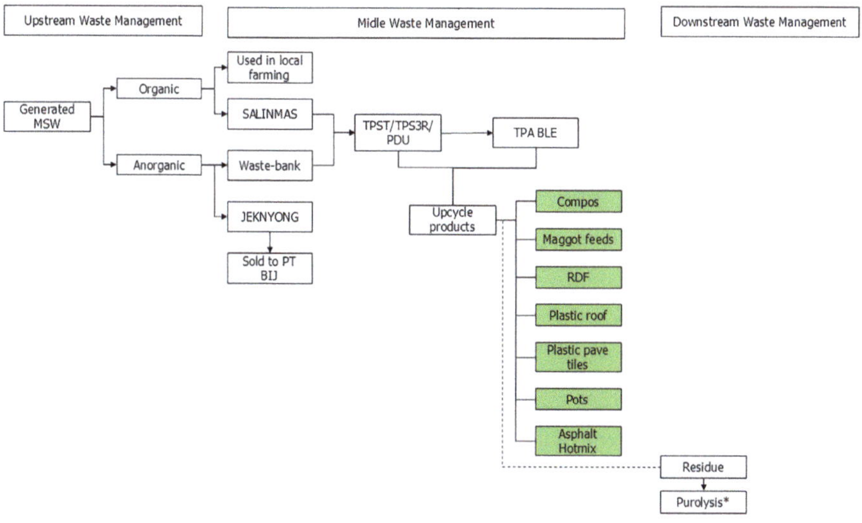

Fig. 1 Organizational chart of waste management in Banyumas Regency

Figure 1 illustrates the comprehensive organizational chart of waste management in Banyumas Regency. The waste management model in Banyumas Regency, as explained, has become a model for other regions in Indonesia, especially small or medium districts/cities. Many regions have visited to learn or replicate the Banyumas waste management scheme.

3.2 *Optimum Service Area and Route for Waste Collection*

In Banyumas Regency, six Waste Technical Implementation Units (Unit Pelaksana Teknis Persampahan/UPKPs) are responsible for waste management, namely UPKP Ajibarang, UPKP Banyumas, UPKP Kembaran, UPKP Purwokerto, UPKP Sumpiuh, and UPKP Wangon. Each UPKP is responsible for managing waste in its surroundings. However, the service area of each UPKP is only allocated to its respective area without direct allocation to existing waste facilities, namely TPST, TPS3R, PDU, and waste bank (Bank Sampah). A distinct allocation of waste facilities for every UPKP would have created a more accountable environment for the waste management system. Such improvement sets clear boundaries between UPKPs' lines of duty and allows for consistent services to the public to be delivered.

Optimizing the transportation system is part of the operations research objectives. In this context, the transportation system of waste collection in Banyumas Regency becomes the optimization target. The system itself is managed under the Environment and Education-based TPA (TPA BLE) unit, which assigns the waste collection task to the most appropriate Waste Management Technical Implementation Unit

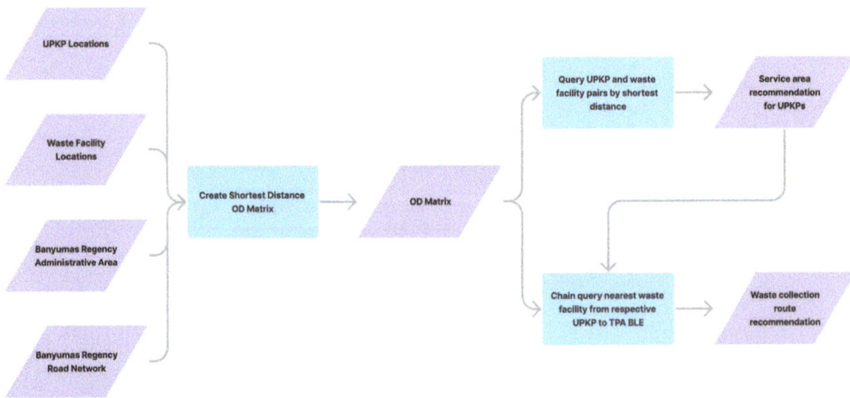

Fig. 2 Methodology to determine the optimum route

(Unit Pelaksana Teknis Persampahan/UPKP) in the Banyumas Regency, depending on the location of the waste collection call. The waste collection vehicle stands by in every UPKP and is conditioned to collect waste as instructed and, at the end of the collection journey, deliver it to the TPA BLE. The waste collection system is arbitrarily conducted in the current situation, where a fixed amount of waste collection vehicles is provided, and the waste facility will coordinate with the TPA BLE unit to call for a collection. Usually, operations research can be delivered to optimize the route. However, the route varies each time a waste collection is carried out. This poses a challenge in measuring the system's efficiency with operations research due to the inconsistent nature of the route.

A service area and waste collection route are proposed to improve the Banyumas Regency waste collection system through operation research using the methodology shown in Fig. 2. UPKP locations, waste facility locations, and Banyumas Regency administrative area and road network are collected as input to the model, all of which were sourced from Banyumas Regency Regional Development Planning Agency (Badan Perencanaan Pembangunan Daerah/BAPPEDA Kabupaten Banyumas) and Geospatial Information Agency (Badan Informasi Geospasial/BIG). Python programming and QGIS software were utilized to carry out the analysis.

An origin-destination (OD) matrix is the foundation of this analysis. An OD matrix is created to get the distance between each of the facilities, namely TPST, TPS3R, PDU, waste bank, and TPA BLE using the QNEAT3 plug-in in QGIS. The information on the distance between each of the facilities enables two things: first, to get the nearest UPKP from each waste facility and determine the service area of each UPKP, and second, from each service area, by assigning UPKP as a start point and TPA BLE as an end point, find the nearest waste facility from the start point (Point A) and find the other nearest waste facility from Point A that is not Point A (Point B), and so on (chain query) until every waste facility is selected as part of the route.

Allocating each waste facility to the nearest UPKP to determine service areas can be classified as a location-allocation (LA) problem. The goal of the LA problem in this research is to find exactly one facility per customer in a way that minimizes the cost of transportation from those facilities (UPKP) to customers (waste facility), which can be modeled with objective function and its constraints as follows [6]:

$$\min \phi = \sum_{i=1}^{m} \sum_{j=1}^{n} Z_{ij}\, r_j\, d(x_i, a_j) \tag{1}$$

subject to

$$\sum_{i=1}^{m} Z_{ij} = 1, \quad j = 1, 2, \ldots, n \tag{2}$$

$$Z_{ij} \in \{0, 1\}, \quad i = 1, 2, \ldots, m; \quad j = 1, 2, \ldots, n \tag{3}$$

Parameters

Φ : Total cost of transportation between facility and customer

n : Number of customers

m : Number of facilities

$$Z_{ij} : \begin{cases} 1 & \text{if customer } j \text{ is assigned to facility } i, \\ 0 & \text{otherwise} \end{cases}$$

r_j : Customer j's demand, $\quad j = 1, 2, \ldots, n$

x_i : Coordinates of facility i, $\quad i = 1, 2, \ldots, m$

a_j : Coordinates of customer j, $\quad j = 1, 2, \ldots, n$

$d(x_i, a_j)$: Distance between customer j and facility i

Equations (1), (2), and (3) reduce the overall cost of transportation, ensure all customer demands are met, and provide standard constraints, respectively [6]. Due to the lack of customer demand data (i.e., the waste capacity of each waste facility), r is assumed to be 1. Since every customer must be assigned to one UPKP, Zij equals 1. The total cost of transportation will only consider the distance between each UPKP and waste facility pair because of data limitations of other possible costs (i.e., speed limits). Therefore, the LA problem for every pair of UPKP and waste facility can be solved by finding a UPKP with the minimum or shortest distance for each waste facility.

Finding the distance between each facility to create an OD matrix can be classified under the all-pairs shortest path (APSP) problem in a graph with a nonnegative edge length [3]. Dijkstra's algorithm [5] is among the most common and fundamental methods to solve this problem by applying the algorithm to each vertex treated as the start point [3, 13, 14]. For each vertex, it is expressed as follows:

Find the shortest path length from the source vertex $(s) \in V$ to each vertex $(v) \in V$ in the directed graph $G = (V, E)$ with edge length $\ell(e) \geq 0$ $(e \in E)$ [13]. Through QGIS software, this algorithm is implemented by QGIS3 to calculate the shortest distance between a pair of points [15]. A QGIS plug-in, QNEAT3, scales this algorithm to create an OD matrix that measures the distance between every UPKP and waste facility [16].

Dijkstra's algorithm is a greedy, node-labeling algorithm. It assumes that any pair of nodes i and j must have a nonnegative distance, denoted as c_{ij}. The labels contain two elements: $d(i), p$. The first component, $d(i)$, represents an upper limit on the length of the shortest path from the source node s to node i, while p denotes the node that comes before node i in the shortest path connecting them [8]. The shortest pathways between s and every other node in the network can be found using the following algorithmic steps [8]:

1. For each node i, assign a number $d(i)$ to indicate the approximate (upper bound) length of the shortest path from s to i that only employs labeled nodes as intermediate nodes. Assume that for every $i \neq s$, $d(s) = 0$ and $d(i) = \infty$. The last labeled node is represented by y. Let $y = s$, and assign the label $\{0, -\}$ to node s.
2. Modify $d(i)$ in the following manner for each unlabeled node i: $d(i) = \min\{d(i), d(y) + c_{yi}\}$. Should $d(i) = \infty$ for every unlabeled vertex i, then stop since there is not a path connecting s to any unlabeled node i that has the lowest $d(i)$ value. Furthermore, let p in the label represent the node that the arc that determined the minimum $d(i)$ came from. Assume that $y = i$.
3. Stop if all nodes have been labeled, as there is a shortest path from s to i for all vertices i that is the unique path of labels $\{d(i), p\}$. If not, go back to Step 2.

The service area of each UPKP is determined using the OD matrix and visualized in Fig. 3. The number of allocated waste facilities for each UPKP is shown in Fig. 4. The proxied actual allocation is collected from the daily waste collection data from each UPKP. Due to the daily variation of waste collection, the entire data could only reflect part of the waste facilities served by each UPKP. Hence, the first week in June 2024 is randomly chosen as a sample to compare with the findings. Additionally, UPKP Purwokerto has the most allocated waste facilities, and UPKP Kembaran has none assigned to it. UPKP Purwokerto is located in a densely populated urban area surrounded by numerous waste facilities. UPKP Kembaran is located near UPKP Purwokerto but is too far from any waste facility, resulting in no allocation from the model that only considers distance as the cost.

Owing to the lack of data availability, the model has no sensitivity to the capacity of each UPKP and waste facility. Thus, the allocation yielded from the model could overwhelm or underwhelm each UPKP, resulting in the inability to serve the waste

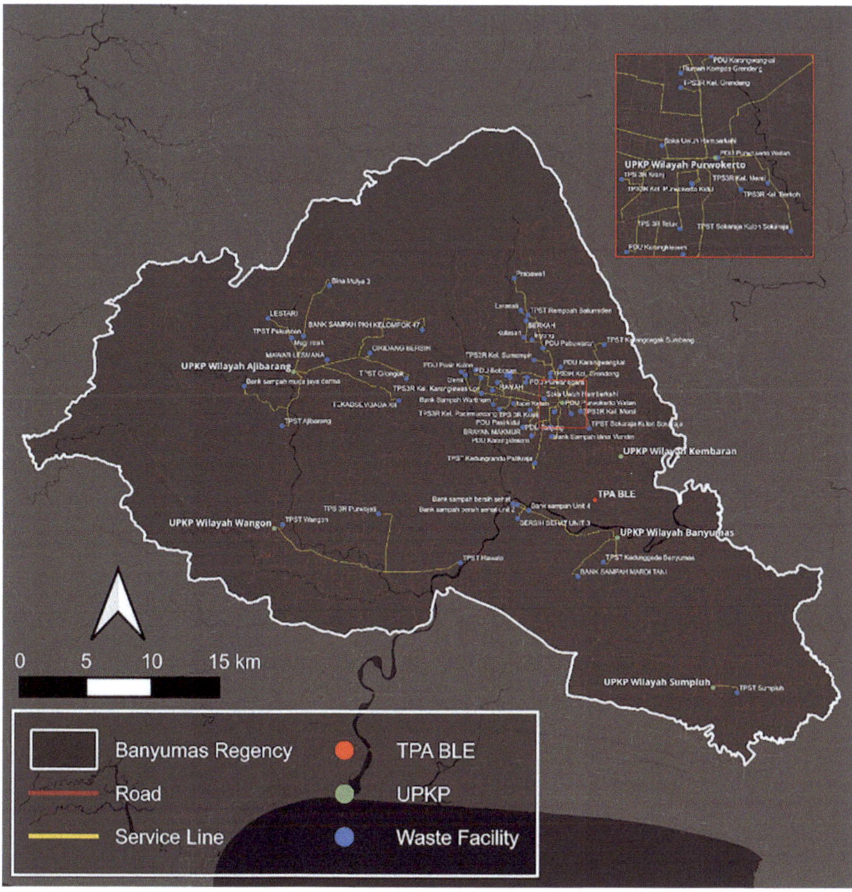

Fig. 3 Service area of UPKPs in Banyumas Regency

facility at optimum capacity in either case. The sampled daily waste collection data from each UPKP is utilized once again to validate the service area that the model outputs, as shown in Fig. 5. Indeed, UPKP Purwokerto relatively collects much more waste due to the high demand from the Purwokerto area; therefore, the finding that it has the most allocation of waste facility matches. From the same data, it was noted that UPKP Kembaran does collect waste from several waste facilities in the Purwokerto area, which are subject to UPKP Purwokerto according to the model. The mismatch between the model output and the actual situation of UPKP Kembaran indicates an opportunity to evaluate its effectiveness and improve the overall cost efficiency of waste collection systems.

The optimum route recommendation and the length for each UPKP are shown in Figs. 6 and 7, respectively. It was found that UPKP Purwokerto has the longest

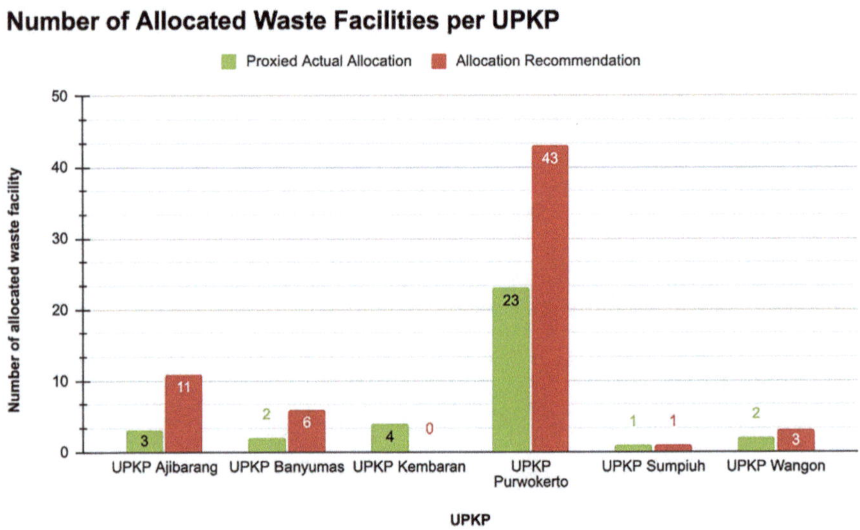

Fig. 4 Number of allocated waste facilities per UPKP

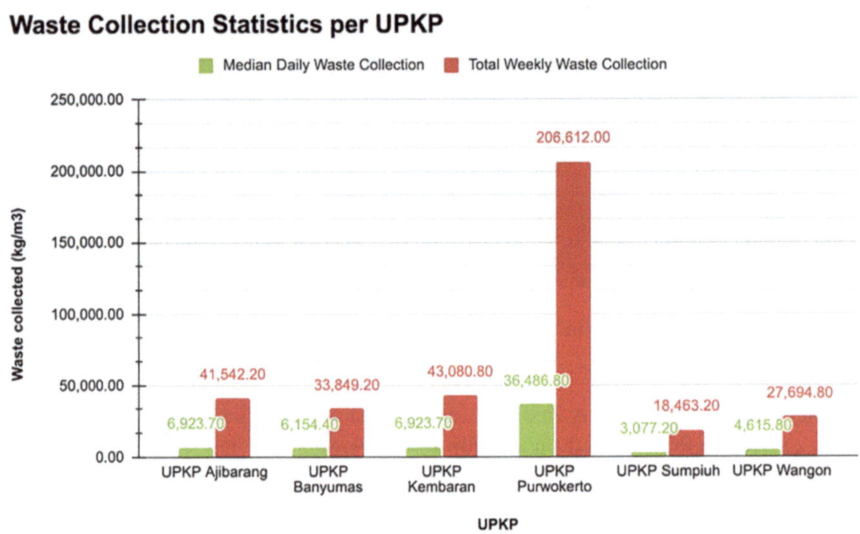

Fig. 5 Sampled waste collection of UPKPs in Banyumas Regency

route due to the amount of waste facility it needs to serve. UPKP Ajibarang serves relatively low waste facilities, but the area is far from the TPA BLE, resulting in a long route close to UPKP Purwokerto's length. UPKP Kembaran has zero route length since no waste facilities are allocated from the model. The actual daily waste collection route varies according to the assignment from the TPA BLE, so the model's output cannot be validated since the varying routes (actual situation) and

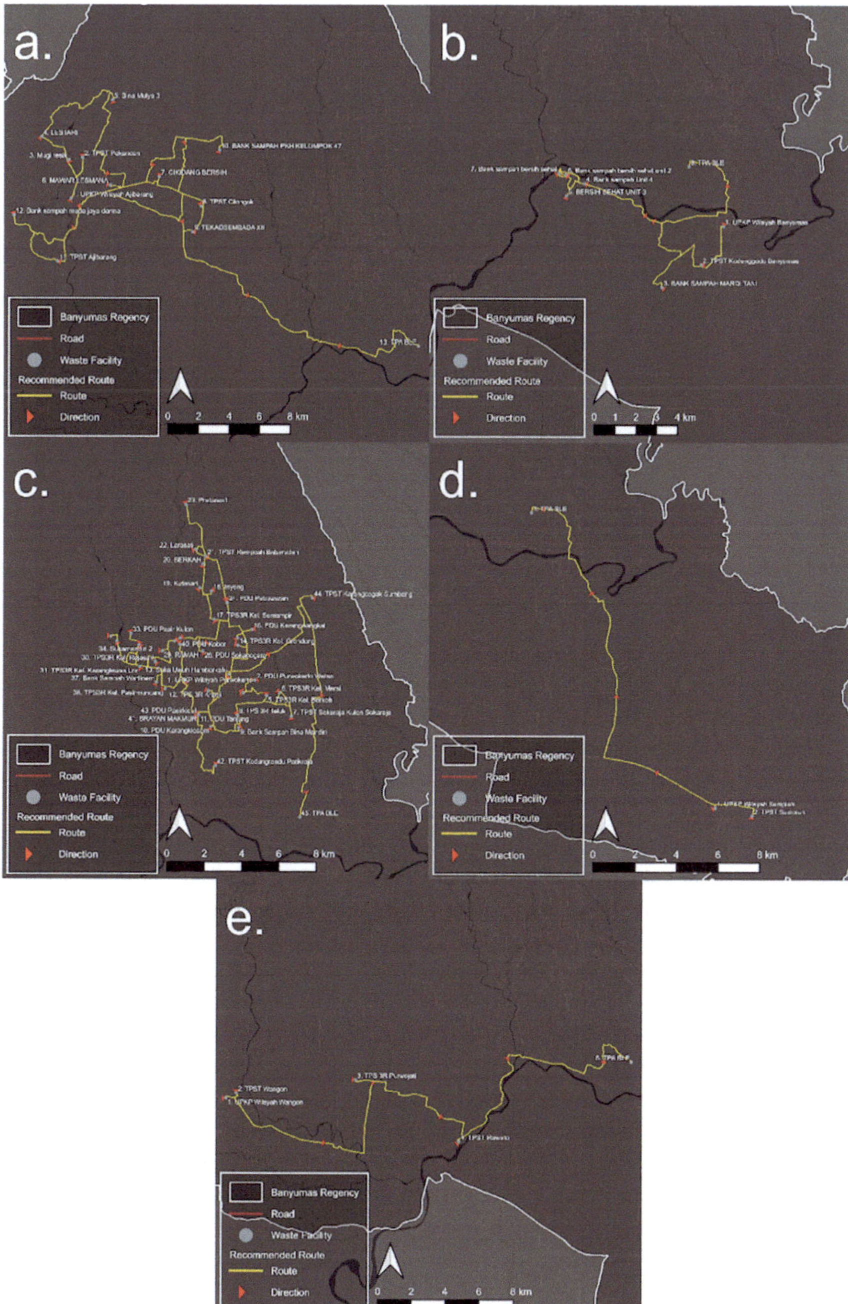

Fig. 6 Recommended routes for (**a**) UPKP Ajibarang, (**b**) UPKP Banyumas, (**c**) UPKP Purwokerto, (**d**) UPKP Sumpiuh, and (**e**) UPKP Wangon

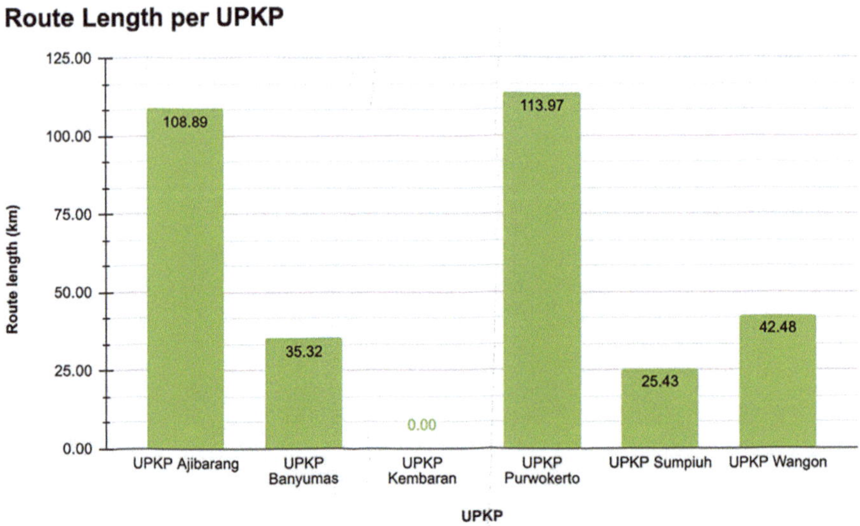

Fig. 7 Route length of UPKPs in Banyumas Regency

the fixed routes (model's output) are incomparable. Therefore, the result of the route recommendation is subject to further discussion with stakeholders from Banyumas Regency responsible for waste management.

The waste collection data sample was also utilized to capture the context of each UPKP's waste capacity and its waste collection (trucks) capacity. Each truck can carry 6 m^3 or 1,318.8 kg of waste [12]. The performance of each UPKP can be assessed by comparing the average daily waste collection and the total waste collection (trucks) capacity, shown in Fig. 8. This assessment shows three UPKPs with significant gaps: UPKP Banyumas, UPKP Kembaran, and UPKP Purwokerto. One of the ways this gap could be filled is by adding more waste collection capacity to each UPKP (i.e., more waste collection vehicles, vehicles with larger waste capacity, and so on). This gap might also exist due to the lack of data since we only know the number of trucks from each UPKP from the waste collection data sample, as there might be other means of waste collection other than trucks.

To refine the route recommendation, the average amount of waste that each truck carries and its route length from each UPKP are considered. For each UPKP, the total route length is multiplied by the ratio of average waste per truck and the average waste to get the number of segments. Afterward, each truck will be assigned to each segment and its respective waste facilities along the segment. This route segmentation can be found in Fig. 9. Although every truck must originate from its UPKP, not every segment starts from UPKP, as shown in Fig. 9. This is deliberately done to better visualize each segment because adding UPKP as a starting point to every segment will saturate the visualization. The route segmentation therefore serves as a recommendation for each UPKP to improve the management of its fleet of trucks.

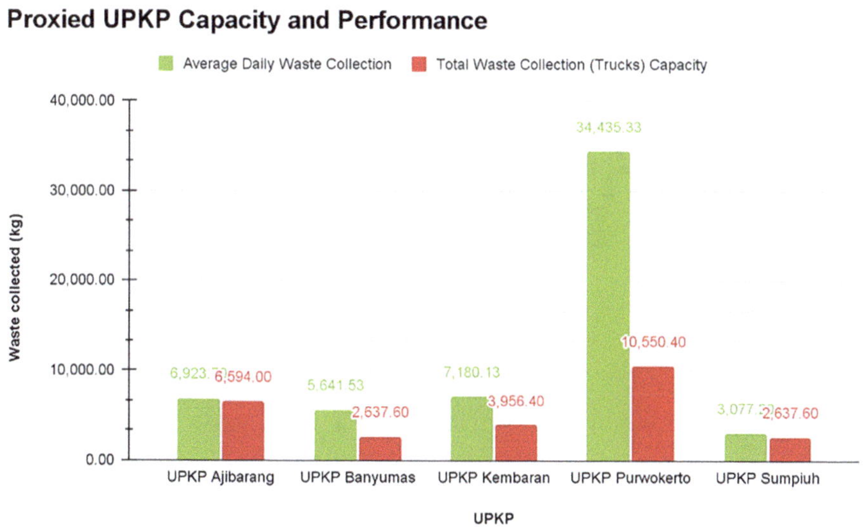

Fig. 8 Proxied UPKP capacity and performance

The waste collection route for each UPKP is therefore listed as follows (by collection order of each UPKP):

1. UPKP Ajibarang, TPST Pekuncen, Mugi Resik, Lestari, Bina Mulya 3, Mawar Lesmana, Cikidang Bersih, TPST Cilongok, TEKADSEMBADA XII, Bank Sampah PKH Kelompok 47, TPST Ajibarang, Bank Sampah Muda Jaya Darma, and TPA BLE
2. UPKP Banyumas, TPST Kedunggede Banyumas, Bank Sampah Mardi Tani, Bank Sampah Unit 4, Bersih Sehat Unit 3, Bank Sampah Bersih Sehat Unit 2, Bank Sampah Bersih Sehat, and TPA BLE
3. UPKP Purwokerto, PDU Purwokerto Wetan, TPS3R Kel. Purwokerto Kulon, TPS3R Kel. Purwokerto Kidul, TPS3R Kel. Berkoh, TPS3R Kel. Mersi, TPST Sokaraja Kulon Sokaraja, TPS 3R Teluk, Bank Sampah Bina Mandiri, PDU Karangklesem, PDU Tanjung, TPS 3R Kranji, Soka Uwuh Hamberkahi, TPS3R Kel. Grendeng, Rumah Kompos Grendeng, PDU Karangwangkal, TPS3R Kel. Sumampir, Inyong, Kutasari, BERKAH, TPST Rempoah Baturraden, Larasati, Prabawa1, PDU Pabuwaran, PDU Purwanegara, PDU Sokanegara, Srayan Makarya, PDU Bobosan, RAMAH, TPS3R Kel. Rejasari, TPS3R Kel. Karanglewas Lor, Suka Mandiri, PDU Pasir Kulon, Sukamandiri 2, Gemi, Mekar Amanah Jipang karang lewas, Bank Sampah Wartinem, TPS3R Kel. Pasirmuncang, Tape Ketan, PDU Kober, Brayan Makmur, TPST Kedungrandu Patikraja, PDU Pasirkidul, TPST Karangcegak Sumbang, and TPA BLE
4. UPKP Sumpiuh, TPST Sumpiuh, and TPA BLE
5. UPKP Wangon, TPST Wangon, TPS 3R Purwojati, TPST Rawalo, and TPA BLE

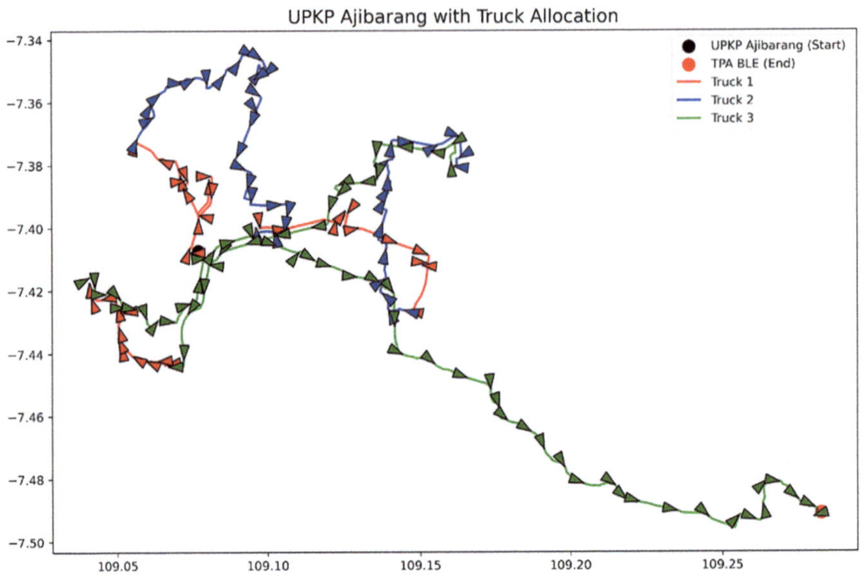

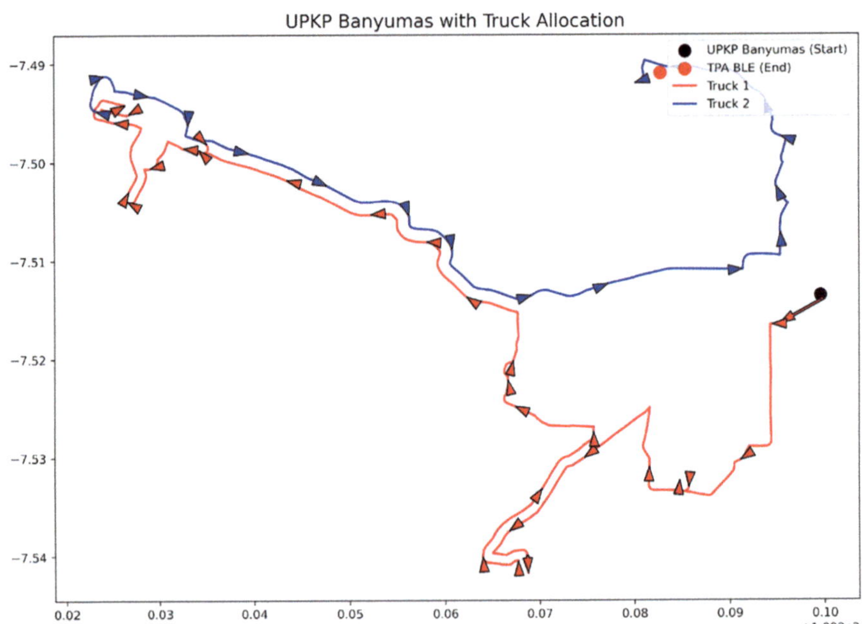

Fig. 9 (continued)

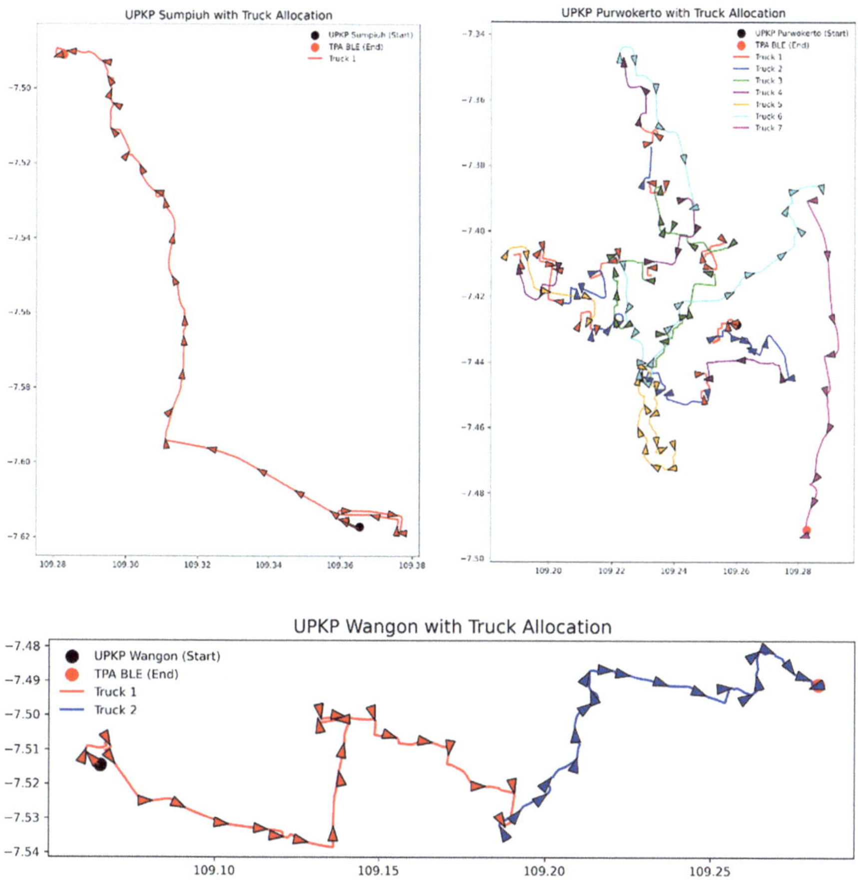

Fig. 9 Proxied UPKP capacity and performance

For each route segmentation, each truck will be allocated to waste facilities; read Table 3 for the complete segmentation.

3.3 Waste Flow Improvement

The Sankey diagram is utilized to analyze the advancements in waste management practices in Banyumas Regency, focusing on data collected from January to May 2023 and 2024. The key objective of this analysis is to examine the flow from sources such as waste collection by the KSMs and Waste Banks to final interventions like RDF production and recycling. The Sankey diagrams visualize the progression of waste through various stages of management, providing insights into the effectiveness of current waste management strategies.

Table 3 Truck assignments by region and facility in Banyumas Regency

UPKP Ajibarang	
Truck 1	UPKP Wilayah Ajibarang
	TPST Cilongok
	Mugi Resik
	UPKP Wilayah Ajibarang
	TPST Pekuncen
	MAWAR LESMANA
	CIKIDANG BERSIH
	TPST Ajibarang
	TPA BLE
Truck 2	UPKP Wilayah Ajibarang
	LESTARI
	Bina Mulya 3
	TEKADSEMBADA XII
	TPA BLE
Truck 3	UPKP Wilayah Ajibarang
	BANK SAMPAH PKH KELOMPOK 47
	Bank Sampah Muda Jaya Darma
	TPA BLE
UPKP Banyumas	
Truck 1	UPKP Wilayah Banyumas
	Bank Sampah Bersih Sehat Unit 2
	Bank Sampah Unit 4
	BERSIH SEHAT UNIT 3
	UPKP Wilayah Banyumas
	TPST Kedunggede Banyumas
	BANK SAMPAH MARDI TANI
	TPA BLE
Truck 2	UPKP Wilayah Banyumas
	Bank Sampah Bersih Sehat
	TPA BLE
UPKP Purwokerto	
Truck 1	UPKP Wilayah Purwokerto
	TPS3R Kel. Purwokerto Kulon
	UPKP Wilayah Purwokerto
	SRAYAN MAKARYA
	PDU Purwanegara
	BERKAH
	TPST Rempoah Baturraden
	Gemi
	PDU Bobosan
	Suka Mandiri

(continued)

Table 3 (continued)

	PDU Pasir Kulon
	Rumah Kompos Grendeng
	Inyong
	PDU Purwokerto Wetan
	TPS3R Kel. Grendeng
	TPS3R Kel. Karanglewas Lor
	Bank Sampah Wartinem
	TPS 3R Teluk
	TPA BLE
Truck 2	UPKP Wilayah Purwokerto
	TPS3R Kel. Rejasari
	Kutasari
	PDU Karangklesem
	TPS3R Kel. Pasirmuncang
	TPS3R Kel. Mersi
	TPS3R Kel. Purwokerto Kidul
	TPS3R Kel. Berkoh
	Bank Sampah Bina Mandiri
	RAMAH
	TPA BLE
Truck 3	UPKP Wilayah Purwokerto
	TPS3R Kel. Sumampir
	PDU Sokanegara
	TPS 3R Kranji
	PDU Tanjung
	Tape Ketan
	Soka Uwuh Hamberkahi
	PDU Karangwangkal
	TPA BLE
Truck 4	UPKP Wilayah Purwokerto
	Sukamandiri 2
	Larasati
	PDU Pabuwaran
	TPST Sokaraja Kulon Sokaraja
	TPA BLE
Truck 5	UPKP Wilayah Purwokerto
	Mekar Amanah Jipang karang lewas
	BRAYAN MAKMUR
	TPST Kedungrandu Patikraja
	TPA BLE

(continued)

Table 3 (continued)

Truck 6	UPKP Wilayah Purwokerto
	PDU Kober
	Prabawa1
	PDU Pasirkidul
	TPA BLE
Truck 7	UPKP Wilayah Purwokerto
	TPST Karangcegak Sumbang
	TPA BLE
UPKP Sumpiuh	
Truck 1	UPKP Wilayah Sumpiuh
	TPST Sumpiuh
	TPA BLE
UPKP Wangon	
Truck 1	UPKP Wilayah Wangon
	TPS 3R Purwojati
	TPST Wangon
	TPA BLE
Truck 2	UPKP Wilayah Wangon
	TPST Rawalo
	TPA BLE

$$Waste\ Generation = Population\ Per\ Capita\ Waste\ Generation\ Factor \tag{4}$$

The total waste generation is calculated using Eq. (4) with the estimated per capita waste generation factor of 0.109, derived from the region's total population of approximately 1.81 million. When multiplied by the population, this factor results in an annual waste generation estimate of 197,758.42 tons. This calculation provides a foundational understanding of the scale of waste produced in Banyumas Regency, forming the basis for evaluating the effectiveness of waste management strategies and initiatives over time. By analyzing trends in waste generation and comparing them with collection rates and management outcomes, policymakers and stakeholders can assess the impact of interventions aimed at reducing waste and promoting sustainable practices.

$$\text{Collection Rate (\%)} = \left(\frac{\text{Total Waste Collected}}{\text{Total Waste Generated}}\right) \times 100 \tag{5}$$

Waste collection rates, crucial for assessing the efficiency of waste management systems, are calculated using Eq. (5) to determine the percentage of total waste generated that is effectively collected for proper management. In 2023, the collection rate was measured at 4.71%, reflecting the portion of waste managed out of the total generated. By 2024, this rate had improved significantly to 7.55%, underscoring enhanced waste collection mechanisms and practices. This substantial increase highlights the region's progress in waste management infrastructure and operational strategies, potentially stemming from improved logistical planning, expanded collection services, and increased public participation in waste separation and disposal practices. Furthermore, the implementation of applications like JEKNYONG and Salinmas may also have contributed to these improvements by facilitating better waste tracking, incentivizing proper disposal practices, and encouraging community engagement in waste management initiatives.

Furthermore, the following sections assess the progress in waste management and explore how applications like JEKNYONG and Salinmas contribute to enhancing waste reduction and recycling practices in the Banyumas Regency. Sankey diagrams for 2023 and 2024 were analyzed using data collected from January to May 2023 and January to May 2024, averaging monthly. The diagrams highlight trends to illustrate improvements in the waste management landscape of Banyumas.

Analysis of 2023 Waste Management Data Through Sankey Diagram

In 2023, waste management efforts in Banyumas Regency were notably effective, with comprehensive waste collection and sorting activities involving various stakeholders, as depicted in Fig. 10. Service providers collected 9,307.33 tons of waste, mainly managed by KSM (9,269.72 tons) and Waste Banks (37.61 tons).

The KSMs played a crucial role in sorting activities, with 4,250.33 tons earmarked for recovery processes. This included 4,167.33 tons for recycling initiatives and 83 tons for producing pavement blocks, promoting sustainable construction practices locally. There was also progress in energy generation from waste, with 3,227.295 tons processed into RDF for alternative fuel and 763.9 tons undergoing pyrolysis, primarily to reduce waste volume.

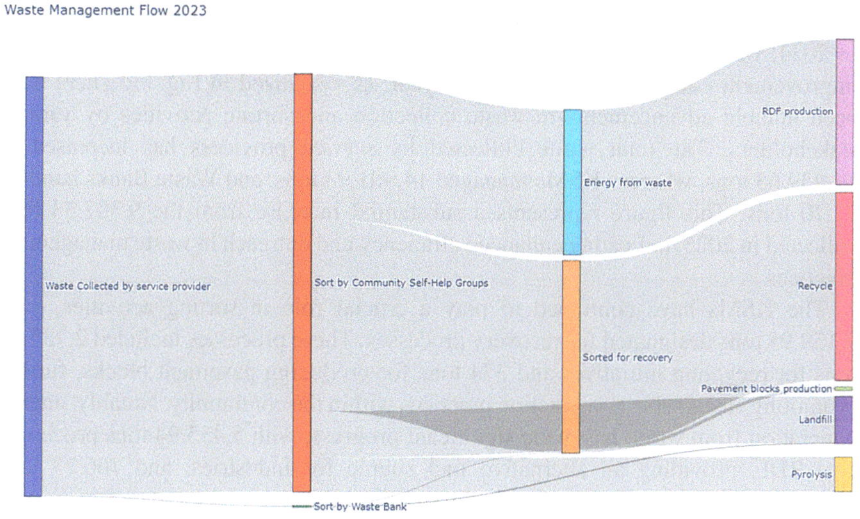

Fig. 10 Sankey diagram for 2023 data

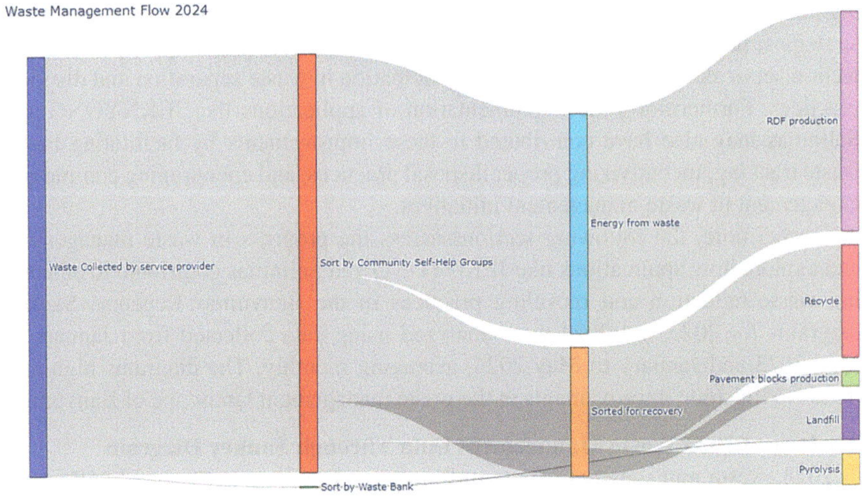

Fig. 11 Sankey diagram for 2024 data

However, the fact that 1,028.2 tons of wastes were still sent to landfills underscores the need to intensify efforts to minimize waste and enhance recycling initiatives. Waste Banks diverted 9.73 tons from landfills and recovered 23.58 tons, showcasing their vital role in local waste management. These initiatives demonstrate Banyumas' commitment to reducing landfill dependence and promoting a circular economy, which contributes to environmental sustainability and resilience in the region.

Analysis of 2024 Waste Management Data Through Sankey Diagram
In 2024, waste management efforts in Banyumas Regency have shown significant improvement compared to the previous year, as visualized in Fig. 11. There have been notable advancements in waste collection and sorting activities by various stakeholders. The total waste collected by service providers has increased to 14,939.63 tons, whereas KSMs managed 14,901.93 tons, and Waste Banks handled 37.70 tons. This figure represents a substantial increase from the 9,307.33 tons collected in 2023, indicating enhanced efficiency and outreach in waste management practices.

The KSMs have continued to play a crucial role in sorting activities, with 3,159.95 tons designated for recovery processes. These processes included 2,785.95 tons for recycling initiatives and 374 tons for producing pavement blocks, further promoting sustainable construction practices within the community. Notably, energy generation from waste has made significant progress, with 5,435.94 tons processed into RDF, providing an alternative fuel source for industries, and 766.23 tons undergoing pyrolysis, primarily to reduce waste volume.

Despite these advancements, challenges in waste disposal remain, although there has been a reduction in landfill dependence. In 2024, 986.05 tons of wastes were sent to landfills—a decrease from 1,028.2 tons in 2023, highlighting ongoing efforts to minimize waste and enhance recycling initiatives. Waste Banks also improved their performance, diverting 4.19 tons from landfills and recovering 30.88 tons through effective sorting practices, underscoring their essential role in the local waste management ecosystem.

These improvements reflect Banyumas' continuous commitment to reducing landfill dependence and advancing circular economy principles, fostering environmental sustainability and resilience in the region.

Comparison of Waste Management in 2023 and 2024

A comparison of waste management efforts in Banyumas Regency between 2023 and 2024 shows significant improvements and changes across various measures. The total waste collected by service providers increased noticeably, rising from 9,307.33 tons in 2023 to 14,939.63 tons in 2024. This growth suggests enhancements in waste collection systems, possibly driven by increased waste generation or improved reporting accuracy.

KSMs have played a crucial role in both years, managing 14,901.93 tons of waste in 2024 compared to 9,269.77 tons in 2023. Similarly, Waste Banks handled 37.69 tons in 2024, slightly up from 37.61 tons the previous year. Despite these increases, the proportion of waste directed to recovery processes by KSMs decreased from 4,250.33 tons in 2023 to 3,159.94 tons in 2024. This decline may indicate challenges or changes in sorting and recovery strategies.

Also, in 2024, there was a significant increase in energy generation from waste. The production of RDF increased from 3,227.29 tons in 2023 to 5,435.93 tons in 2024. This increase was made possible by agreeing to higher production targets with the off-takers from the cement industry, such as PT Solusi Bangun Indonesia and PT Semen Bima. Additionally, Pyrolysis, which is used to reduce waste volume, saw a slight increase from 763.9 tons to 766.22 tons over the same period.

Waste Banks have made significant strides in landfill diversion efforts, reducing disposals from 9.73 tons in 2023 to 4.19 tons in 2024. Simultaneously, sorted recoveries increased from 23.58 tons to 30.88 tons, underscoring their substantial contribution to improving local waste management practices through civil society voluntary initiatives.

Despite these advancements, recycling efforts experienced a decline from 4,167.332 tons in 2023 to 2,785.946 tons in 2024, while RDF increased. However, it is important to note that pavement block production rose significantly from 83 tons to 374 tons, indicating a positive trend in sustainable waste utilization practices.

These improvements highlight Banyumas Regency's ongoing commitment to reducing landfill dependence and advancing circular economy principles. The correlation between these improvements and the implementation of initiatives like Salinmas and JEKNYONG underscores the potential benefits of such applications. Salinmas, for instance, could facilitate better waste collection coordination among stakeholders, improving efficiency and coverage. Meanwhile, JEKNYONG's technology-driven approach might enhance data management and analysis, supporting more informed decision-making in waste management strategies.

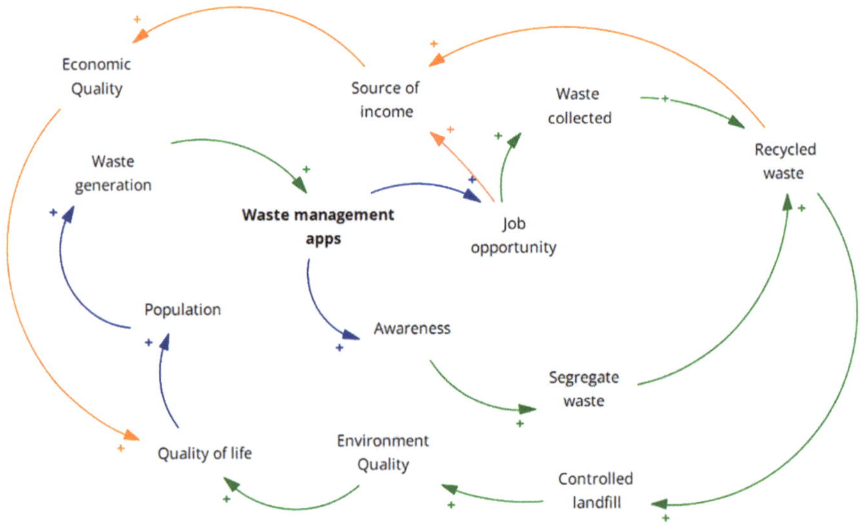

Fig. 12 Causal loop diagram the impact of waste management apps in Banyumas Regency

4 Waste Management Evaluation in Banyumas Regency

This section will further explain the evaluation of JEKNYONG and Salinmas Applications as digital waste management tools in the Banyumas Regency.

The Banyumas Regency is addressing its escalating waste issue with two innovative applications called JEKNYONG and Salinmas. A Causal Loop Diagram (CLD) analysis (Fig. 12) illustrates the positive feedback loop created by these apps. They work to increase public awareness about sorting and recycling organic and inorganic waste, leading to a reduction in the amount of untreated waste that ends up in landfills. By implementing controlled landfills and continuing waste management efforts, environmental conditions in Banyumas are expected to improve.

The benefits go beyond just the environment. The waste management system, supported by these apps, generates new job opportunities, such as waste managers in KSMs and drivers for household waste collection. These opportunities help increase household income and provide financial rewards for app users. Every waste transaction or sale results in additional income, either directly or periodically, through cash or cashless systems. Users can save these earnings within the app for later disbursement. Consequently, these applications contribute to improved socioeconomic conditions and an overall higher quality of life for the Banyumas community. A detailed evaluation of each application is provided in the following subsections.

4.1 *JEKNYONG (Ojeke Inyong)*

JEKNYONG is an innovative waste management tool in the form of an application that can be accessed by the public to sell their inorganic waste. The JEKNYONG application began operating on January 12, 2022. The operationalization of this application is run by a Regional-Owned Enterprise (Badan Usaha Milik Daerah/BUMD), PT Banyumas Investama Jaya (BIJ). This application can only be used by Android users. The implementation of the JEKNYONG application is concentrated in the Banyumas urban area, specifically Purwokerto and its surrounding areas, including Sumbang, Sokaraja, Patikraja, Karang Lewas, and Baturaden districts. In addition to improving the waste management system, the app aims to raise awareness and encourage people to be more aware and responsible for their waste. It has been successful in raising public awareness about sorting inorganic waste, evidenced by the good level of community acceptance and the 6000 users of the application in the Banyumas urban area.

As of 2022, JEKNYONG was developed into JEKNYONG IBMS. Its main feature for waste management is OJEKE NYONG, a transportation service to pick up waste at households. In addition, JEKNYONG also has other features for information in Banyumas Regency, namely VAKSINE NYONG, PANGANANE NYONG, OLEH-OLEHE NYONG, UMKMe NYONG, and WISATANE NYONG. These additional features aim to establish JEKNYONG IBMS as a regional information center and waste service provider in Banyumas Regency. The application workflow is shown in Fig. 13, involving three main actors who play a role in the waste management system: households (waste generators), couriers, or drivers who buy and pick up the household waste and sell it again to PT BIJ as an off-taker

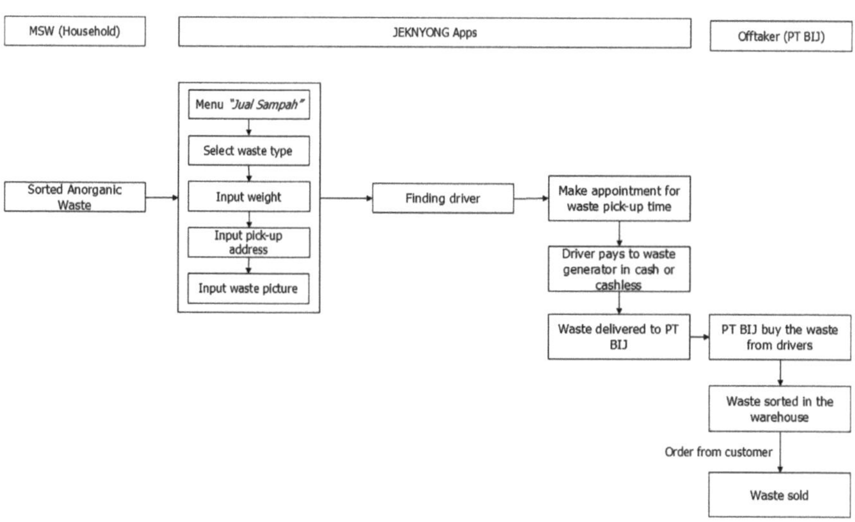

Fig. 13 JEKNYONG IBMS application workflow

of inorganic waste that has been collected through drivers. The types of inorganic waste that can be purchased by PT BIJ are very diverse, including plastic (17 items), aluminum, hollow iron, solid iron, cans, brass, copper, used cooking oil, used oil, various types of paper, glass bottles, and used electronic goods. The existence of such a waste management system has the potential to increase the amount and type of waste recycled in the area.

For its operation, there are four drivers and four fleets that collect waste from households. In the waste pickup process, households that have sorted their waste can upload the weight details of the waste, photos of the waste condition, and the pickup address. Then, the driver will get the order and process it for transportation. The pickup time is agreed upon by the waste generator and the driver, providing flexibility. Upon arrival, the driver will pay the resident according to the price listed on the application in cash or cashless. The waste is then delivered to the warehouse and sold to PT BIJ, where it is stored until a buyer is found.

Economically, this application has succeeded in increasing residents' income by enabling them to sell inorganic waste. The application listed the selling price of each type of waste per kilogram, subject to market price changes. Unfortunately, through this system, waste generators can only sell their inorganic waste to the available driver in their area. Therefore, the market for inorganic waste through this system is limited. It has yet to be opened for the public to buy waste directly from waste generators.

In addition, the application has opened up new job opportunities as a courier or driver of the garbage pickup fleet. The drivers benefited from the waste sales, earning more as they transported more waste to PT BIJ. Transactions could be done in cash or cashless using a coin system in the application, which could be withdrawn at any time. This system provided a convenient way for users and drivers to track transaction history.

Despite these advantages, the application could only manage 5–10% of the inorganic waste in Banyumas. Unfortunately, JEKNYONG ceased operations in 2024 due to the withdrawal of the transportation fleet and warehouse facilities by the local government. Until now, there is no plan to continue or reoperate JEKNYONG due to the budget constraint to provide the necessary facilities such as transportation fleets and warehouses. This has led to a flood of complaints from the community as they are unable to access the app, and their inorganic waste is not being collected, forcing them to find alternative waste management solutions. Table 4 shows the overview of JEKNYONG features availability. By completing these criteria, JEKNYONG is hoped to be a better waste management application.

4.2 Salinmas (Sampah Online Banyumas)

Salinmas (Sampah Online Banyumas, meaning Banyumas' Online Waste in English) is a waste management application specifically for organic waste in urban areas of the Banyumas Regency. This application was launched in November 2019

Table 4 Multi-criteria analysis for the features of JEKNYONG

Criteria	JEKNYONG
Easy to use	Yes
Offline feature	No
Map feature	Yes, only for tagging pickup location
Live tracking	No
Users must have a smartphone to make an account	Yes, website is not available
Educational Resources	Yes
Digital payment	Yes, cashless transactions using saldo in the app

Source: Author Analysis, 2024

and is designed to overcome the problem of limited land in managing organic waste in urban areas. The Salinmas application can be downloaded via Google Playstore and used only with Internet access.

The implementation process begins with socialization in residential areas, particularly targeting mothers in the neighborhood. At this early stage, residents were also given tutorials on sorting household waste. The response to this initiative was so good that residents were willing to participate. Initially, each household will be given one 60L barrel to store the organic waste. However, as the program progressed, this size was deemed too large for one household, so now one barrel is used for one Dasa Wisma (the group consists of neighboring households) consisting of 10–15 households. This approach has increased the awareness of urban communities to manage organic waste.

In the implementation of Salinmas, residents act as waste generators, while KSMs collect, weigh, transport, and verify the organic waste generated by each Dasa Wisma. The waste collected and transported by the KSMs is recorded through the Salinmas app and purchased by the Banyumas Regency Environmental Agency (Dinas Lingkungan Hidup Kabupaten Banyumas/DLH). The purchased organic waste is then used for maggot feed or compost, which has a positive impact on the environment by increasing the amount of managed organic waste and the selling value of the organic waste management results as illustrated in Fig. 14.

The proceeds from the sale of waste by each resident will be combined to create income for Dasa Wisma as an incentive. The amount of incentive earned will be shown in the application and can be saved or cashed out as desired. The amount of incentive earned by each depends on the amount of waste collected. For example, 14 Dasa Wismas in Sokanegara Subdistrict, Banyumas, earned six million rupiahs in 6 months by collecting organic waste. They used this incentive to finance community activities in their neighborhood. The Salinmas program has positively impacted the collective income of residents.

Unfortunately, the implementation of the Salinmas has experienced several obstacles. The Salinmas application update in 2022, known as Salinmas IBMS (Ide Baru Memilah Sampah or New Idea for Sorting Waste), has caused difficulties for residents as some devices are no longer compatible. The application is only

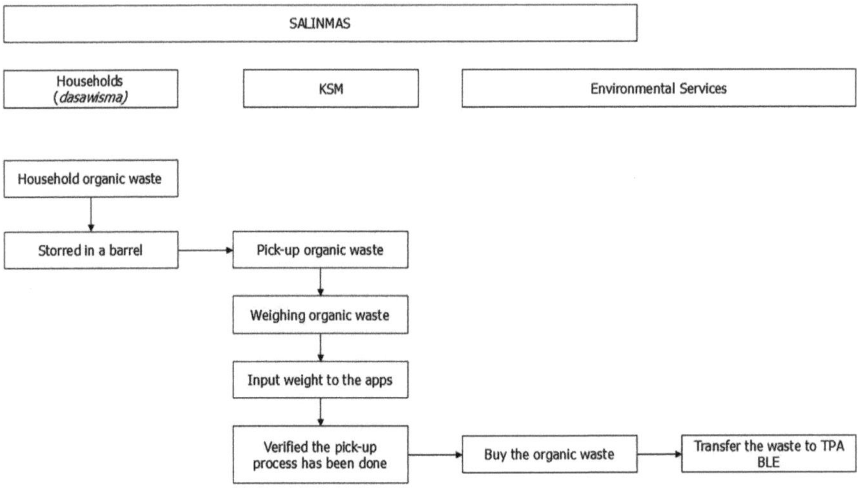

Fig. 14 Salinmas application workflow

accessible from Android devices. Therefore, residents are currently using the collection link or web-based application to input, verify, and view balances in Salinmas. Problems also arise when residents want to create an account in Salinmas. Creating a Salinmas account is limited to residents living in areas with registered and verified KSMs by DLH. Manual verification takes several days. As a result, the reach of Salinmas implementation is limited and time-consuming.

Currently, some KMSs have shown declining commitment to implementing Salinmas, leading to a buildup of uncontrolled organic waste. To overcome this problem, DLH is looking for alternative off-takers who need organic waste (food waste) and are willing to collect it from households such as worm, fish, chicken, and maggot farmers. For instance, in Karanglewas Lor Village, a local youth organization managed a fish farm that purchases organic food waste from households. This ensures that organic waste is managed and does not pollute the environment. Despite the fact that the price per bucket (about 20 kg) of organic waste is IDR 10,000, this scheme eliminates Salinmas' role in the organic waste supply chain. The profit obtained by residents is higher than the scheme purchased by DLH through KSMs. Meanwhile, in neighborhoods where KSMs are still actively running Salinmas, they usually have a special committee to collect and sell organic waste from households. This division of tasks within the KSMs can ensure the continuity of Salinmas in the Banyumas Regency.

Table 5 shows the overview of Salinmas features availability. By completing these criteria, Salinmas is hoped to be improved as a waste management application.

Table 5 Multi-criteria analysis for the features of Salinmas

Criteria	Salinmas
Easy to use	No
Offline feature	No
Map feature	No
Live tracking	No
Users must have a smartphone to make an account	No, the website link is available and more reliable
Educational resources	Yes, but only in the initial year of implementation
Digital payment	Yes, cashless transactions using saldo in the apps

Source: Author Analysis, 2024

5 Conclusion and Recommendation

This chapter explored how circular economy principles can optimize waste management at the regency level in Indonesia, with a focus on the Banyumas Regency as a case study. This chapter examined waste transportation, flow, and innovative waste management applications. Our thorough analysis led to the following conclusions:

1. A fixed route for waste collection by each UPKP could serve as a foundation for continuous and measured evaluation of the waste management system in the Banyumas Regency.
2. Compared to 2023, waste management efforts in Banyumas Regency in 2024 have shown significant improvements and changes that reduce landfill dependency and implement circular economy principles.
3. Digital waste management tools like JEKNYONG and Salinmas have had positive impacts on improving the social, economic, and environmental conditions in the Banyumas Regency, as illustrated in Fig. 12.

Based on the conducted analysis, several recommendations for improving waste management in Banyumas Regency are as follows:

- The effectiveness of UPKP Kembaran needs to be evaluated because although it is located close to the Purwokerto area, it is not close enough to any waste facility within it. Further investigation is needed since there could be an important reason for the specific location of UPKP Kembaran. If no important reason were found, the suggested evaluation could lead to, for instance, the replacement of UPKP Kembaran with a facility located closer to the Purwokerto area or its termination to save the maintenance and operational cost of the overall waste management system in Banyumas Regency.
- A fixed optimum route for waste collection from every UPKP in the Banyumas Regency could save transportation operational costs since every route is optimized. The route could be further segmented according to the total route length and average waste carried by trucks available in each UPKP. However, validation is key in implementing the recommendation, because there could be

an important reason why a varying route is a better implementation in the context of waste collection in the Banyumas Regency. For instance, daily collection is not required for each waste facility in the area. This favors the implementation of various routes since the vehicle will not have to travel to every waste facility, thus reducing operational costs.

- Implement standardized data collection and reporting procedures to ensure reliable waste management data, utilize advanced data analytics (including Sankey diagrams) for continuous monitoring and optimization of waste flow, and establish short-term, mid-term, and long-term targets aligned with circular economy goals to increase waste diversion rates and enhance RDF production efficiency over time.
- Based on the results of the evaluation of the two waste management applications in Banyumas Regency, the study found several areas for improvement. These include integrating organic and inorganic waste management into a single application to ensure that each feature of the JEKNYONG and Salinmas applications complements the other, thereby improving overall waste management. This study also recommends expanding the application's operational area beyond the Banyumas urban area, creating job opportunities as drivers or couriers for waste collection, expanding the waste market from Banyumas Regency to the public, and incorporating educational content within the apps to raise public awareness.

References

1. Abdullah N, Alwesabi OA, Abdullah R (2019) IoT-based smart waste management system in a smart city. Adv Intell Syst Comput 843:364–371. https://doi.org/10.1007/978-3-319-99007-1_35
2. Asian Development Bank (2022) Battambang smart solid waste financial management option analysis report. Retrieved from https://events.development.asia/system/files/materials/2022/08/202208-battambang-smart-solid-waste-financial-management-option-analysis-report.pdf/
3. Chapuis G, Djidjev H, Lavenier D, Andonov R (2017) GPU-accelerated shortest paths computations for planar graphs. In: Advances in GPU research and practice. Elsevier, Amsterdam, pp 273–306. https://doi.org/10.1016/B978-0-12-803738-6.00011-2
4. Circular Action (n.d.) Kolekt. Circular Action. Retrieved June 10 2024, from https://circular-action.com/kolekt
5. Dijkstra EW (2022) A note on two problems in Connexion with graphs. In: Apt KR, Hoare T (eds) Edsger Wybe Dijkstra, 1st ed., pp 287–290. ACM, New York. https://doi.org/10.1145/3544585.3544600
6. Farahani RZ, Hekmatfar M (2009). Facility location: concepts, models, algorithms and case studies. Springer Science & Business Media, Berlin. https://books.google.com/books?hl=en&lr=&id=a7JhwTwHveMC&oi=fnd&pg=PP6&dq=Facility+Location+farahani&ots=iHFavfPasS&sig=r5ROOr1A9RFSOHv8266C-wL-lnE
7. Fatimah YA, Govindan K, Murniningsih R, Setiawan A (2020) Industry 4.0 based sustainable circular economy approach for smart waste management system to achieve sustainable development goals: a case study of Indonesia. J Clean Prod 269:122263. https://doi.org/10.1016/j.jclepro.2020.122263
8. Gass SI, Fu MC (eds) (2013) Encyclopedia of operations research and management science. Springer US, New York. https://doi.org/10.1007/978-1-4419-1153-7

9. IPCC (2022) Climate Change 2022: Impacts, Adaptation, and Vulnerability. Contribution of Working Group II to the Sixth Assessment Report of the Intergovernmental Panel on Climate Change. In Pörtner H-O, Roberts DC, Tignor M, Poloczanska ES, Mintenbeck K, Alegría A, Craig M, Langsdorf S, Löschke S, Möller V, Okem A, Rama B (eds). Cambridge University Press, Cambridge, 3056 pp. https://doi.org/10.1017/9781009325844
10. IPCC (2023) Sections. In Climate Change 2023: Synthesis Report. Contribution of Working Groups I, II and III to the Sixth Assessment Report of the Intergovernmental Panel on Climate Change [Core Writing Team, Lee H, Romero J (eds)]. IPCC, Geneva, pp 35–115. https://doi.org/10.59327/IPCC/AR6-9789291691647
11. ISWA (2019) How Industry 4.0 transforms the waste sector. International Solid Waste Association, New York
12. Munggiarti, A, et al (2022) Kajian Evaluasi Sistem Persampahan Kabupaten Banyumas. Pemerintah Daerah Kabupaten Banyumas.
13. Murota K, Shioura A (2014) Dijkstra's algorithm and L-concave function maximization. Math Program 145(1–2):163–177. https://doi.org/10.1007/s10107-013-0643-2
14. Ojekudo NA, Akpan NP (2017) An application of Dijkstra's Algorithm to shortest route problem. IOSR J Math (IOSR-JM) 13(3):14
15. QGIS (2024) 19. Network analysis library—QGIS Documentation. https://docs.qgis.org/3.34/en/docs/pyqgis_developer_cookbook/network_analysis.html
16. Raffler C (2024) Root676/QNEAT3 [Python]. https://github.com/root676/QNEAT3

Part II
Data Science Applications in Waste Management

The Price and Attitude to Recycling: Manufacturers' and Consumers' Points of View

Anastasia Y. Nikitaeva and Olga I. Dolgova

Abstract This chapter proves that a significant factor limiting the volume of recycling is its high price. For businesses, this price is connected to the high investments required to operate circular models. For consumers, the price is to pay a higher fee for the product. With this in mind, the expansion of recycling is associated with the development of a positive attitude toward it. This chapter suggests using semantic analysis of materials from the most cited newspapers and the social network of Russia to identify attitudes toward recycling. To determine the specifics of the image of a recyclable product broadcast through the mass media of the Russian Federation, an analysis of publications in the most popular Russian mass media and social media was carried out. A semantic assessment of the tonality of the selected publications using the multilingual sentiment method showed that most publications are negative in content, while the emotional coloring is more pronounced in informal publications on social networks. It is proposed to pay more attention to the publication of materials on successful cases and effective examples of recycling to expand the positive image of recycling and the circular economy as a whole.

Keywords Circular economy · Recycling · Industrial development · Semantic analysis · Attitudes toward recycling · Mass media · Sustainable development · Waste revolution · Price of recycling · Industry 4.0 · Recyclable products

1 Introduction

Currently, the importance of putting the concept into practice and achieving the sustainable development goals is beyond doubt among researchers and experts. One of the most significant components of sustainable development within the framework of the achieved level of the economic and technological progress of

A. Y. Nikitaeva (✉) · O. I. Dolgova
Southern Federal University, Rostov-on-Don, Russia
e-mail: aunikitaeva@sfedu.ru; oldolgova@sfedu.ru

B. Jarboui et al. (eds.), *Data-Driven Waste Revolution*, Computational Intelligence Methods and Applications, https://doi.org/10.1007/978-981-96-5890-9_7

society is the sphere of waste disposal and recycling. In the scientific field, this issue is characterized by high scientometric indicators that have maintained the growth dynamics in recent years. By searching for the keywords “recycling of waste,” 305 744 articles were found in the ScienceDirect publication database of the Elsevier ecosystem (and in 2014, 8150, and in 9 months of 2024, 36104) [22]; in the system Lens.org, with a similar request, according to data for September 2024, 203 848 papers were found with an exponential increase in their number in recent years [12].

However, in practice, the share of recycled waste in industrial economies remains quite low. In 2022, 40.8% of waste was recycled and 30.2% landfilled in the European Union [26]. The percentage of total electronic waste that is recycled varied in 2019 from 1.3 percent in Latin America to 31.81 percent in Europe and North America (the average share in the world was 17.5%) [19]. The overall recycling rate shows little progress in general, even in Europe, which is leading in terms of relevant indicators in the world (the total share of recyclable waste is less than 50 percent) [25]. In Russia, the share of disposed and neutralized production and consumption waste in the total volume of generated production and consumption waste amounted to 45.7% in the country in 2022. At the same time, there is a noticeable differentiation of indicators by regions of the country, where they range from 0% to values exceeding 100 percent [17].

Insufficient (relative to the target settings) growth rates of volumes and the share of recycling are largely determined by the price that producers and consumers of products must pay for this.

For manufacturers, switching to the use of recycled raw materials is an expensive and risky investment. Because of this, they are often not ready to invest in the production of recyclable goods until a steady demand for such goods is formed, whereas the formation of such demand is related to the attitude toward recycling. Another option is the occupation of dominant positions by noneconomic goals of enterprises, the presence of a conscious need to introduce circular models. In this case, we are talking about a change in attitude toward recycling.

For consumers in turn, paying a higher price for products made from recycled raw materials can be justified enough only if there is a positive motivation for processing. This motivation is closely associated with the attitude toward recycling. At the same time, it is the consumer who is the final person making the decision to accept or reject circular behavior.

In other words, the price of recycling and using products from recycled materials is higher compared to the traditional linear economy. Under these conditions, the attitude toward recycling in the logic of sustainable development and the circular economy is a driver or limiting factor in increasing its volumes in absolute and relative terms. In this regard, a research question arises related to how the attitude of producers and consumers toward recycling can be assessed. How much do price issues sound in relation to recycling in the media space. The relevance of this issue is determined by the fact that the attitude toward recycling is a procyclical indicator in relation to recycling itself, acting with a certain lag. This research is aimed at finding an answer to this research question.

2 Literature Review

Since the recycling fee largely determines its scale and the interest of all stakeholders in the relevant processes (product manufacturers, recycling enterprises, authorities financing relevant programs, consumers, etc.), researchers study the cost issues of recycling from different angles.

Malak Anshassi and Timothy G. Townsend note that in the period from 2019 to 2024 there has been a notable increase in recycling processing costs at material recovery facilities, escalating from an approximate 50*pertontomorethan*150 per ton [1]. As a result, some recycling programs may be curtailed or stopped altogether. As a solution, the scholars proposed an evaluation model, the analysis of the results of which showed that a targeted recycling program, focusing on high-value materials, reduced both environmental impacts and system costs.

For some types of products, such as used photovoltaic modules, industrial recycling is not typical precisely because of economic barriers [5]. Such operations are very complex and require large investments for the safe recycling of used photovoltaic modules [6, 21, 28]. Another example is related to the fact that although recycled concrete is considered a viable alternative to natural concrete, this product has not been widely used in the construction industry due to insufficient price competitiveness [14].

At the same time, the high price of recycling may decrease with an increase in its volume due to economies of scale. Although a number of studies show that, taking into account the changing interests of the supply and demand sides, there is a nonlinear relationship between the price and the volume of recycling [8].

It should be noted that many entities participate in the recycling processes, including manufacturers, retailers, third parties, and various combinations of them. Given the information asymmetry, it is difficult for participants in these processes to maximize efficiency and profit. Li Q., Zhang H., and Liu K., with this in mind, confirm the important role of the state reward-penalty mechanism in increasing the level of recycling of used products and the overall profit of a closed supply chain [13].

Chen et al. noted that even with financial government support for recycling (for example, through a subsidy), the operation of a circular value chain still requires huge investments, such as the purchase of waste recycling equipment. Successful recycling companies invest heavily in various components of their respective processes. As a result of the study of the problems of pricing, greening, and waste disposal in a closed supply chain with limited capital investments, researchers emphasize the importance of government subsidies for waste recycling, institutional conditions, and increasing consumer awareness of environmentally friendly consumption [4].

Studies aimed at research of consumers' attitudes toward recycling and various aspects of their adoption of circular models have also appeared in recent years. Kirchherr et al. noted that in the European Union the main barriers to the spread of the circular economy are cultural ones. Such barriers cover the lack of interest/awareness of consumers and indecision in the corporate culture [10]. Camacho-Otero et al. also pay significant attention to the study of the role of consumers in the development of a closed-loop economy and the adoption of recycling [3].

Further enhancing the consumer perspective, Hazen B.T., Mollenkopf D.A., and Wang, Y. note that for the stability of the circular economy, consumers must not only return goods after use but also purchase refurbished products. However, studies show that consumers have a low opinion of reconstituted products and, as a rule, are not ready to use them [7].

Significant attention is paid to the study of the consumer side in the context of circularity by Bucker C., Geissdoerfer M., and Kumar M., covering the concepts of circular consumer behavior, circular consumer adoption, circular consumer acceptance, and impact factors, as well as highlighting factors and practices of consumer behavior [2]. In the conceptual framework of circular consumer behavior, researchers show the relationship of concepts, where a logical transition from consumer awareness to attitude, intention, and behavior is traced [2].

According to the literature analysis, it can be concluded that while recognizing the importance of the attitude of producers and consumers to recycling in order to overcome price issues, insufficient attention is paid to the analysis of such an attitude in a mass context. Important research based on case studies and literary analysis is present in scientific publications. Based on this, it seems relevant to analyze the image of a recyclable product broadcast through the mass media.

3 Material and Methods

In this chapter, the analysis of attitudes toward recycling is carried out using the example of the Russian Federation. In Russia, firstly, there is a high share of the industrial sector in the economy, which seriously actualizes the problem of recycling. Secondly, in the National Development Goals for Russia, appropriate strategic guidelines and indicators of their achievement are fixed. Thirdly, technological solutions in the field of recycling are accumulating more and more. Fourthly, in recent years, the country has initiated a series of national programs in the field of closed-loop economic development, primarily in terms of waste recycling. Fifth, there is currently limited research reflecting attitudes toward recycling, which can act as a driver or limiting factor in the development of a circular economy.

To determine the specifics of the image of a recyclable product in the mass media of the Russian Federation, an analysis of publications in the most popular Russian media was conducted (according to the rating of Russian media for August 2024 [15, 16]). To collect a large amount of information about publication in various media, a parser program was written in Python, automating the process of data collection and systematization on the Internet. To ensure comprehensive coverage of various types of media, three most popular (in terms of citation) newspapers were selected: Izvestia, Kommersant, and Rossiyskaya Gazeta, and one most cited news agency,

RIA Novosti [9, 11, 18, 20]. These resources are official (formal) media. Also, in order to include more informal media in the analysis, in which information is more accessible and which users themselves can post, data were collected on the largest Russian social network in terms of coverage among the population—VKontakte [24]. To ensure uniformity of data on the subject, a time period of 3 years was taken in the analysis of official media (from September 1, 2021 to September 1, 2024).

The selected time interval is due to the launch of the Federal project "Circular Economy" in January 2022. This project includes six main areas: reducing waste generation, creating infrastructure for recycling, stimulating the transition to the use of recyclable materials, limiting the turnover of non-environmentally friendly packaging, developing a system for traceability of waste movement, as well as environmental education. Taking into account the latter direction, it was suggested that during the implementation of the project and several months before its launch, the media will pay more attention to waste recycling issues in order to encourage the population to switch to responsible consumption and businesses to responsible production.

On the one hand, the VKontakte social network does not have the functionality to select publications for certain years, and on the other hand, it places a huge number of information posts per day: For three analyzed days, the number of posts in it almost equaled the total number of publications in all three major newspapers (although the volume of posts is much less than the volume of articles). These factors led to the choice of a 3-day time interval in September 2024, which allows reflecting the current situation with the perception of waste recycling. After that, all selected publications were analyzed using semantic text analysis tools in the Orange Data Mining program (semantic text analysis, hierarchical cluster analysis, word cloud, bag of words). Biases in determining the tonality of publications were eliminated due to the use of artificial intelligence in assessing the tonality. The multilingual sentiment method of semantic text analysis was used in the Orange software. The evaluation results were interpreted in a standard way: A zero score meant a neutral tone of the article, more than zero was positive, and less than zero was negative. The research methodology is schematically presented in Fig. 1.

Rossiyskaya Gazeta is the official print organ of the government of the Russian Federation, and business news is often published in the Kommersant newspaper. The target audience of these publications includes many business representatives, including senior executives (TOP management). The information published in the Izvestia newspaper, RIA Novosti, and the VKontakte social network includes a wider range of readers and is more aimed at the target audience of possible buyers of goods (consumers) from recycled materials.

4 Results

The semantic evaluation of the tonality of the selected publications using the multilingual sentiment method of semantic text analysis (Fig. 2) indicates that the

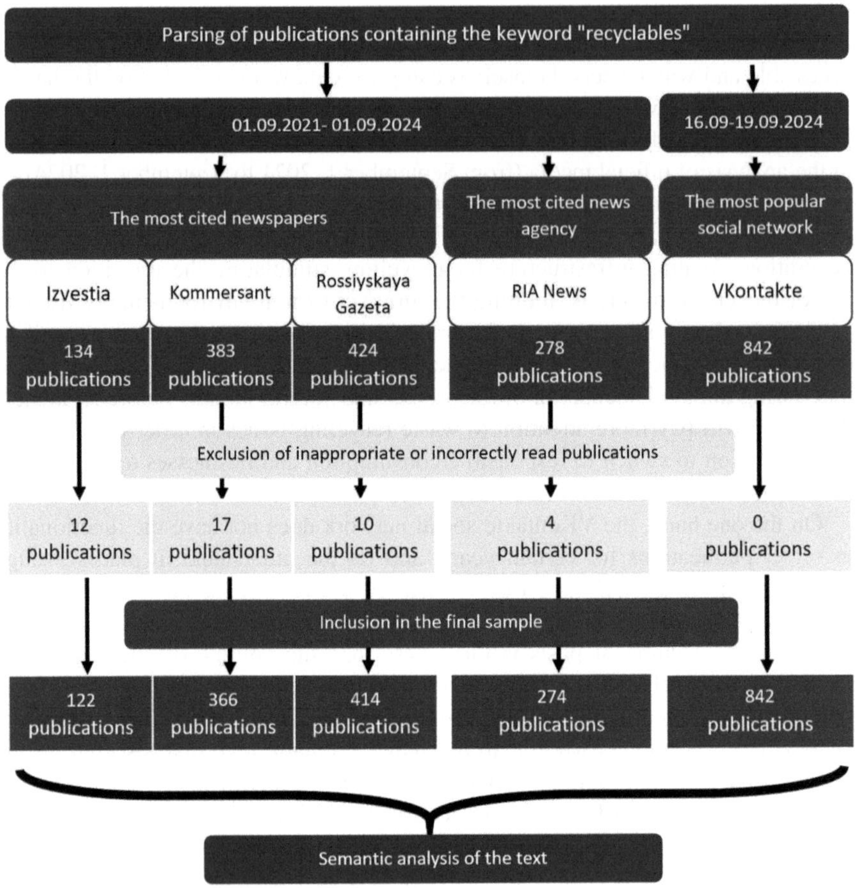

Fig. 1 Methodology overview

vast majority of publications are negative, while the emotional coloring is much more pronounced in informal publications posted on social networks. The higher the evaluation of the tonality of the text, the more positive it is. A zero grade corresponds to neutral publications in which emotionally colored vocabulary is practically not used. Texts with a score below zero contain words perceived as negative. In most sources, publications have a weakly expressed positive connotation; nevertheless, the proportion of articles with a weakly expressed negative connotation is about 25%. One of the main factors influencing this is the general negative connotation of the terms "waste" and "garbage." At the same time, publications on the use of recyclables often include descriptions of various negative consequences of environmental degradation, which should "push" the population to switch to more environmentally friendly products. The most pronounced positive character (20 points) is observed in the posts of the VKontakte social network; the most "positive" post has a tonality score of 20, while 26 publications have a pronounced positive

	Izvestia	Kommersant	Rossiyskaya Gazeta	RIA Novosti	VKontakte
from 15 to 20	0	0	0	0	1
from 10 to 15	0	0	0	0	2
from 5 to 10	1	0	1	1	26
from 0 to 5	81	255	287	180	466
equal to 0	12	38	32	16	150
from -5 to 0	28	73	94	77	190
from -10 to -5	0	0	0	0	6
from -15 to -10	0	0	0	0	1
Maximum	5,26	3,75	5,56	6,36	20,0
Minimum	-2,55	-3,61	-3,14	-2,27	-14,29
Total	122	366	414	274	842

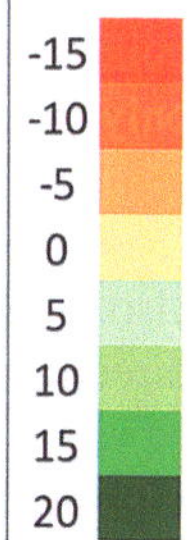

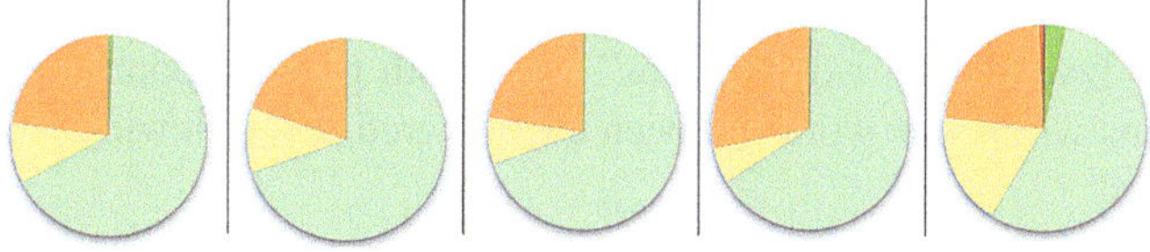

Fig. 2 Semantic coloring of publications based on analyzed sources [18, 20, 24]

coloring. At the same time, this mass media has the largest number of neutral publications compared to other analyzed sources of information. Based on this, it can be concluded that some of the publications are informational (neutrally colored), while other posts reflect the personal position of the author, which increases the emotionality of the text.

Kommersant newspaper has one of the highest proportions of neutral articles. Among the publications of RIA Novosti, the highest proportion of "negative" publications. This may be due to the nature of the news presented and the agency's focus on promptly reflecting the consequences of various emergencies. The semantic profile of Rossiyskaya Gazeta and Izvestia is quite similar; however, the articles of Rossiyskaya Gazeta are characterized by a slightly brighter emotional coloring.

A cloud of words contained in publications from all types of sources was created. Among them, the most popular are recyclables, waste, garbage, collection (including separate), solid municipal waste, recycling, packaging, plastic, as well as words reflecting the territorial nature of the publication—Russia, the region. An interesting feature is the perspective orientation of the publication; most of the popular verbs are presented in the future tense: "can," "need," and "plan." Words related to units of measurement, names of social networks, emails, and some of the most popular pronouns were included in the stop list as insignificant (because they did not allow us to draw a conclusion about the subject of the publication but interfered with visual perception). Words related to price or value were not among the most popular either among the general pool of articles on all sources or among publications of certain media.

A more in-depth analysis of the content of publications was also carried out, which allows us to identify the categories with which most publications in various media containing references to recyclables are associated. Word clouds were also built for each source; however, since the texts of the articles are written in Russian, only the most popular words in text form will be presented in this study.

In the Izvestia newspaper, the most popular words are "recyclables," "waste," "garbage," "goods," and "packaging," while the words "RF" and "Russia" are also popular, which indicates a focus on events inside the country. The words directly related to "price" were not among the most popular to reflect the issues of value. But it was decided to analyze the publications in which the currency was mentioned. The official currency in the Russian Federation is the ruble. Unlike other world currencies, it has entered the TOP of the most frequently used words. Then, separately for all considered media, a selection of publications containing references to the national currency was made, and their main topics were analyzed. So, in the newspaper Izvestia, the word "rubles" was in 8th place in popularity. Among the publications where this word has been encountered, the following main topics can be identified:

- Issues related to the prevention of crimes in the field of recycling. This topic affects both manufacturers (in December 2021, Samsung was fined almost 90 million rubles for fake recycling,[1] and Norilsk Nickel paid a fine of 146.2 billion rubles in the same year for a fuel spill near Norilsk[2]) and consumers (who will have to pay up to 15 thousand rubles for improper disposal of garbage).[3]
- Problems related to waste recycling and solutions. For example, there is a shortage of Rosprirodnadzor inspectors caused by the low level of prestige of the profession and low wages (34 thousand rubles per month)[4] or attracting the population to separate garbage collection by exempting individuals from paying income tax on the sale of recyclables.[5]
- Informing about various environmental trends. For example, about sustainable fashion, the transition to which is also supported within the framework of the federal project "Closed-loop Economy."

---

[1] Pertseva E. Full scrap: garbage processors will be monitored with the help of energy sensors. Izvestia. [in Russian]. 22.02.2022. Available via https://iz.ru/1294717/evgeniia-pertceva/polnyi-util-za-pererabotchikami-musora-proslediat-s-pomoshchiu-energodatchikov.

[2] Lezhneva L. Stocks will never run out. Izvestia. [in Russian]. Available via https://iz.ru/1270166/liubov-lezhneva/zapasy-ne-konchatsia-nikogda.

[3] The State Duma deputy recalled the fine for burning garbage in the cottage. Izvestia. [in Russian]. 29.03.2024. Available via https://iz.ru/1673709/2024-03-29/deputat-gosdumy-napomnil-o-shtrafe-za-szhiganie-musora-na-dache.

[4] Platonova A. Now only 7% of waste is recycled. Izvestia. [in Russian]. 14.02.2023. Available via https://iz.ru/1469278/anastasiia-platonova/seichas-na-pererabotku-idet-lish-7-otkhodov.

[5] Perevoshchikova M. Resource status: in Russia personal income tax from recycling is wanted to be abolished. Izvestia. [in Russian]. 18.01.2022. Available via https://iz.ru/1278283/mariia-perevoshchikova/resursnoe-sostoianie-v-rossii-khotiat-otmenit-ndfl-ot-sdachi-vtorsyria.

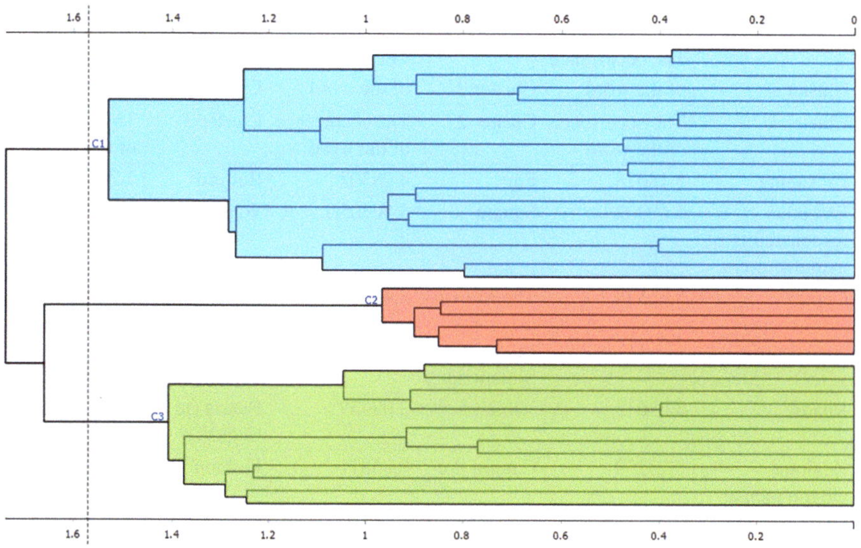

Fig. 3 Hierarchical cluster analysis of publications in the Izvestia newspaper

- Local news affecting the sphere of secondary use of resources. As an example, we can cite an article that the government of the Primorsky Territory terminated the agreement with LLC Polygon on the construction of a recycling complex for the management of municipal waste in the settlement of Gorodechny. One of the main reasons is the numerous complaints of local residents who opposed the construction.[6] This situation clearly demonstrates the "recycling price." Despite the declared course for sustainable development, citizens are not ready to accept waste recycling enterprises located near their homes.

To determine the main categories of publications, all the words used in the selected articles were analyzed using a bag of words using the IDF method (inverse frequency of document use). The peculiarity of this method is that the more often a word is used in documents, the lower the score it is assigned. Based on this, publications on similar specific topics were identified using hierarchical cluster analysis. Figure 3 shows a hierarchical cluster analysis of publications in the Izvestia newspaper. Cluster analysis revealed three main clusters. After that, an analysis of the most common words was carried out for each of them; its results are presented in Table 1. The publications of the first cluster are mainly related to the use of recyclables in packaging, extended responsibility of manufacturers, construction, and recycling.

[6] The authorities of Primorye refused to build a landfill. Izvestia. [in Russian]. 04.03.2024 https://iz.ru/1659684/2024-03-04/vlasti-primoria-otkazalis-ot-stroitelstva-musornogo-poligona.

Table 1 TOP 10 most popular words by clusters in the Izvestia newspaper

Cluster 1	The occurrence of the word	Cluster 2	The occurrence of the word	Cluster 3	The occurrence of the word
Cluster 1	The occurrence of the word	Cluster 2	The occurrence of the word	Cluster 3	The occurrence of the word
Packaging	5.18028	Fire	4.08302	Rewards	1.24266
Extended responsibility of manufacturers	4.60302	Square	3.50303	Waste paper	1.11477
Goods	3.66574	The Ministry of Emergency Situations	3.44258	Russia	0.919173
Roads	3.53613	Vehicles	3.09359	Russia (in English)	0.8877
Regional environmental operator	3.23259	Caught fire	2.85811	Fashion	0.886076
Construction	3.04523	Warehouse	2.85811	Leather	0.835482
Recycling	2.89021	Occurred	2.44981	Car	0.83337
Said	2.83014	Fires	2.44123	Putin	0.81076
Prohibition	2.82458	Burning	2.44123	Garbage	0.785346
News	2.8121	Open	2.44123	Environment	0.753031

The second cluster includes publications related to fires and the work of special services (this is the smallest cluster).

The articles of the third cluster are mainly devoted to various initiatives in the field of ecology and environmental protection (including various awards), many of which were created as part of the initiatives of Russian President Vladimir Putin.

Among the most popular topics in the publications of the Kommersant newspaper are: solid municipal waste, various projects, regional publications, separate garbage collection, and recycling issues.

Since the target audience of Kommersant is largely representatives of business, it is strongly interested in monetary issues; therefore, the word “rub” (short for “rubles”) is in second place in popularity. Among the main topics of articles that touch on finance, we can highlight the following:

- Various promising investment projects at different stages of implementation. Basically, these projects are ecotechnoparks; for example, ecotechnopark, whose developer and investor is Ecocenter LLC, will require 3.5 billion rubles.[7]
- Informing about the progress of initiatives on separate garbage collection in the regions of Russia. For example, the authorities of the Samara region decided

[7] Volkodav M. Only large operators remain in the industry, ready to work for a long time. Kommersant. [in Russian]. 30.08.2024. Available via https://www.kommersant.ru/doc/6919629.

to start with the separation of waste according to the principle of dry and wet, which will lead to the separation of garbage into packaging and food residues. To achieve this goal, the Samara Region received 28 million rubles from the federal budget, which were used to purchase 1,800 containers for separate garbage collection.[8]

- Prospects for the use of recyclables of various types. For example, information about the shortage of recycled plastic bottles, which are a valuable resource in the recyclable materials market, the demand for which exceeds supply, so in Russia the cost of a pressed PET bottle can reach up to 60 thousand rubles per ton, while in other countries their cost is $80-350 per ton, which means several times less[9]
- Environmental actions and initiatives for business. For example, Sberbank, the largest Russian bank, has developed a green credit project in which the interest rate depends on the borrower's fulfillment of certain ESG indicators (such as environmental responsibility, reuse, ESG reporting, etc.) [10]
- Various business problems related to the new environmental legislation. This was how the problems related to the adoption in 2023 of the law on expanded producer responsibility, according to which responsibility for the disposal of packaging from manufacturers of goods to those who produce containers, were discussed. The Executive Director of the Association of Manufacturers of Glass Containers notes that the adoption of this law will increase the cost of production of glass containers by 20–30%, depending on the type, while increasing the share of recyclables in packaging manufacturers is hampered by strict technological requirements and a lack of capacity for waste collection and processing. According to business representatives, this law may lead to an increase in prices for consumer goods, a decrease in the competitiveness of glass containers, and the "shutdown of glass factories".[11]

Hierarchical cluster analysis using the Ward method demonstrated a clear division of articles containing information about recyclables by subject (Fig. 4). In total, five clusters can be distinguished, characterized by a greater or lesser degree of similarity:

The first cluster includes publications related to the coverage of various events, part of which is the collection of recyclables. This also includes events directly devoted to the issue of the use of recyclables, for example, the festival of reasonable consumption or the ecological holiday "Ecodvor." As part of such events, there is

---

[8] Recycling of waste in the Samara region: dual sorting is one of the key steps to reduce the burden on the environment. Kommersant. [in Russian]. 23.12.2021. Available via https://www.kommersant.ru/doc/5143021.

[9] SPIEF is going green. Kommersant. [in Russian]. 07.06.2024. Available via https://www.kommersant.ru/doc/6745621.

[10] Smirnova O. Green transformation. Kommersant. [in Russian]. 22.02.2022. Available via https://www.kommersant.ru/doc/5227531.

[11] Shapovalov A. Money from can.Kommersant. [in Russian]. 31.05.2022. Available via https://www.kommersant.ru/doc/5380332.

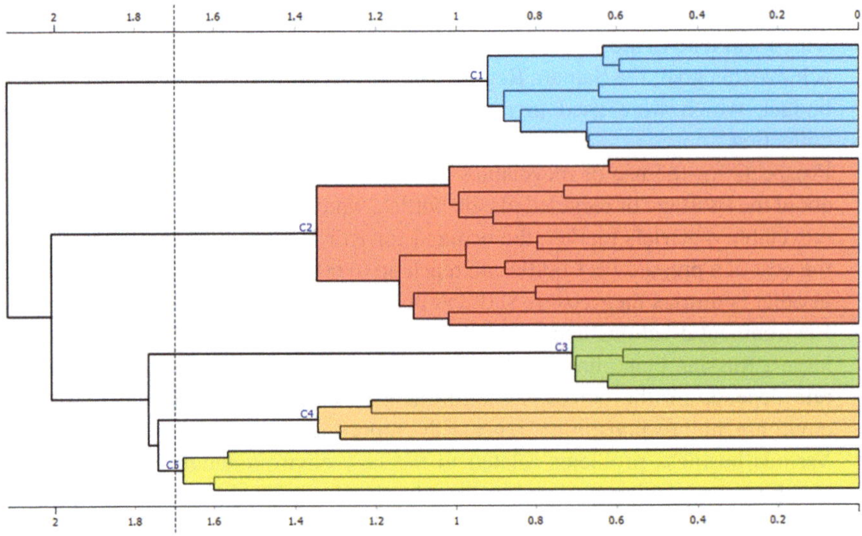

Fig. 4 Hierarchical cluster analysis of publications in the Kommersant newspaper

an opportunity to hand over recyclables for recycling, exchange things and books, attend a master class on the second life of things, or see the full cycle of recycling of different types of recyclables.

The second cluster includes texts dealing with issues of eco-catchments for producers, the responsibility of enterprises for environmental damage, as well as various projects to reduce it. The issues of environmentally friendly packaging and disposal of hazardous substances (for example, lead) are also being considered.

The articles of the third cluster touch upon environmental initiatives related to recycling at one of the largest agricultural enterprises in the south of Russia, Agroholding "STEPPE."

The fourth cluster presents news reflecting the activities of the Russian Environmental Operator and regional environmental operators within the framework of waste management projects.

The fifth cluster includes articles dealing with the issues of sustainable development of regions and the specifics of the implementation of environmental initiatives at enterprises located in the territories of advanced socioeconomic development.

In Rossiyskaya Gazeta, among the most popular words are the following: recyclables, waste, collection, garbage, solid municipal waste, recycling, percent production, and waste paper. The word "rubles" appeared on the 7th place in the TOP of the most common words and among the publications where it occurs, the following areas can be distinguished:

- Projects of recycling enterprises, including industrial parks. For example, in the Kemerovo Region, it is planned to create the Prigorodny industrial park, in which the investor (the Ecoimpulse organization) plans to invest 650 million rubles. The implementation of this project will create conditions for the work of enterprises engaged in the collection, sorting, and disposal of MSW, as well as using such waste as recyclable materials (the total amount of investments in these enterprises is estimated at 10 billion rubles).[12]

- Green investments of industrial enterprises. For example, the management of the Volga Abrasive Plant has spent about 500 million rubles over the past five years to reduce emissions of sulfur-containing substances, and the management of the Volgograd Aluminum Plant in 2017–2020 invested over 30 million rubles in environmental measures.[13]
- Increase the efficiency of separate waste collection. This topic is quite diverse and includes various solutions aimed at both consumers and producers: control over spontaneous landfills (by installing surveillance cameras and container sites for garbage collection in such places), monetization of separate garbage collection (in Syktyvkar, they pay for certain types of recyclables, and some residents at the level of a multistorey building organize its collection, for example, in one of the nine-storey buildings in the first half of 2021, residents collected about 10 tons of recyclable materials and earned about 118 thousand rubles on this[14]), improvement of infrastructure for the collection of recyclables (installation of fandomats, which are robotic kiosks for receiving recyclable packaging in exchange for remuneration), and for municipal authorities (in Novorossiysk, within the framework of the eco-marathon "Hand over waste paper—save a tree," the municipality, in which more than two tons of waste paper were handed over, received a monetary reward[15]).

Figure 5 shows a hierarchical analysis of articles in this newspaper by category. There are three main clusters in total:

The articles of the first cluster are devoted to the implementation of environmental initiatives on garbage collection initiatives of Pyotr Biryukov, Deputy Mayor of Moscow.

The second cluster includes publications related to various services for the removal of unnecessary items (including bulky ones).

[12] Potapova Yu. An industrial park for the processing of MSW will be created in Kuzbass. Rossiyskaya Gazeta. [in Russian]. 14.09.2023. Available via https://rg.ru/2023/09/14/reg-sibfo/v-kuzbasse-sozdadut-industrialnyj-park-dlia-pererabotki-tko.html.

[13] Merzlyakov R. Plants in Russia will reduce their carbon footprint. Rossiyskaya Gazeta. [in Russian]. 19.09.2021. Available via https://rg.ru/2021/09/19/reg-ufo/zavody-v-rossii-budut-sokrashchat-uglerodnyj-sled.html.

[14] Mikhailov A. The design of garbage chutes in new buildings was proposed to be banned. Rossiyskaya Gazeta. [in Russian]. 23.05.2023. Available via https://rg.ru/2023/05/23/reg-szfo/proektirovanie-musoroprovodov-v-novostrojkah-predlozhili-zapretit.html.

[15] Mikhailova T. The participants of the "BumBatl" campaign saved more than 1.5 million trees in a month. Rossiyskaya Gazeta. [in Russian]. 15.11.2023. Available via https://rg.ru/2023/11/15/reg-pfo/uchastniki-akcii-bumbatl-za-mesiac-spasli-bolee-15-milliona-derevev.html.

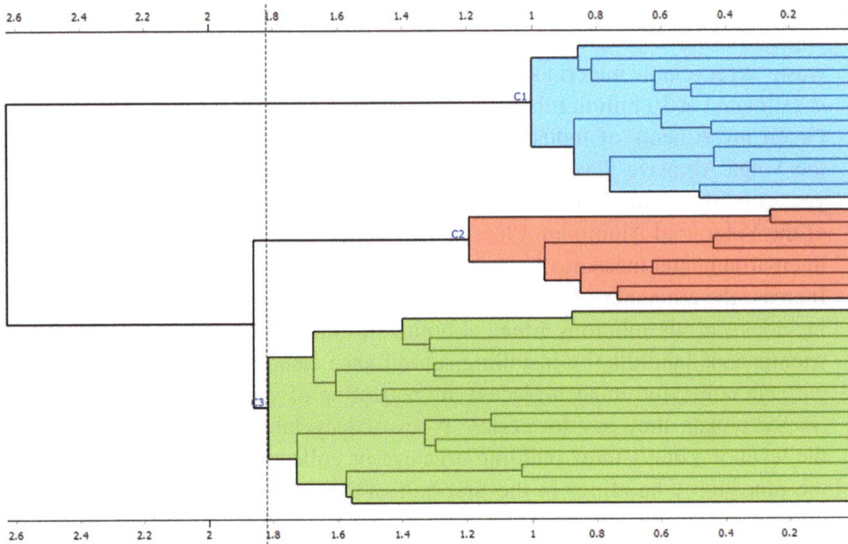

Fig. 5 Hierarchical cluster analysis of publications in the newspaper Rossiyskaya Gazeta

The third cluster includes articles on the recycling of waste paper (including the availability and location of reception points) and the manufacture of environmentally friendly packaging from it.

Among the most frequently used words in the RIA Novosti news agency, recyclables, waste, garbage, and recycling also occupy leading positions, but terms such as news, plastic, and recycling are also added. The specification of financial values in currency is used less frequently in this source—the word "rubles" is in 41st place in the TOP of the most popular words. Among the main topics in these publications are the following:

- Promising projects in the field of recycling. For example, the Yugra company "Taiga" received a government loan in the amount of 115 million rubles for 5 years at 3% for the construction of a pellet production plant.[16]
- Informing about environmental initiatives of companies. For example, Avito plans to double its investments in social and environmental initiatives by 2024, while attracting employees and users of the platform to raise funds.[17]
- Solutions for separate garbage collection. For example, in the city of Khimki, Moscow region, a recycling collection point "Megabak" was opened, where you

[16] Sukhorukov A. The Yugra Development Fund will finance the production of pellets in Yugorsk. RIA Novosti. [in Russian] 03.07.2024. Available via https://ria.ru/20240703/yugorsk-1957171367.html.

[17] Avito has launched a new digital project for social assistance. RIA Novosti. [in Russian] 28.04.2023. Available via https://ria.ru/20230428/avito-1868346991.html.

can hand over recyclables for money, and collected more than 10 tons of tires, plastic, glass, and other waste in it.[18]

- Informing about changes in the recycling infrastructure. For example, in the year, the news was posted that from January 1, 2024, access to the "Electronic trading platform for the purchase and sale of recyclables" posted on the REO website will be paid for all users (five thousand per month). These funds are required to support and develop the platform and should not become a barrier for professional market participants.[19]
- Promotion of an eco-friendly lifestyle. Such articles talk about various ways to organize activities more environmentally friendly. For example, citizens are advised to switch to using bags and shoppers instead of plastic bags, which is not only more environmentally friendly, but also more environmentally friendly, since one package costs an average of 10 rubles.[20]

The subject matter of publications is somewhat more diverse than in Rossiyskaya Gazeta; four clusters are clearly expressed (Fig. 6). The first cluster also includes, as in Rossiyskaya Gazeta, texts dedicated to the allocation of funds on the initiative of Pyotr Biryukov for projects to improve the landscaping and ecology of Moscow.

The second cluster consists of publications containing information about a large fire on February 1 at a warehouse with recycled plastic in the city of Krasnogorsk near Moscow. The fire area was 1,400 square meters.

The third cluster includes news about the functioning of the municipal economy complex in terms of household waste disposal (in particular, the organization of containers for separate garbage collection).

The fourth cluster included articles on waste management culture, in particular, plastic recycling and solid municipal waste disposal.

On the VKontakte social network, the most popular topics turned out to be posts calling for participation in various recycling and collection campaigns for garbage, waste paper, and bottles. Interestingly, it was in this medium that the verb "show" was often used. The topic of money in this social network turned out to be the least popular among all the studied sources. The word "rubles" was ranked 75th in the TOP of the most popular words. Among the main topics of publications related to the mention of the Russian currency, the following can be identified:

- A large number of vacancies related to work at enterprises for the collection, disposal, or recycling of materials. Some of the vacancies relate directly to recycling activities (a recyclable material sorter) and some related activities (for example, a driver).

---

[18] An updated Megabak separate waste collection point has been opened in Khimki. RIA Novosti. [in Russian] 15.11.2023. Available via https://ria.ru/20231115/sbor_musora-1909692133.html.

[19] exchange. RIA Novosti. [in Russian] 30.09.2023 Available via https://ria.ru/20230930/podpiska-1899581110.html.

[20] Belousov V. It is easier than easy: how to master separate garbage collection and make it a habit. RIA Novosti. [in Russian] 06.03.2023. Available via https://realty.ria.ru/20230306/sbor-1856058379.html.

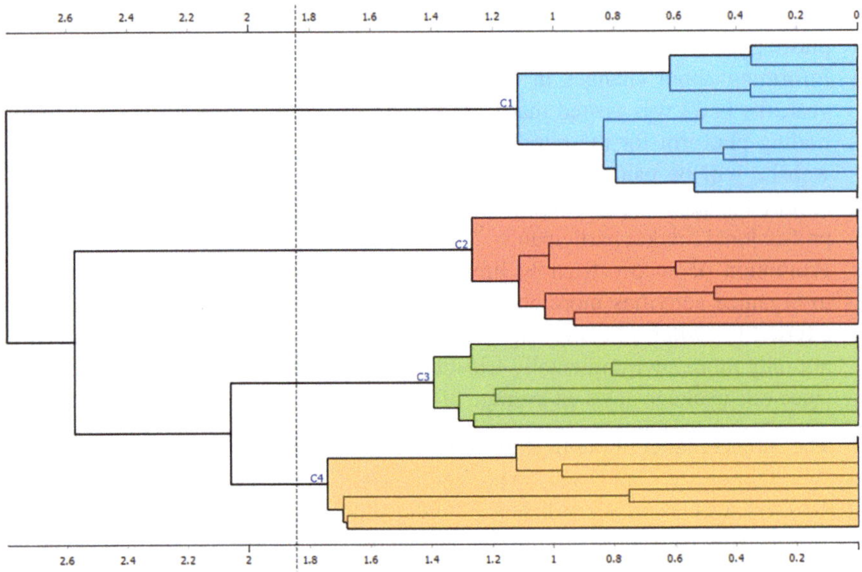

Fig. 6 Hierarchical cluster analysis of RIA Novosti publications

- Informing about the implemented initiatives related to the decriminalization of the recyclable materials market, for example, increasing the transparency of the recycling market. It is assumed that this project will bring an additional 50 billion rubles to the budget in 2025 from previously unpaid taxes and fees.[21]
- Advertising of various promotions and charity auctions related to the collection of recyclables. In most cases, such publications contain calls to join; often the authors are representatives of educational institutions, among them, for example, the BOU "Gymnasium No. 9" of the city of Omsk, whose pupils hand over waste paper for recycling, and the regional environmental operator allocates money to replace windows at school (since the beginning of participation in the action, the educational institution has collected and delivered 188 kg of plastic and 37683 kg of waste paper for 411083 rubles).[22]
- Informing about the possibility of receiving money for recycling. For example, in the eco-houses of Udmurtia for 3.5 years of existence, more than five thousand tons of energy resources were accepted, for which residents were paid over 22 million rubles.[23] However, the amounts paid for the same types of waste vary depending on the region: So in the Novosibirsk region, a girl received 30 rubles

[21] The LDPR will seek to increase transparency in the handling of ferrous and nonferrous scrap. VKontakte. [in Russian]. 17.09.2024. Available via https://vk.com/wall-433349_722611.

[22] BOU Omsk "Gymnasium No. 9." Waste paper news. VKontakte. [in Russian]. 18.09.2024. Available via https://vk.com/wall-198405523_5182.

[23] Tanishuv Aloqa Suhbatlar. The jubilee check was repulsed at a point on Baranova Street in Izhevsk. VKontakte. [in Russian]. 18.09.2024. Available via https://vk.com/wall-170221658_2202.

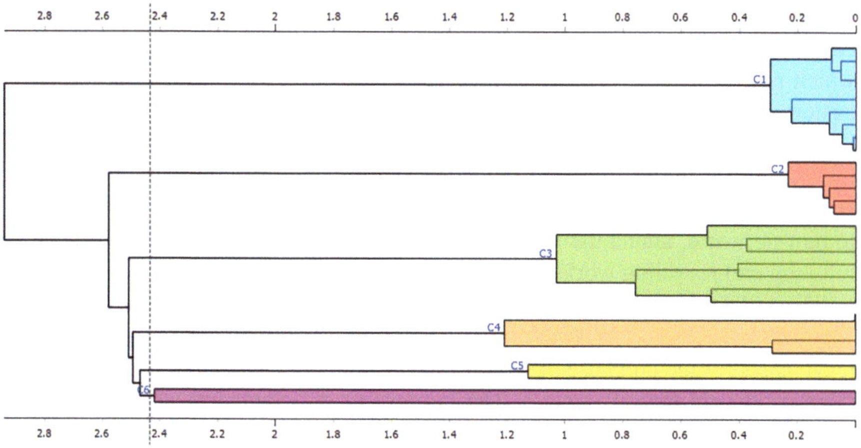

Fig. 7 Hierarchical cluster analysis of publications on the VKontakte social network

(5 rubles per kg) for recycling 6 kg of bottles,[24] while in Udmurtia they pay 17 rubles for 1 kg of plastic bottles.

- Notification of the population by various companies about their investments in improving the environmental friendliness of production. For example, Rosatom State Corporation is modernizing equipment and annually allocates hundreds of millions of rubles for environmental protection measures.[25]

Figure 7 shows clusters of publications on the VKontakte social network. The most popular words from the publications of the first cluster sound very specific and include such words as "rainbow," "Peter," "mushroom," "eco-friendly," "Mikhail Grigorievich," "winner," "children," and "spawn." In fact, this group of posts is an advertising campaign for the environmental video contest "Rejoice—eco-friendly!" in which families with children should tell their way of eco-friendly organization of the holiday. And Mikhail Grigoryevich mushroom, bred in the SPAWN laboratory [23], was chosen as the expert who was supposed to announce the results of voting for the best video. This laboratory is engaged in the recycling of cellulose-containing composite waste using mushrooms into building materials, furniture, decorative items, and other things.

The posts of the second cluster are aimed at informing residents of the city of Ussuriysk about the placement of additional orange metal containers for recyclables. Addresses of recycling collection points can be found on the website of the Primorsky Environmental Operator (spzv.ru).

[24] Drach E. I took a trial batch of bottles for recycling. VKontakte. [in Russian]. 18.09.2024. Available via https://vk.com/wall67113283_1932.

[25] Akhmetov A. It's serious... One of the priorities is ecology.VKontakte. [in Russian]. 17.09.2024. Available via https://vk.com/wall42776011_5318.

The third cluster also contains posts reflecting the advertising of the Chinese enterprise Wenzhou Keshu Cotton Textile Co. (Wenzhou Keshu Cotton Textile Co.), which produces yarn for knitting from recycled cotton and polyester according to customer parameters.

The fourth cluster consists of posts on temporary closure for the improvement of the territory of the ecocenter in the city of Kirov. The main purpose of this center is environmental education of the population. On its territory, it will be possible to recycle waste, attend thematic events, as well as listen to lectures by experts. A museum and recycling workshop will also be opened.

The fifth cluster includes publications reflecting information about the holding by the regional movement "Circle of Life" of an autumn campaign to collect recyclables for the purchase of seedlings, which, upon completion of the collection, were planted in the city square. The purpose of the campaign was to strengthen environmental thinking.

The sixth cluster includes the most general publications related to various regional garbage collection and recycling campaigns.

Publications in VKontakte not only have a brighter emotional coloring but also more often cover the same news with minor changes. This is due to the fact that the VKontakte social network is essentially a platform for posting news from individual communities and users who have their own separate audience of users. Also, VKontakte, unlike the official media, has quite a lot of advertising publications that are posted in different communities.

5 Conclusion

The analyzed media reflects several key areas related to recycling price and attitude.

First covers promising projects of ecotechnoparks and individual processing enterprises. Such projects require significant financial costs and are largely implemented thanks to government support.

The second is connected to the implementation of separate garbage collection in the regions of the Russian Federation. These publications are mainly devoted to the successful experience of individual cities and regions in the field of waste collection and recycling. One of the most effective and often mentioned methods of attracting residents to recycle is a monetary reward for these actions. Actions of various formats also serve to educate and involve the population in participating in environmental projects aimed at collecting or using recyclables.

Informing about the implementation of environmental initiatives in various companies. On the one hand, this creates a positive image of the organization and attracts potential investors, and on the other hand, it serves as a guideline and motivation for other enterprises in the industry.

Since the words "waste" and "garbage" in general have a negative emotional connotation, one of the possible recommendations is to popularize the terms "reuse" and replace the term "recyclable materials" in the semantic construction of "recyclable products" with the name of a specific product (for example, furniture made of recycled plastic, packaging made of recycled cardboard, etc.). Such a solution can not only remove negative associations from potential customers but also motivate them to participate in separate garbage collection programs because instead of the abstract term "recyclable materials," they will see concrete examples of the use of recycled waste.

For manufacturers, switching to using recycled raw materials is an expensive and risky investment. Because of this, they are not ready to invest in the production of recyclable goods until a steady demand for such goods is formed.

With this in mind, it seems advisable to pay more attention to the publication of materials on successful cases and effective (in environmental, economic, and social aspects) examples of recycling in order to expand the positive image of recycling and the circular economy as a whole.

References

1. Anshassi M, Townsend TG (2024) Residential recycling in Florida: a case study on costs, environmental impacts, and improvement strategies. Resour Conserv Recycl 206:107627. https://doi.org/10.1016/j.resconrec.2024.107627
2. Bücker C, Geissdoerfer M, Kumar M (2022) 100 practices to foster consumer acceptance in the circular economy. In Apollo—University of Cambridge Repository. https://doi.org/10.17863/CAM.80149
3. Camacho-Otero J, Boks C, Pettersen IN (2018). Consumption in the circular economy: a literature review. Sustainability 10:2758. https://doi.org/10.3390/su10082758
4. Chen J, Tian Y, Chan FTS, Tang H, Hou Che P (2024) Pricing, greening, and recycling decisions of capital-constrained closed-loop supply chain with government subsidies under financing strategies. J Clean Prod 438:140797. https://doi.org/10.1016/j.jclepro.2024.140797
5. Deng R, Chang NL, Ouyang Z, Chong CM (2019) A techno-economic review of silicon photovoltaic module recycling. Renew Sust Energ Rev 109:532–550. https://doi.org/10.1016/j.rser.2019.04.020
6. Faircloth CC, Wagner KH, Woodward KE, Rakkwamsuk P, Gheewala SH (2019) The environmental and eco-nomic impacts of photovoltaic waste management in Thailand. Resour Conserv Recycl 143:260–272. https://doi.org/10.1016/j.resconrec.2019.01.008.44421
7. Hazen BT, Mollenkopf DA, Wang Y (2017) Remanufacturing for the circular economy: an examination of consumer switching behavior. Bus Strategy Environ 26:451–464. https://doi.org/10.1002/bse.1929
8. He H, Zhang C, Wang S, Sun J, Ma F, Sun Q (2024). Dynamic optimization of battery recycling e-platforms under non-equalizing supply and demand: recycling price and service commissions. Waste Manag 177:266–277. https://doi.org/10.1016/j.wasman.2024.02.013
9. Izvestia. Daily broadsheet newspaper in Russia. Available via Iz.ru. https://iz.ru/. Cited 23 September 2024
10. Kirchherr J, Piscicelli L, Bour, R, Kostense-Smit E, Muller J, Huibrechtse-Truijens A, Hekkert M (2018) Barriers to the circular economy: evidence from the European Union (EU). Ecol Econ 150:264–272. https://doi.org/10.1016/j.ecolecon.2018.04.028
11. Kommersant. Nationally distributed daily newspaper. Available via Kommersant.ru. https://www.kommersant.ru/. Cited 25 September 2024
12. Lens. Scholarly Works by keywords "recycling AND (of AND waste)" Available via Lens.org. https://www.lens.org/lens/search/scholar/analysis?q=recycling%20of%20waste. Cited 2 October 2024

13. Li Q, Zhang H, Liu K (2021) Research on closed-loop supply chain decision-making in different cooperation modes with government's Reward-Penalty mechanism. Sustainability 13:6425. https://doi.org/10.3390/su13116425
14. Ma M, Tam VWY, Le KN, Osei-Kyei R (2023) Analysing the impacts of key factors on the price of recycled concrete: a system dynamics model. J Build Eng 80:108123. https://doi.org/10.1016/j.jobe.2023.108123
15. Medialogy. TOP 10 Russian media—August 2024. Available via Mlg.ru. https://www.mlg.ru/ratings/media/?ysclid=m1w7zwxqfw886423237. Cited 20 September 2024
16. Mediascope Cross Web. Internet ratings. August 2024. Available via Mediascope.net. https://mediascope.net/data/. Cited 20 September 2024
17. National set of SDG indicators. Available via Rosstat.gov.ru. https://rosstat.gov.ru/sdg/national. Cited 1 October 2024
18. RIA Novosti. Russian state-owned domestic news agency. Available via Ria.ru. https://ria.ru/. Cited 20 September 2024
19. Ritchie H (2023) Waste Management. Available via Our-WorldinData.org. https://ourworldindata.org/waste-management. Cited 1 October 2024
20. Rossiyskaya Gazeta. Russian newspaper published by the Government of Russia. Available via Rg.ru. https://rg.ru/. Cited 21 September 2024
21. Salim HK, Stewart RA, Sahin O, Dudley M (2019) Drivers, barriers and enablers to end-of-life management of solar photovoltaic and battery energy storage systems: a systematic literature review. J Clean Prod 211:537–554. https://doi.org/10.1016/j.jclepro.2018.11.229.44421
22. Sciencedirect. Articles with terms "recycling of waste". Available via Sciencedirect.com. https://www.sciencedirect.com/search?qs=recycling%20of%20waste. Cited 2 October 2024
23. SPAWN is a company that grows products from waste. Available via Spawntech.ru. http://spawntech.ru. Cited 20 September 2024
24. VKontakte. Russian online social media and social networking service. Available via Vk.com. http://vk.com. Cited 19 September 2024
25. Waste recycling in Europe. European Environment Agency. Published 19 Dec 2023. Available via Eea.europa.eu. https://www.eea.europa.eu/en/analysis/indicators/waste-recycling-in-europe. Cited 2 October 2024
26. Waste statistics. Eurostat. Available via Ec.europa.eu. https://ec.europa.eu/eurostat/statistics-explained/index.php?title=Waste_statistics. Cited 30 September 2024
27. Yu H, Mi J, Xu N (2024) Recycling, pricing decisions, and coordination in the dual-channel closed-loop supply chain: a dynamic perspective. J Clean Prod 456:142297. https://doi.org/10.1016/j.jclepro.2024.142297
28. Zhang Q, Liu C, Zheng S (2023) Investment and pricing in solar photovoltaic waste recycling with government intervention: a supply chain perspective. Comput Ind Eng 177:109044. https://doi.org/10.1016/j.cie.2023.109044.44421

Transition Pathway from Linear to Circular Economy in Waste Management in Indonesia: A Study of Two Cities

Elisabeth Rianawati, Maya Larasati, Firdha Laila Amalia, Almira Rahma Hinanto, Yudhistira Jalu Waskitha, Nabila Safitri, Akhirul Insan, Febrianty Hasanah, Pradnya Rahmani, and Fikri Aswan

Abstract Waste management in Indonesia faces the challenges of increasing waste generation and has traditionally relied on a linear system focused on waste collection and transportation. Shifting toward a circular economy is imminent as the National Development Plan of 2020–2024 emphasizes integrating circularity in waste management. This chapter describes the transition process in two Indonesian cities, Semarang and Makassar, through case-study analysis primarily using qualitative methodology, including desk reviews, focus group discussions, interviews, and field surveys. The discussions explored the dynamics between actors in Semarang's waste management system while examining the operational challenges faced in Makassar. Both cases underscored the potential of decentralized waste management through community involvement to enhance waste circularity. Key findings indicate that both cities face common challenges of limited funding and lack of waste segregation, while robust community initiatives and digital platforms could potentially help. In addressing challenges in transitioning into a circular economy, policy and governance reform, community programs, and stakeholder collaboration are the key factors dictating success. Good practices in the waste management activities in Semarang and Makassar include fostering digital initiatives to improve community participation, engaging the informal sector, and enacting the "one neighborhood, one waste bank" policy.

E. Rianawati (✉) · M. Larasati · F. Laila Amalia · A. Rahma Hinanto · Y. Jalu Waskitha · N. Safitri · A. Insan · F. Hasanah
Resilience Development Initiative, Bandung, Indonesia
e-mail: maya.larasati@rdi.or.id; firdhalaila@rdi.or.id; almira.hinanto@rdi.or.id; yudhistira.waskitha@rdi.or.id; nabila.safitri@rdi.or.id; akhirul.insan@rdi.or.id; febrianty.hasanah@rdi.or.id

P. Rahmani · F. Aswan
Clean Cities Blue Ocean, Jakarta, Indonesia
e-mail: PRADNYA.RAHMANI@tetratech.com; fikri.aswan@cleancitiesblueocean.org

B. Jarboui et al. (eds.), *Data-Driven Waste Revolution*, Computational Intelligence Methods and Applications, https://doi.org/10.1007/978-981-96-5890-9_8

Keywords Waste management · Waste collection and transportation · Waste circularity

1 Introduction

Semarang and Makassar, two major port cities in Indonesia, are strategically located—Semarang in Central Java and Makassar in Southern Sulawesi, with Makassar serving as a key port for eastern Indonesia. As bustling coastal hubs, both cities have significant economic, cultural, and environmental importance. However, their coastal locations also make them vulnerable to marine waste pollution, particularly from mismanaged land-based waste like plastic, which can easily travel through rivers and waterways into the sea. This accumulation of waste severely impacts coral reefs, mangroves, and fisheries, threatening the livelihoods of communities that depend on marine resources.

One of the primary causes of this mismanagement is inadequate waste handling infrastructure and poor practices in the waste management system, including a lack of waste minimization efforts. The transition to the circular economy is crucial to address this issue. A circular economy model shifts away from the traditional "take, make, dispose" practice to a regenerative system that treats waste as a resource. In this model, materials like plastics, metals, and glass are continuously cycled back into the system through closed-loop recycling, requiring effective waste segregation at the source. By maximizing waste recycling and maintaining recovered materials' quality, coastal cities can reduce the waste entering water bodies and contribute to more sustainable waste management practices, preserving both the environment and marine-based livelihoods.

This chapter describes the transition process in two key Indonesian coastal cities, Semarang and Makassar, using a case-study approach. Although cities in Indonesia have diverse characteristics, the two cities exemplify the common waste management scene prevalent across the country, including low waste handling rate, undocumented informal sectors, sporadic community initiatives, emerging digital platforms, and insufficient government support in terms of technical assistance and governance. Both cases unravel the flaw in the Indonesian waste management system while highlighting the potential of decentralized waste management through community involvement to promote waste circularity in Indonesia.

While much of the existing literature discusses the general waste management landscape in Indonesian cities, limited attention has been paid to the potential of decentralized waste management systems like TPS3R and community-based waste banks. Studies often highlight the importance of community participation in waste management but fail to critically examine the long-term sustainability and effectiveness of decentralized solutions. Our study seeks to fill this gap by offering a comparative analysis of how decentralized waste management initiatives are implemented in Semarang and Makassar, shedding light on their potential scalability and impact on national waste reduction goals.

2 Waste Management in Indonesian Cities

Indonesia's waste management policy is governed by Law No. 18 of 2008 on Waste Management, which defines solid waste or trash as unwanted solid products derived from human and/or natural activities. The law categorizes solid waste into domestic waste, domestic waste equivalents, and specific waste [1]. Municipal Solid Waste (MSW), which originates from public daily activities, falls under the purview of local governments and municipalities for its management [2, 3]. To foster responsible waste management practices at both national and municipal levels, the Indonesian government established a national waste management roadmap and targets called Jakstranas (*Kebijakan dan Strategi Nasional*/National Strategy Policy) through Presidential Regulation Number No. 97 of 2017. The goals of Jakstranas are to reduce waste at its source by 30% and to manage and process up to 70% of waste by 2025. Following the Jakstranas roadmap and targets, local governments have to formulate and establish a regional strategy policy at the provincial and city levels. This comprehensive approach aims to decentralize waste management, promoting greater local responsibility while ensuring alignment with national objectives. By involving a wide range of stakeholders, Indonesia's waste management policy seeks to create a more sustainable and effective system for handling the country's waste management [4].

In practice, Indonesia categorizes waste management into the following sectors [5, 6]:

- Nongovernmental sector: In this sector, waste management is normally carried out by the neighborhood unit (*Rukun Tetangga/Rukun Warga*/RT/RW), which collects waste from garbage bins at waste sources, such as households, and transports it to a temporary transfer point provided by the community before later transporting it into a temporary storage station (*Tempat Penampungan Sementara*/TPS) and landfill.
- Formal sector: This sector involves local government or other government-issued institutions to manage the waste. Waste management in this category begins with collection at the sources, moving it to the waste transfer station/depot/container, and then to the TPS. At the TPS, waste is then transported by state-owned fleets to the waste processing facility (*Tempat Pengolahan Sampah Terpadu*/TPST) or disposed of at a landfill. Regional environment agencies oversee the overall management, which includes program development and monitoring, mentoring programs, and financial and infrastructure aspects.
- Informal sector: The informal sector includes individuals, groups, and small firms that are not incorporated or do not follow standard operating procedures. The informal sector arose from the low-income bracket's economic drive, leveraging waste's economic value by selling recyclables. Scavengers, junkmen, intermediaries, and the recycling business all engage in this cycle of trading activities, contributing significantly to municipal waste management despite not being documented. Generally, in informal waste management schemes, scavengers and

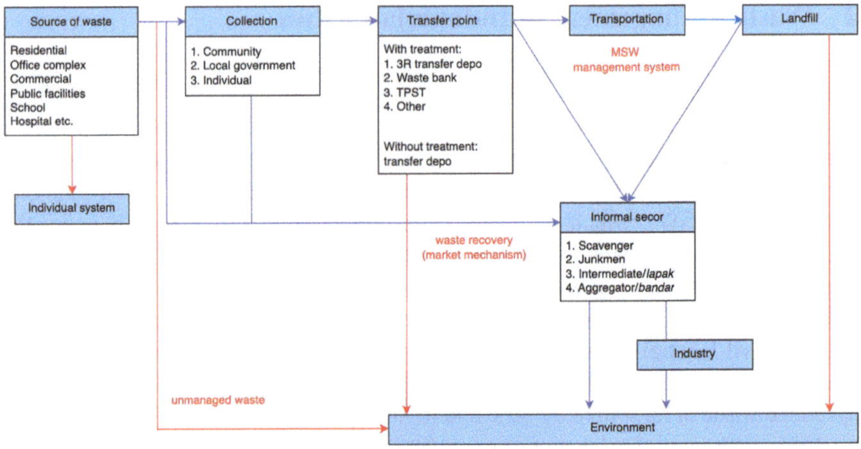

Fig. 1 Path flow of MSW recovery and actors involved [9]

junk shops collect the waste and then sell it to intermediaries. The latter then sells the waste to recycling facilities [7–9].

- Semiformal sector: This is carried out by the community on their own initiative with initial support from the government in the form of waste banks, TPS3R (*Tempat Pengolahan Sampah Reuse-Reduce-Recycle*—a small-scale material recovery facility that serves as a temporary waste storage station and facilitates waste sorting, recycling, and composting at the community level), and other community-based activities.

In Indonesia, waste management actors include waste generators, operators, collectors, recyclers, and end users, comprised of local governments, private, and informal sectors. In a larger sense, the waste management cycle includes numerous players managing various activities at each stage of the cycle. Figure 1 illustrates the flow/process of MSW and the involvement of each actor in waste management.

2.1 *Semarang City*

Semarang city, with its complex waste management challenges, is actively working toward improving waste segregation and recycling rates. Community-driven waste banks and digital platforms support the city's initiatives. This chapter dissects the waste management practices in Semarang, unfolding each actor's role in the system.

2.1.1 Waste Management in Semarang

Semarang is the capital and largest city in Central Java province, a major port city since the Dutch colonial era. As a coastal city, Semarang faces the issue of marine plastic waste, with a total of 920.2 kg of waste and a density of around 5.2 kg/m2, discovered on seven beaches across Semarang in July 2020 [10]. The waste found on the beaches consisted of plastic, wood, cloth, rubber, metal, plastic foam, glass and ceramics, paper and cardboard, and other materials, with plastic as the majority. Beach clean-up actions conducted by various actors see increasing amounts of waste in coastal areas. A total of 1,758 kg of waste was collected from Tirang Beach alone in 2023 [11], and 937.4 kg of waste was collected at Mangunharjo Beach in 2024 [12]. Waste found on the beach comes from land waste entering the water bodies way through to the sea or being washed up from another region along with sea current.

According to the National Information System of Waste Management (Sistem Informasi Pengelolaan Sampah Nasional/SIPSN), Semarang waste generation in 2021 reached 430,749.75 tons/year [13] (Fig. 2). The amount of waste handled in 2021 was 74%, and the percentage of waste reduction was 24%. Meanwhile, the waste recycling rate in Semarang was only 6% in 2021. MSW in Semarang primarily consists of organic waste (61%) and plastic (17%).

Similar to common practice in Indonesia, most households in Semarang do not sort their waste. Residents put their waste outside their houses to be collected to TPS using three-wheeled motorcycles (*triseda*) by waste officers appointed by the neighborhood association (RT/RW). The waste is then transported to Jatibarang Landfill using dump trucks and arm-roll trucks. In the process, the informal sector actors also actively collect waste from the households or TPS. The city is also implementing integrated solid waste management to improve waste management in Semarang by operationalizing TPST to reduce waste disposed at Jatibarang Landfill [14, 15].

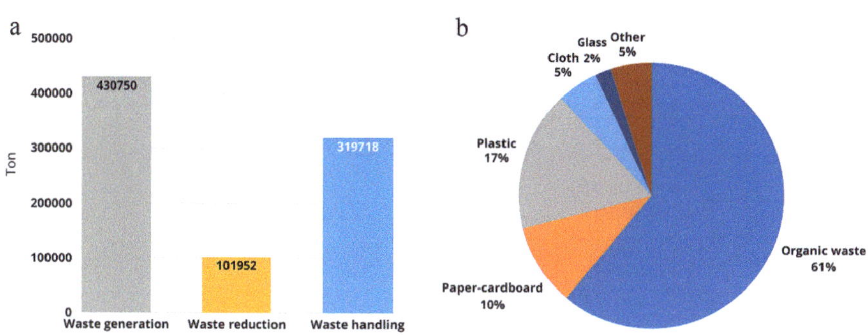

Fig. 2 Waste management performance and waste composition of Semarang. (**a**) Waste management performance. (**b**) Waste composition of Semarang

2.1.2 Involved Actors

In the case of Semarang's waste management, numerous actors from different backgrounds are involved in managing the waste. Figure 3 illustrates the waste management cycle in Semarang. Informal sector actors, scavengers, junkmen, and intermediaries channel the waste to the recycling industry. The waste flows and actors as depicted in Fig. 3 are common schemes in the waste management system in Indonesian cities.

Informal Waste Collectors (Scavengers and Junkmen)

Informal waste collectors such as scavengers (*pemulung*) and junkmen play a significant role in the waste management cycle, especially in the initial collection phase, by collecting recyclable waste from streets, residential areas, commercial areas, and TPS. Another type of scavenger, known as *penceker*, primarily collects valuable waste from TPS and occasionally assists in loading the waste onto trucks for transport. Another type of informal waste collector is junkmen, or *perosok*, who purchase recyclable waste from offices, factories, or households on a call-for-service basis. These actors earn fluctuating income from selling the valuable waste to third parties or intermediaries. Waste collectors form a patron-client relationship with intermediaries as observed at Jatibarang Landfill, where most scavengers have a

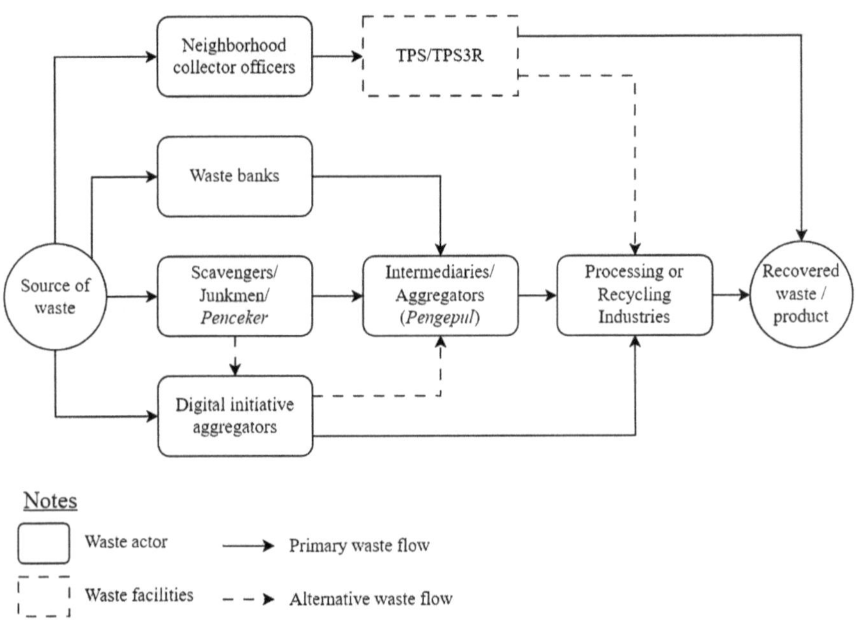

Fig. 3 Flow diagram of waste management in Semarang

particular intermediary with whom they trade, and rarely change. This unspoken rule was performed to reduce potential conflict, and the bond is heavily influenced by trust between involved parties. Despite their significance in the waste management system, informal waste collectors receive minimal assistance and recognition from the government. The Social Agency and Environment Agency faced challenges in working with informal actors, mainly due to the lack of structured data vectorization and negative stereotypes about creating slums. However, unless there is a radical change in the waste management system, scavengers and junkmen will continue to be key players in the cycle.

Intermediaries

Intermediaries/aggregators, or pengepul, serve as key hubs in the waste management system. They purchase waste at a profitable margin from various sources, including scavengers, junkmen, waste officers, and institutions generating large amounts of waste or even from smaller intermediaries. After sorting and arranging the waste, intermediaries trade the waste to recycling industries. Most intermediaries focus on specific recyclable waste desired by recycling industries, such as PET bottles, cardboard, iron, and aluminum. Intermediaries often form associations to coordinate efforts and maintain price stability. Intermediaries face limited facilities, resources, and capital challenges in running the activity. Legal issues also arise at times, necessitating business registration for interactions with recycling industries and the government.

Recycling Industries

Recycling industries from private sectors purchase waste from scavengers and intermediaries and then process it into finished or semifinished products sold to subsequent industries or consumers. As the products of recycling industries are no longer considered waste, the market determines the price. This pricing system impacts the value of the type of waste and the purchase price—from suppliers (scavengers, intermediaries, etc.). As the only end buyer in the waste market scene, recycling industries hold power in setting the price of waste. In addition to the competitiveness of the market, this dynamic results in fluctuation of the waste price. To counteract the high bargaining power of recycling industries, several examples have shown that forming collectives, such as waste banks and other associations, helps suppliers negotiate for better prices.

Due to the nature of the recycling industries and the scale of their operation, some of the recycling factories face challenges in legitimizing their business. The presence of national waste recycling associations helps tackle legal issues and support waste management business by improving coordination and worker welfare.

Waste Banks

Another significant component of waste management in Semarang is the Waste Banks, which are community organizations supported by the city Environment Agency. Like scavengers and intermediaries, waste banks act as entry points into Semarang's waste management cycle. A distinct point differentiating them is the focus and motivation of the activity. Unlike the latter, which are counted as business activity, waste banks are mainly formed as community efforts to support environmental sustainability, hence gaining less profit from the waste trading. Due to their similarities and differing motivations behind the act, waste banks and scavengers share a quite competitive relationship.

Waste banks are typically established through grassroots initiatives and neighborhood activity, with many originating from women's associations or community gatherings. Thus, the service is mostly bound by area. Waste banks purchase sorted recyclable waste from residents within administrative boundaries such as RT/RW or subdistricts. The collected waste is then sold to intermediaries or recycling activities. Most waste banks only accept inorganic and recyclable waste, with a few also accepting organic waste and used cooking oil. Profits from waste sales are distributed to customers and fund waste bank operations. Approximately 180 waste banks are registered in Semarang, with 95% of them still active.

While waste banks are supposed to run a profit-sharing scheme, in practice, the majority of waste banks run on a charitable basis relying on community or association support. This then resulted in challenges in funding the facilities. To help coordinate waste bank operations across the region, the Waste Bank Association connects waste bank managers to allow knowledge transfer, collaborations with stakeholders, companies, and Corporate Social Responsibility (CSR) initiatives, as well as facilitating access to funding programs.

Digital Initiatives

Widespread use of digital technology is starting to expand to the waste management sector, and Semarang is no exception. Digital initiatives create platforms that act as hubs and information systems for waste management actors, potentially improving management quality in the process. Several digital programs were initiated in Semarang. Notable ones are "Ambilin," a mobile application used to connect junkmen and intermediaries, and "Rapel," a digital platform that collects and buys sorted recyclables from households. Given that several digital initiatives share a similar drive with informal actors in the pursuit of monetary profit, collaboration between them might be mutually beneficial. However, not all models have proven to be sustainable or long-lasting. Despite the challenges in creating viable business models, digital initiatives show promise for modernizing and improving the current structure of informal waste management.

Neighborhood Waste Collection and TPS/TPS3R Facilities

The neighborhood's waste collection officers are individuals affiliated with, or endorsed by the neighborhood association to collect the waste from designated residential areas and transport it to the TPS, mainly using *triseda*. Households pay retribution for this service, ranging from USD 1.3 to 3.4 per month. While collection officers typically do not segregate waste, at times they pick up recyclable materials in good condition to be sold for personal gain. In most cases, the non-valuable waste is taken directly to the TPS, while the valuable one goes to TPS3R or is collected by scavengers. Residual waste from TPS and TPS3R will then be transported to the landfill after being processed.

The number of TPS across Semarang reached 145 units, excluding informal waste transfer points spread throughout the city. In addition, there are 23 units of TPS3Rs in the city, each serving around 300–400 households on average, below their capacity of 600 households. Unfortunately, many TPS3R are operating below their intended outcome. Some TPS3R even have similar functions as conventional TPS, in that it is only being used to contain waste temporarily. The role of waste processing is not carried out properly in many TPS3R due to several reasons. One major issue is inconsistent community participation, which contributes to resource limitations. Additionally, a lack of monetary incentives to encourage waste segregation and low compliance with tipping fee payments have been observed.

Government Institutions

The Environmental Agency of Semarang plays a vital role in coordinating and facilitating waste management within the city. However, the agency insufficiently monitors and assists waste banks in the city and gives inadequate programs and support to address issues. For instance, there is no formal coordination forum between the agency and waste management actors. In managing waste management in the city, the Environmental Agency faces challenges that hinder their outcome. Programs aimed at waste actors like scavengers, and TPS3R operators, are constrained by limited funding and resources. Despite all this, they try to remain committed to supporting community-based initiatives and welcome collaboration to enhance waste management in Semarang city.

Actors in the waste management system faced numerous challenges in addressing issues in their respective roles. Informal waste collectors, who operate without formal recognition, face social stigmas and financial instability despite their crucial role in the initial collection of recyclables. While promoting environmental sustainability, waste banks struggle with maintaining operations due to limited resources and community participation. Intermediaries, essential for streamlining the recycling process, face logistical challenges and difficulty in accessing financial support. These difficulties underscore the necessity for stronger coordination.

2.2 *Makassar City*

Makassar city faces significant problems regarding its waste management, ranging from landfill overcapacity, and low handling rate, to inefficiency in waste collection and transport system. To address these issues, the local government enacted programs involving formal and informal sectors. This section discusses the waste management practices encountered in Makassar.

2.2.1 Waste Management in Makassar

According to Local Government Regulation of Makassar Number 4 of 2011, the local government of Makassar is responsible for providing and managing waste services in Makassar city. The responsibility consists of segregation, collection, transportation, processing, and disposal of solid waste. The process includes waste segregation at the source, and the provision of collection facilities for segregated waste in every household, settlement, commercial area, industrial area, and other facilities. Segregated waste from the source is then collected to TPS, and the residue then transported to a waste processing facility or landfill [16].

In 2022, waste generated in Makassar was around 365,924 metric tons, with only 281,678 tons (76.98%) being managed or transported. This number indicates that the waste collection and transportation system in Makassar is insufficient to manage all solid waste in the city. As for waste composition, food waste (65%) and plastics (16%) are the dominant types. Waste banks and TPS3R reduce only 6,766 tons (1.85%) of all generated waste at the source, and 1,313 tons (0.36%) of the waste is recycled (Fig. 4). In contrast, the rest of the handled waste (around 75% of all waste) goes directly to Tamangapa landfill [16]. The Tamangapa landfill serves as the primary waste disposal site for the entire city and is one of the largest landfills in Eastern Indonesia. The landfill employs a semi-controlled method of disposal [17]. For a large urban area such as Makassar, Indonesian Regulation

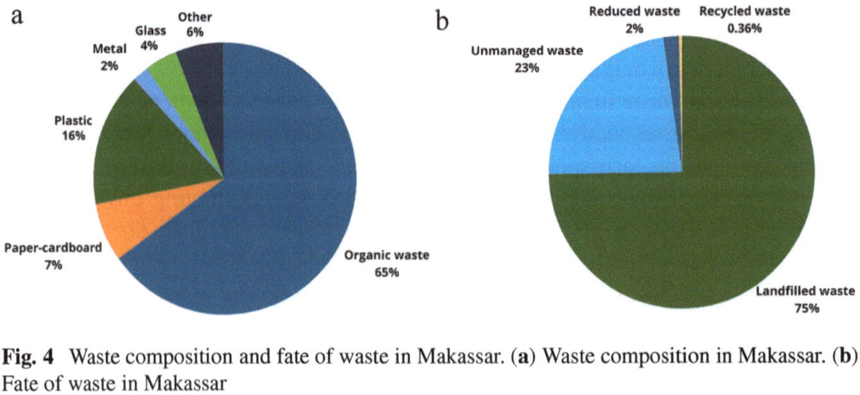

Fig. 4 Waste composition and fate of waste in Makassar. (**a**) Waste composition in Makassar. (**b**) Fate of waste in Makassar

Number 18 of 2008 regarding Waste Management mandates the adoption of sanitary landfilling techniques to mitigate the adverse impacts of landfills on the population and the environment. However, inadequate funding hinders sanitary landfilling, which prolongs the practice of controlled landfill methods. Like most landfills in Indonesia, Tamangapa Landfill also faces overcapacity problems. The landfill is estimated to have already exceeded its capacity limit in 2020 [17]. Temporary measures, such as landfill expansion, have been implemented to address the issue; however, financial constraints underscore the need for the government to adopt more sustainable and innovative long-term solutions.

2.2.2 Waste Collection and Transportation in Biringkanaya District, Makassar

In Makassar city, each district regulates their own waste collection and transportation system, while the Environmental Agency provides facilities and infrastructure for the process, including the provision of fleets. In the Biringkanaya district, this arrangement falls under the authority of the district cleaning division. Waste collection and transport mechanism in Biringkanaya District, as the study area, is divided into several activities: (1) Collection of waste from source to collection fleets, (2) Transfer of waste collected to transportation fleets at the transfer point, and (3) Transporting waste from transfer points to landfill. There are four types of fleets operating in the Biringkanaya district for waste collection and transportation purposes, namely:

- Three-wheeled motorcycle with a container (*triseda*) collects waste from households and transfers it to a transfer point, where it is then transferred into trucks for waste transportation. The capacity of the *triseda* is around 1.5 m^3.
- Arm roll, a truck-type vehicle with a detachable container to hold the collected waste. This vehicle can directly attach a full container of waste to the truck. The capacity of arm roll is around 6–7 m^3.
- Dump truck, a truck-type vehicle that can directly dump the waste from the container by tilting the container. This vehicle is transporting collected waste at the transfer point to the landfill. The capacity of dump trucks ranges from 6 to 7 m^3.
- *Tangkasaki* is a truck-type vehicle equipped with a built-in waste compactor. *Tangkasaki* does not collect waste at transfer points and instead collects and transports waste from main roads and cluster housing directly to the landfill. The capacity of the *tangkasaki* truck ranges from 6 to 7 m^3 (Fig. 5).

Each fleet has designated service areas and routes which are bound by the administrative district. The starting points for *triseda* are near the collection area, while the truck starts near the Tamangapa landfill outside Biringkanaya. *Triseda* are deployed in high-density areas with narrow streets, whilst trucks are deployed in main roads. *Triseda* collected waste directly from houses, brought it to the transfer point, and started another cycle to serve different routes or service areas. Meanwhile,

Fig. 5 Waste collection and transport fleet in Biringkanaya district, Makassar. (**a**) *Triseda*. (**b**) Arm roll. (**c**) Dump truck. (**d**) *Tangkasaki*

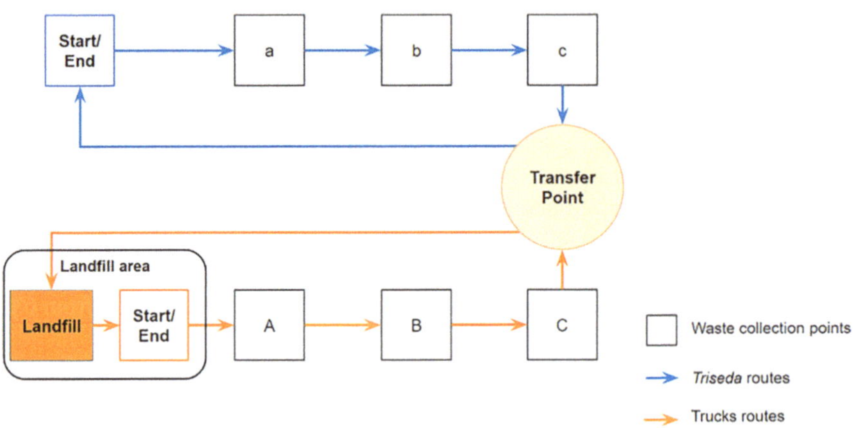

Fig. 6 Waste collection and transportation flow in Biringkanaya district, Makassar

the trucks collected solid waste from transfer points, industry, and commercial areas along the main road to later transport it to landfill. Figure 6 shows the scheme for waste collection and transportation.

Waste collection and transportation activities in the Biringkanaya district are carried out every day in the early morning. *Triseda*, operated by one to two officers, start operating at 6 or 7 AM to align the collection time with the truck schedule, for all waste collection and transport activities to be done by noon. Due to space limitations, there are no TPS in Makassar. Instead, the waste was transferred directly, without storing, into the trucks at transfer points, which are mainly located on the roadside. In this kind of waste transportation scheme, the waiting time for *triseda* could vary between 2 and 8 hours, depending on the arrival time of the trucks. The whole operational duration of the *triseda* is approximately 2–4 hours, including break time and time for transferring the waste into the trucks, which averages around 20–40 minutes.

As for the trucks, most of the fleet start operating around 6–8 AM by collecting waste from the main roads and continue heading to transfer points. The operational duration of trucks depends on the type of the fleet. *Tangkasaki* trucks typically operate for 6–9 hours a day for one trip to the landfill, while dump trucks, which transported waste from both transfer points and main roads, operated for 4–10 hours a day for a trip. As for arm roll, the operational time is only 2–5 hours for two trips to the landfills in a day. The short duration and more trips for arm roll are due to differences in operation, in which the arm roll trucks collected waste that had already been gathered in the container at transfer points. In contrast, other trucks need to load the waste from *triseda* at transfer points. Besides time spent in service areas and transfer points, trucks operational duration is also determined by the road traffic and the queue in the landfill. The traffic occurs in peak hours around 11–14 AM due to trucks from various districts in Makassar heading to the landfill simultaneously, in addition to daily peak hours in Makassar around 11–13 AM (Fig. 7).

In the landfill, arm roll trucks are unloaded manually or mechanically with the help of an excavator, taking around 5 minutes to unload, taking similar duration to dump trucks which are equipped with hydraulics systems. On the other hand, manual unloading of trucks could take about 40 minutes to 2 hours to complete. For *tangkasaki*, the trucks are divided into two types: medium-capacity trucks with no hydraulic system and large capacity ($30\,m^3$) with a hydraulic system. Large *tangkasaki* trucks take around 10–30 minutes to unload, while medium ones need around 60–90 minutes. This practice shows that manual unloading significantly takes up more operational time. Road traffic, queues at the landfill area, and unloading time in both transfer points and the landfill make it difficult for the fleet to make more than one trip to the landfill daily.

Aside from lengthy duration, waste transportation in Makassar also faces an overcapacity issue. Almost 70% out of 55 fleets in the Biringkanaya district collected waste exceeding its container capacity. Overcapacity can be seen from the height of the waste pile which exceeds the height of the container, or when the waste is being hung on the sides of the container. Although the existing waste collection system can still operate normally in a state of overcapacity, there are concerns that the system could not match the increasing waste generation resulting in more waste unhandled or mismanaged (Fig. 8).

Fig. 7 Collected waste exceeding *triseda* container's capacity

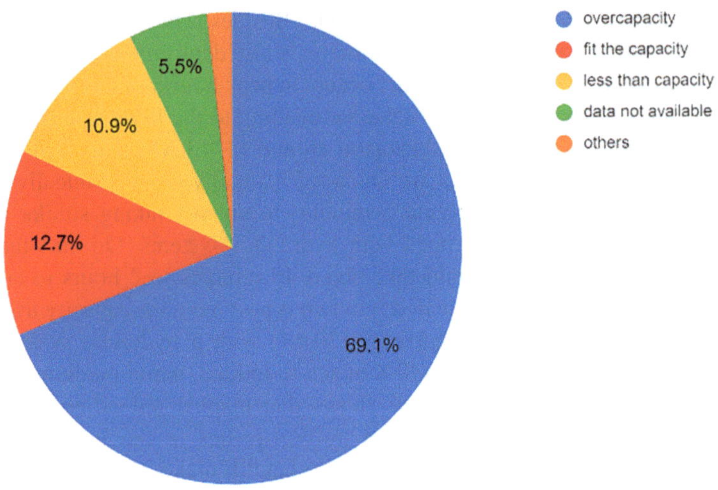

Fig. 8 Waste container at the end of collection process

In short, the current volume of collected waste in Makassar is overwhelming for the existing waste facilities, both the collection fleets and the landfill. To effectively manage all waste produced, it is crucial to increase the number of waste collection facilities while also implementing strategies to reduce the amount of waste sent to landfill.

2.2.3 Waste Collection and Transport Optimization and Potential to Improve Waste Circularity in Makassar

Observations and discussions in the waste collection and transportation process in Makassar reveal that the lack of permanent waste storage points influences the main constraints. Currently, to transfer waste, *triseda* and truck must meet at specific times at transfer points. Often, these meetings are not synchronized, causing long waiting times for one of the fleets and reducing the time available for additional collection rounds. Since most transfer points are located roadside, collected waste cannot be left unattended, necessitating that transfer and loading occur only when a truck is present.

Building permanent waste storage locations like TPS or TPS3R facilities in Makassar is advisable to eliminate waiting times. Additionally, TPS3R would decrease the amount of waste that needs to be transported to landfills by sorting, processing, and recycling the collected waste. The concept of TPS3R is widely recognized as an effective method of integrating community involvement into the waste management system. This aligns with the local government's goal in pursuing the provision of waste banks and TPS3R.

Currently, there are nine units of TPS3R in Makassar; this amount falls short of the city's target of having at least one TPS3R facility in each district. The record shows that waste circularity in Makassar has the potential to generate profit as high as USD 74,200 from waste turnover in 2022, highlighting the economic value of waste and providing an effective way of boosting waste reduction. Three of the nine TPS3R in Makassar are situated in the Biringkanaya district, namely in Pai, Untia, and Bulurokeng subdistricts, which are all managed by the community. Though beneficial, the facilities face many challenges in managing daily operations, like what happened in TPS3R Bulurokeng, which was inactive due to labor and facility issues.

Lack of waste segregation at the source and during waste collection adds to the workload for TPS3R. Suppose constraints can be overcome, TPS3R will be an effective measure to reduce waste entering the landfill and promote circularity, as in the case of TPS3R Untia and TPS3R Darul Aman, Pai subdistrict. TPS3R Untia mainly processed recyclable waste, achieving a waste reduction percentage of 3.7% of the total waste in 2022. Aside from processing recyclable waste, these two TPS3R compost waste from food and other biodegradable waste. The facilities receive around 231 tons of waste and produce around 85 tons of compost and 65 tons of recyclables. Other TPS3R in Makassar, TPS3R Barrang Lompo, and Rusunawa Panambungan also achieved a high number of waste reduction, around 8 and 21 tons in 2023.

Apart from TPS3R, the informal sectors and local waste banks play a big role in reducing waste and promoting waste circularity in Makassar city. Waste banks, in particular, are growing rapidly. In 2018, there were 544 waste banks in Makassar, and by 2023, the number had increased to 1,077 units, spreading across the 15 districts. It is estimated that up to 21,000 people use the service of waste banks and have managed around 943 tons of waste [18]. Implementation of extensive "one neighborhood, one waste bank" campaigns by the local government, a policy initiated in 2015, resulted in a twofold increase in the number of waste banks. To accomplish this level of success, the government appointed a "motivator" for each district to promote community waste reduction and recycling. The policies were followed by the establishment of the central waste bank of Makassar city in 2016, built and managed by the local government through the Regional Technical Implementation Unit *(Unit Pelaksana Teknis Daerah*/UPTD). This central waste bank aims to facilitate the trade between community waste banks and the recycling activity/off-taker. The facility is equipped with a large waste storage yard and crushing equipment to sort and process recyclable materials. It also provides on-call waste collection services for waste banks. The government uses the central waste bank to analyze and issue reference prices for recyclable waste and to ensure fairness in recyclables trading.

In addition to on-the-ground waste reduction activities, Makassar implemented approaches to prevent waste generation by issuing regulations. The city has enacted a ban on single-use plastics through Mayor Regulation Number 21 of 2023. The regulation encourages the substitution of plastic bags with more eco-friendly alternative materials.

Significant efforts are also shown by the private sector in Makassar. For example, waste recycling and upcycling efforts, such as those done by Rappo Indonesia, contribute to the goal of waste circularity. These efforts address the city's waste management issues and create economic opportunities for residents by encouraging the local community to take part in craftsmanship workshops and training for running the recycling business independently. With the current rising awareness of waste reduction and environmental issues, upcycled products are gaining popularity in the markets, providing a potential source of income for many residents.

3 Lesson Learned

As shown in Semarang and Makassar, the transition from linear to circular economy in waste management systems faces significant challenges. These difficulties often stem from a lack of understanding among policymakers and officials regarding the principles of circular economy and its potential to address waste management issues. Overcoming these obstacles requires robust policy reforms, increased funding for waste management infrastructure, and extensive public education campaigns.

This section discusses the lessons learned from the studies of Makassar and Semarang, which are supported by successful cases from other countries. While the scope of the study limits the depth and specificity of contributing factors in waste management systems, the results indicate that broader adoption of circular economy principles can significantly improve overall waste management.

3.1 Decentralizing Waste Management Through TPS3R and Waste Bank

The enactment of new regulations emphasizing the establishment of TPS3R facilities and waste banks to reduce waste at the source indicates a shift toward a decentralized approach to waste management in Indonesia. The shift is primarily driven by the increasingly complex problems in centralized waste management systems, such as low waste handling and recovery rates, landfill overcapacity, and high material handling costs. Implementation of TPS3R and waste banks as decentralized waste facilities proved beneficial in reducing the waste, as observed in Semarang and Makassar, as well as other cities such as Bogor, Depok, and South Tangerang, in which the facilities can reduce up to 2–4% of the city's waste [19].

A report from the Commission on Audit of the Philippines highlights a similar community-centered model being used in the Philippines. The Barangay Waste Management Program involves local communities to develop waste management systems that include waste segregation, composting, and recycling under government support. The program has improved waste handling, closing 1,232 illegal dumpsites by 2021 [20]. This demonstrates the importance of strong cooperation between the government and local communities in achieving waste management goals [21].

From the technical perspective, decentralization of waste management comes with advantages as well as disadvantages. In the case of Makassar, the lack of proper temporary waste storage facilities contributes to inefficiency in the duration of waste collection and transportation, ultimately increasing the cost of transportation. It is estimated that up to 50% of the waste management budget is allocated for waste collection and transportation services in conventional waste management [22]. As a result, minimal resources are left for waste processing, which is essential to improve the circularity of the waste management system. Decentralization of waste facilities could be beneficial in lowering transportation costs, reducing the volume of waste sent to landfills, and increasing efficiency for waste processing. However, it can also be argued that decentralization efforts may not fully capture the benefits of economies of scale. In optimizing service area and waste processing scale, harmony should be achieved in a balanced allocation of transportation costs (more efficient in decentralized systems) and treatment costs (more favorable in centralized systems). Gujarat State in India offers a strong example of inter-municipal cooperation to optimize waste management scale. In 2005, the Gujarat Urban Development Company Limited (GUDCL) initiated a statewide solid waste management program across 159 urban local bodies (ULBs). GUDCL found that decentralized treatment facilities and 36 regional landfill sites could lower waste management costs from 25 USD/ton to 9.4 USD/ton. By 2015, 93 vermicomposting plants and seven regional landfills had been built to serve multiple ULBs [23]. Future decentralization efforts should focus on addressing this scale optimization to improve cost efficiency.

3.2 Challenges of TPS3R and Waste Bank Development

As a community-based waste management program, the performance of TPS3R and waste banks is highly dependent on the voluntary participation of the local communities. Most TPS3R would require households to pay retribution fees to support the operation. In Makassar, despite community agreements on retribution fees, some households refuse to pay, resulting in disrupted waste services. This situation could lead to social strain and complaints from other community members. For TPS3R and waste banks to become sustainable, it is important to segregate waste at the source to improve the value of waste and reduce the operational costs associated with waste segregation. The challenge lies in the current practice of most regions in Indonesia where collected waste is mixed, thus resulting in poor-quality recyclables that cannot compete with virgin materials. Although awareness of the importance of waste management continues to grow, many community members struggle to adopt consistent waste segregation practices or engage in recycling programs due to factors such as limited access to facilities, information, or motivation [24]. Similarly, a feasibility study on decentralized waste management in rural Lebanon reveals that the primary reason respondents refrain from segregating their waste at the source is the perceived redundancy, as the waste is ultimately mixed after collection [25]. The introduction of new TPS3R facilities or waste banks also carries the risk of disrupting the livelihoods of existing informal sectors and establishments. Since TPS3R and waste banks operate upstream, they may reduce the availability of the supply of recyclable materials, negatively impacting the recycling revenue downstream, the area mainly handled by informal sectors. In Semarang, scavengers have a complicated relationship with waste banks, though they both serve as the initial entry point for waste management. People deem scavengers lowering the commercial worth of waste banks and neglecting environmental concerns in their practice. These potential conflicts need to be mitigated to establish mutual support and collaboration between actors.

The government needs to bridge the communication between different waste actors to prevent social conflicts regarding the equity of the waste management practice. In Makassar, the government convinced the various waste management actors to collect different types of wastes. Scavengers collected more valuable types of wastes, such as wire cables and electric appliances, while waste banks collected materials like plastics, metal, paper, and glass. This initiative has been deemed successful in minimizing livelihood disruption in Makassar. Efforts are also being made to address this issue by facilitating the informal sector to feel more included. In Semarang, *Yayasan Kesejahteraan Keluarga Soegijapranata* (YKKS) links waste banks, scavengers, and junkmen to various institutions that may benefit them. Similarly, *Yayasan Bina Karta Lestari* (Bintari), through funding from USAID Clean Cities Blue Ocean, supported informal sectors to improve their working condition and social recognition by providing work aids and protective gear. Fostering harmony between formal and informal actors is essential for the equitable development of waste management.

3.3 Ensuring Sustainability in Decentralized Waste Management in Indonesia

Prospective analysis on the sustainability of TPS3R identifies regulation and stakeholder collaboration as the most influential variables determining the sustainability of the operations [26]. The study also highlights the significant contributions of the private sector (companies through CSR programs) and public sectors (government, NGOs) to TPS3R sustainability. The role of local government is particularly pivotal, serving not only as the party responsible for the implementation of overall waste management activity but also as the primary regulatory institution.

The existing regulatory framework in Indonesia provides a good starting point for achieving sustainable waste management systems, but further work is needed to address the complex problems in the implementation of the policy. For example, the performance of most waste management systems in Indonesia is heavily dependent on political leaders. In Makassar, the change of the city's mayor raised concerns among residents and waste bank organizers about the continuity of the waste bank program in the city. The political disruption is linked to decreased community participation in waste banks. The former mayor is then appointed as a caretaker of the program to establish the government's commitment to supporting the local waste bank.

Financial sustainability is a crucial aspect that most facilities fail to address properly. In the development of TPS3R or waste banks, capital investments are often provided by either the government or the private sector, while operational expenses are funded through tipping fees and revenue from the facility's waste products. However, due to the lack of a market for recyclable products and the low value of waste, most TPS3Rs operate on a deficit. This leads to facilities operating under their intended outcomes. Improvement to the revenue sources is needed for facilities to sustain operations. The main revenue sources considered by the government include retribution fees, waste valorization, and private sector co-funding [27]. Each revenue source faces problems that require innovative policies and regulations to address effectively. For example, the current retribution fees (1) do not reflect the true cost of waste handling, (2) are deemed expensive compared to illegal waste burning and dumping, and (3) lack financial transparency. In terms of waste valorization, the process is hindered by difficulties in creating a circular economy for waste materials. Low-value flexible plastics dominate the waste stream, indicating insufficient efforts by producers to design products that are easier to recycle. Small-scale waste recyclers often struggle to secure competitive materials prices and meet the off-taker's minimum quantity requirements. Addressing these challenges requires empowering communities through educational initiatives, such as training on waste segregation and composting, which can enhance the quality of recyclables and reduce contamination [28].

However, incremental improvements alone will not suffice; a more comprehensive strategy is necessary to boost the circularity of recyclables. This approach must involve stronger collaboration between producers, recyclers, and policymakers, with a focus on creating economic incentives and improving the infrastructure needed to support a more robust recycling ecosystem.

Regarding private sector involvement, Indonesia already has two mechanisms to incorporate private funding into the waste management system: private-public partnership/PPP (*Kerjasama Pemerintah dengan Badan Usaha*/KPBU) and CSR. Currently, many CSR programs aim to establish waste banks within local communities. However, most of the funding is allocated solely for facilities and equipment, while subsequent daily operations are often not properly sustained and monitored. These programs are also largely uncoordinated and sporadic, and their impact is not tracked. This is due to a lack of coordination between the provider of CSR programs and the local governments.

On the other hand, PPP schemes are relatively more robust in terms of sustainability and impact in the long term. One notable example is the PPP between the local governments of Surabaya and PT Sumber Organik for a landfill gas facility. They involve structured agreements between the government and private entities, ensuring a shared responsibility in financing, managing, and operating waste management projects. However, despite their potential, the effectiveness of PPP schemes is sometimes hindered by bureaucratic complexity, lack of transparency, and coordination issues between the involved parties. Developing policies to address these challenges is important to improve the circularity of waste and sustainability of TPS3R and waste banks.

3.4 Optimizing Community Involvement Within TPS3R and Waste Bank

TPS3R and waste banks were both initiated by the government as top-down programs, often perceived as a "gift" for communities. As a result, the operation requires active community involvement to improve operational effectiveness. In Semarang, this dependency on the community is partly mitigated through collaboration with waste banks or NGOs and CSOs, supporting behavioral change of the community.

An example of a lesson learned based on the study in Semarang is the collaboration between the NGOs and the local community to improve involvement through capacity building in waste management and business management. TPS3R also has significant potential for collaboration with local waste banks and digital aggregators, as shown in a TPS3R in Semarang. The facilities experience raised worker welfare by paying facilities officers monthly salaries up to IDR 5,000,000 (USD 337)—with profit sharing of 20% for officers—in light of effective management and collaboration with various institutions, including waste banks, digital aggregators, and successful product diversification.

The involvement of the government in Makassar's waste reduction efforts can serve as a model for other cities. Implementing policies such as establishing TPS3R in each district, promoting the concept of "one neighborhood, one waste bank," and appointing a motivator for each district have demonstrably increased community participation in local waste management. This active engagement helps establish and secure future political commitment to city waste management. Although it may seem trivial, local officials' political will can significantly influence a city's commitment to waste management, especially in a politically driven and closely knit community like Makassar.

The need for an active engagement and awareness campaign is also highlighted in lessons learned from India by the Thiruvananthapuram Municipal Corporations. The program utilizes massive awareness campaigns such as house-to-house surveys, sensitization activities, awareness events, as well as social media campaigns [29]. Studies reported that the program has made progressive changes in behavioral patterns of the residents, increasing the participation of households. This success led to 60–70% of biodegradable wastes being treated at the household level [30].

Vietnam's Material Recovery Facilities (MRF) in Cu Lao Cham represent a comparable model to TPS3R. While TPS3R has faced challenges with community involvement, Vietnam's MRFs benefited from an early introduction of waste segregation at the source (WSAS) in 2009. However, WSAS's effectiveness has waned over time, highlighting the need for continuous community engagement. In response, the local government and NGOs in Cu Lao Cham implemented a pilot MRF that operates as a community-based waste management model. With the support of the "Mitigating Marine Plastic Debris in Vietnam" Project, the MRF now processes organic waste more efficiently using advanced technology, such as waste crushers, to increase productivity. The MRF in Cu Lao Cham has successfully recycled nearly 50% of the waste generated by the community, demonstrating the positive potential of community-driven waste management systems when backed by technological innovation and continuous stakeholder engagement [31, 32].

In Manila, the Ecowaste Coalition and Plastic Bank have implemented community-based waste management models. The Ecowaste Coalition promotes zero-waste practices at the barangay level, supporting recycling and social entrepreneurship aligned with circular economy principles. Plastic Bank's Social Recycling model empowers informal waste collectors, particularly women, to improve livelihoods and address social challenges. In addition, waste management in the Philippines follows a waste hierarchy. Waste generators and barangays are responsible for avoiding, reducing, reusing, recycling, and recovering waste. Barangays operate MRFs for sorting and recycling before sending residuals to landfills. Municipal and provincial authorities oversee compliance and manage waste recovery and treatment, involving informal collectors, recycling facilities, landfills, and off-takers [33, 34].

In Thailand, the collaboration between WWF Thailand and the Surat Thani Municipality to support the Baan Bang Lha community offers another example of effective grassroots waste management. Baan Bang Lha, a community of 315 households built on narrow pathways over the Tapee River, faced significant waste management challenges, including waste contamination into the river due to the lack of formal waste services. Motivated community leaders initiated a waste collection system funded by a monthly fee of 50 baht per household. WWF Thailand provided segregation bins, training, and coordination with local junk shops for recyclable waste sales. As of November 2023, the community has sold 1,195 kilograms of recyclable waste, including 405.2 kg of plastic, earning 7,015 baht [35]. This initiative highlights the effectiveness of community-led efforts, supported by NGOs and local governments, in improving waste management practices and generating income for community development [36].

Drawing lessons from domestic and international models is essential to optimizing community participation in TPS3R and waste banks. The initiatives in Vietnam, the Philippines, and Thailand demonstrate the potential of community-led waste management systems and underscore the importance of sustained community engagement, innovation, and partnerships with multiple stakeholders, including NGOs and the private sector.

4 Recommendation for Pushing Waste Circularity

4.1 Rearranging Waste Actors Roles and Integrating Active Participation for an Optimized Waste Management System

To promote waste circularity, all key players, including the public, must shift their perspective and start seeing waste as a resource. Both cities in the studies demonstrate a strong sense of community, requiring full cooperation from every sector. Champions may be needed to motivate others to build a circular waste management ecosystem. Figure 9 illustrates a recommendation scheme for effective collaboration, ensuring a smoother transition of waste circularity.

Waste sources, typically divided into household and non-household, should segregate their waste into inorganic and organic categories to minimize segregation efforts. Cooperation and collaboration among waste management actors are essential to reduce competition and assign clear roles. For instance, digital aggregators can partner with scavengers or junkmen to collect inorganic waste directly from households and businesses. Digital aggregators can list registered residents and businesses needing waste collection, while scavengers and junkmen can collect, re-sort, and weigh the waste. Households can also opt to give their inorganic waste to the nearest waste bank, if available in their community. Waste banks and digital aggregators also play a crucial role in motivating people to promote behavior change toward waste segregation at the source. Waste banks can provide financial incentives or community benefits, while digital aggregators can offer streamlined collection services and real-time updates. Integrating digital initiatives with existing waste management frameworks offers an innovative approach to modernize and optimize waste management and expand their customer base. The collected and segregated inorganic waste can then be sold to intermediaries or directly to recycling industries.

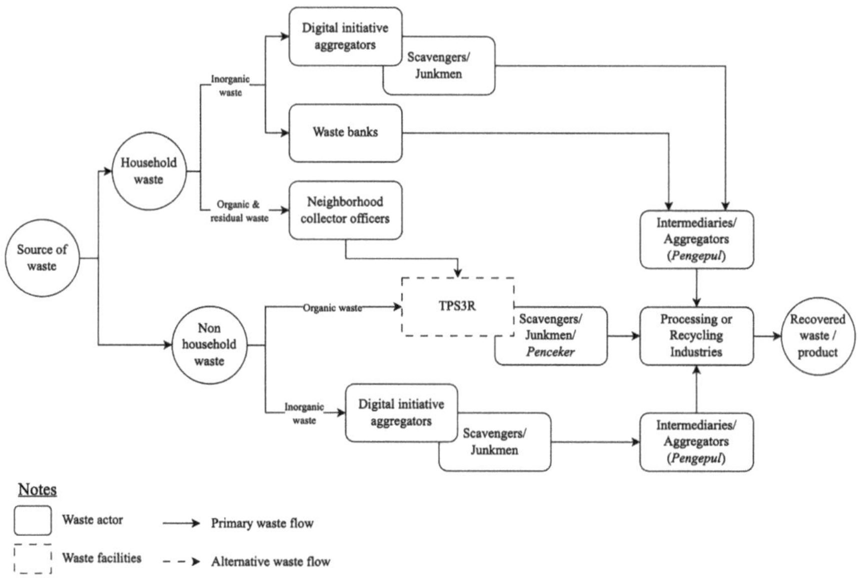

Fig. 9 Recommendation for optimizing the role of waste management actors

This shift allows TPS3R to manage waste more efficiently, focusing on processing organic and residual waste instead of sorting mixed waste at TPS3R. Scavengers and junkmen can also play a role at TPS3R as workers, helping to sort and process waste. Therefore, establishing and reactivating TPS3R functions is essential to handling waste effectively. Effective engagement and education of the community are crucial for optimizing the benefits of TPS3R and waste banks. The government, especially the Environment Agency, should be involved in empowering community-level actors such as informal waste collectors, waste banks, and TPS3R through training, capital provision, facilities, and active monitoring. Waste-derived products or semifinished goods can be distributed to off-takers or processing industries for further use.

The waste reduction rate could potentially be higher with the implementation of waste segregation at the source and better waste processing. A comprehensive, multi-stakeholder approach that involves enforcement of the Extended Producer Responsibility (EPR) policy, promotion of household waste segregation, and partnership with local and private actors should be implemented to tackle the problem. Hence, the relationship between waste actors and the community is vital for the system's success, as they complement each other's roles in Indonesia's waste management.

The inclusion of off-takers in the waste management system will help create a sustainable business ecosystem that balances the supply and demand of recovered waste and builds a circular economy environment. Identifying the types of waste-derived products in demand within the industry is crucial. By understanding industry needs, the waste recovery and processing methods can be tailored accordingly. Moreover, developing niche markets for specialized waste products, such as organic cleaning products (EcoEnzyme), Black Soldier Fly (BSF) Larvae Products, compost and organic fertilizer, biochar, biodegradable products, eco-bricks, and plastic waste furniture products, is necessary to make the most of waste recovery efforts. This comprehensive approach ensures that all types of wastes are efficiently processed and converted into valuable resources.

Structured division of responsibilities in the Philippines waste management system—from households to community and higher government levels—can be taken as an example to ensure clarity in roles and accountability. It offers a model that could be replicated in Indonesia by establishing clear responsibility lines for waste management, ensuring communities and government have distinct, coordinated roles. The integration of informal waste collectors and the establishment of MRFs and TPS3Rs can also be replicated to enhance waste recovery and recycling efforts at the local level. Such a system could drive better waste management, reduce landfill dependency, and promote circular economy practices across Indonesia's municipalities and provinces.

4.2 *Policies and Governance Reforms to Support a Transition to the Circular Economy*

However, as explained in the previous section, the government needs to play a more substantial role in implementing an integrated circular waste management system with active role sharing. Comprehensive planning, thorough study, and stakeholder mapping are essential to setting up the proposed system. A pilot project is recommended to be started, and if successful, it can be scaled up or replicated in other cities.

For starters, governments should minimize reliance on the motivation of individual leaders and instead adopt a more institutionalized approach to development. Local governments could take more significant actions if directed by national mandates and propose plans and budgets. The national government has outlined initiatives for a circular economy, including the Circular Economy Roadmap and Action Plan for 2025–2050, as established by the Ministry of National Development Planning. Policymakers can advance this action plan by enforcing stricter regulations on EPR, applicable waste reduction measures, and clear responsibility lines for waste management actors. They could require communal waste segregation systems in newly developed residential areas and the establishment of TPS3R or MRF facilities for commercial and industrial areas as the business' responsibility in mitigating the environmental impact of their activity.

Implementing plans for waste circularity adoption will require a concise plan and increased funding for local Environmental Agencies. Such funding would enable these local agencies to better execute and expand their waste management efforts, reducing their reliance on retribution fees and private sectors for providing and operating waste management services. Therefore, governments must commit to allocating a significant portion of the budget for waste management by issuing a secured policy, as currently, the budget for waste management in most cities is less than 1%. The government often struggles to sustain these facilities independently, as waste management is not yet seen as a priority compared to basic needs like health, education, and infrastructure. The government must recognize that the long-term effects of mismanaged waste can lead to serious health issues. Neglecting waste management not only exacerbates environmental pollution but also increases the risk of disease outbreaks and imposes heavy burdens on public health systems. Policymakers should enforce waste management policies and integrate them with public health initiatives to enhance awareness and encourage more proactive efforts.

Fostering community-based initiatives and active participation may significantly improve the implementation of role sharing and the proper upkeep of both existing and future waste facilities, allowing them to become self-sustaining through collective efforts. The community-based waste management (CBWM) approach highlights that solving waste issues requires technical solutions, government policies, behavioral changes, and active community involvement. It promotes participation from all demographics, ensuring that knowledge and opportunities are shared across the community. The approach must be tailored to the local social, economic, and cultural context [37]. However, shifting community behavior takes time, in addition to ensuring financial sustainability. Strengthening partnerships and developing innovative business models are key to making waste management beneficial. Through coordinated actions, the government can link TPS3R and the community with NGOs or CSOs interested in endorsing sustainable waste management and circularity activities to design programs that provide immediate support and foster the long-term sustainability of CBWM. For example, programs could include training workshops, resource-sharing networks, and ongoing mentorship to empower waste actors to continue developing and managing their operations effectively after external support ends.

5 Conclusion

As Semarang and Makassar exemplified, the transition to a circular economy in waste management in Indonesia highlights significant potential and challenges. The national government's Medium-Term Development Plan (2020–2024) and Circular Economy Roadmap and Action Plan (2025–2050) set a strategic framework for shifting toward waste circularity, underscoring the importance of strong policy enforcement. For this vision to be realized, local implementation of a circular waste management system in Indonesia must be supported by well-designed policies, thorough planning, adequate funding, and strong collaboration across all stakeholders. In addition, a community-based approach focused on capacity building and promoting behavioral change is needed to ensure the long-term success of decentralized waste management systems.

References

1. Law of the Republic of Indonesia Number 18 of 2008 on Waste Management
2. Damanhuri E, Handoko W, Padmi T (2013) Municipal solid waste management in Indonesia. Environ Sci Eng 139–155. https://doi.org/10.1007/978-981-4451-73-4_8
3. Damanhuri E, Tri P (2016) Pengelolaan Sampah Terpadu. ITB Press, New York
4. Presidential Regulation Number 97 of 2017 on Indonesian National Strategy Policy on Management of Household Waste and Household-like Waste
5. Putra HP, Damanhuri E, Sembiring E (2019) Sektor Baru Pengelolaan Sampah Di Indonesia (Studi Kasus Di City Yogyakarta, Kabupaten Sleman dan Bantul). Jurnal Sains dan Teknologi Lingkungan 11(1):11–24
6. Samadikun BP, Rezagama A, Ramadan BS, Andarani P, Rumanti ED (2020) Understanding informal actors of plastic waste recycling in Semarang City. Jurnal Ilmu Lingkungan 18(1):162–170
7. Amelia D, Putra HP (2018) Potensi pengelolaan sampah yang dilakukan oleh sektor informal di wilayah City Yogyakarta [Undergraduate thesis, Universitas Islam Indonesia]. Universitas Islam Indonesia Repository. https://dspace.uii.ac.id/handle/123456789/13103?show=full
8. Damanhuri E (2011) Typology of informal recycling (Waste management—waste recycling activity in Indonesia) [PowerPoint slides]. In 5th Workshop of Asia Circulation Policy Research
9. Damanhuri E, Padmi T (2018) Integrated Solid Waste Management. Bandung. Penerbit ITB, Indonesia
10. Pratiwi AH, Budiyono, Dewanti NAY (2021) Identification types of the Marine Debris and Factors Related them in Semarang City. Jurnal Presipitasi 18(1):64–72
11. Dinas Lingkungan Hidup Kota Semarang (2023) Gandeng DLH, Pelindo bersihkan Pantai Tirang. Dinas Lingkungan Hidup Kota Semarang. https://dlh.semarangkota.go.id/gandeng-dlh-pelindo-bersihkan-pantai-tirang/
12. Saputra IY (2023) Hari Lingkungan Hidup, PLN pungut 1 ton sampah dari Pantai Mangunharjo Semarang. Solopos. https://jateng.solopos.com/hari-lingkungan-hidup-pln-pungut-1-ton-sampah-dari-pantai-mangunharjo-semarang-1935678
13. Ministry of Environment and Forestry (n.d.) Sistem Informasi Pengelolaan Sampah Nasional: SIPSN. Retrieved July 29, 2024, from https://sipsn.menlhk.go.id/
14. Yuliastuti N, Saraswati N (2014) Environmental quality in urban settlement: the role of local community association in East Semarang Sub-district. In Procedia—Social and Behavioral Sciences 135:31–35. https://doi.org/10.1016/j.sbspro.2014.07.321
15. Kustyardhi R, Suwitri S, Djumiarti T (2013) Pengelolaan Sampah Terpadu di Kota Semarang. J Publ Policy Manag Rev 2(2):1–8. https://doi.org/10.14710/jppmr.v2i2.2514
16. Regional Regulation of Makassar City Number 4 of 2011 on Waste Management
17. Juhaidah S, Usman F, Subagiyo A (2019) Pengelolaan Sampah TPA Tamangapa Kota Makassar. Planning for Urban Region and Environment 8(3):133–142
18. Kubota R, Horita M, Tasaki T (2020) Integration of community-based waste bank programs with the municipal solid-waste-management policy in Makassar. Indonesia. Journal of Material Cycles and Waste Management 22(3):928–937. https://doi.org/10.1007/s10163-020-00969-9
19. Putri MOA, Zakianis S, Djaja IM., Agustina H, Fauzia S (2020) Solid waste reduction through 3R-based waste management unit and Waste Bank in Indonesia in 2018. Indian J Public Health Res Dev 11(4):1003–1008
20. Commission on Audit (2023) Performance audit report: PAO-2023-01, solid waste management program: progress in the achievement of the goals of the Ecological Solid Waste Management Act needs stronger support and the cohesive efforts and strategies of all stakeholders
21. Shu Y, Booker A, Karetny J, O'Keefe K, Rees K, Schroder L, Roe BE (2023) Evaluation of a community-based food waste campaign using a national control group. Waste Manag 160:101–111. https://doi.org/10.1016/j.wasman.2023.02.011

22. Zafira AD, Damanhuri E (2019) TPS 3R Sustainability strategy analysis for Waste Transport Minimation to Landfill (Case study: 3R Bandung district, West Java). Jurnal Teknik Lingkungan 25(2):33–52
23. Kojima M (2019) Regionalization of Solid Waste Management in Asia: Benefits and Challenges. In Toward Regional Cooperation of Local Governments in ASEAN. ERIA Collaborative/Support Research Report. IDE-JETRO
24. Pamuji K, Rosyadi S, Nasihuddin AA (2023) The legal institutional model of community-based waste management to reinforce multi-stakeholder collaboration in Indonesia. Kasetsart J Soc Sci 44(1):73–82. https://so04.tci-thaijo.org/index.php/kjss/article/view/264451
25. Ahad MaA, Chalak A, Fares S, Mardigian P, Habib RR (2020) Decentralization of solid waste management services in rural Lebanon: barriers and opportunities. Waste Manage Res J Sustainable Circular Economy 38(6):639–648. https://doi.org/10.1177/0734242x20905115
26. Nopriani M, Fauzi A, Nuva N (2022) Analisis Prospektif untuk Keberlanjutan Pengelolaan TPS3R di Kota Pangkalpinang. Jurnal Pendidikan Tambusait 6(3):13791–13808
27. SYSTEMIQ (2023) Building robust governance and securing sufficient funding to achieve Indonesia's waste management targets. SYSTEMIQ. https://www.systemiq.earth/wp-content/uploads/2021/11/Building-Robust-Waste-System-Governance-and-Securing-Sufficient-Funding_Final-Report_26Nov2021.pdf
28. Ramang R, Dirawan GD, Ruslan R (2023) Domestic waste management study of community based in Kupang City. IOP Conf Ser: Earth Environ Sci 1209(1):012020. https://doi.org/10.1088/1755-1315/1209/1/012020
29. Menon VJ, Palackal A (2022) Centralized and decentralized approaches to solid waste management—a case study. Holistic Approach Environ 12(4):155–164. https://doi.org/10.33765/thate.12.4.3
30. Véron R, Fernando N, Narayanan NC, Upreti B, Ambat B, Pallawala R, Rajbhandari S, Dhananka SR, Zurbrügg C (2018) Social processes in post-crisis municipal solid waste management innovations: a proposal for research and knowledge exchange in South Asia. Res Ideas Outcomes. https://doi.org/10.3897/rio.4.e31430
31. Setyono JS, Anas N, Putri EM, Ma'rif S (2018) Karakteristik Pengelolaan Sampah Rumah Tangga Skala Komunitas di City Semarang. Riptek, 12(2):119–230
32. Plasticsmart Cities (2023) Vietnam community-based municipal solid waste. https://plasticsmartcities.org/vietnam-community-based-municipal-solid-waste
33. Plasticsmart Cities (2023) Vietnam material recovery facility initiative. https://plasticsmartcities.org/vietnam-material-recovery-facility-initiative/
34. Plasticsmart Cities (2023) Manila City launches innovative solutions with social entrepreneurs to tackle plastic pollution. https://plasticsmartcities.org/manila-city-launches-innovative-solutions-with-social-entrepreneurs-to-tackle-plastic-pollution
35. Plasticsmart Cities (2023) Community Generates Funds from Selling Recyclables through a Waste Bank Initiative. https://plasticsmartcities.org/wp-content/uploads/2023/12/Case-Study-Thailand-Community-Generates-Funds-from-Selling-Recyclables-through-a-Waste-Bank-Initiative.pdf
36. Environmental Management Bureau, Department of Environment and Natural Resources, Philippines (2021) The Ecological Solid Waste Management, Act of 2000. https://emb.gov.ph/wp-content/uploads/2021/03/ESWM.pdf
37. Parinduri RY, Zsazsa CSKM, Yusup M (2024) Optimizing community-based waste management: a review of the literature. J Community Dedication **4**(2):354–367

Multicriteria Sustainability Index (MSI) in the Chico Mendes Extractive Reserve, in the State of Acre, Brazil

Ademar Ribeiro Romeiro, Lucas Ferreira Lima, Raimundo Cláudio Gomes Maciel, João Alfredo Carvalho Mangabeira, Sérgio Gomes Tôsto, Mitali Daian Alves Maciel, Oscar Sarcinelli, Oleides Francisca Oliveira, and Lauro Charlet Pereira

Abstract Several initiatives have been developed to conserve the Amazon Forest, valuing and promoting the livelihood of populations to maintain their social reproduction. Conservation units, such as the Chico Mendes Extractive Reserve (RESEX), emerge as a way of solving problems related to the struggle for land ownership in the region and environmental issues arising from unsustainable activities, such as extensive livestock farming. This research evaluates multicriteria sustainability, based on the Social Carbon concept, at Chico Mendes RESEX, Acre, in 2023. It works with the Social Carbon methodology, instrumented from the Multicriteria Decision Support Analysis (MCDA), through field research comprising workshops and interviews with regional strategic actors. The sustainability criteria evaluated were Governance, Agronomic, Environmental, Economic, and Social. The research findings indicate that the Chico Mendes RESEX Multicriteria Sustainability Index (IMS) was 0.73, a proxy for a sustainable scale. In the field research, the IMS was validated with the Porongaba Seringal, whose value was 0.66, below the RESEX proxy, represented by the moderate sustainability scale for four of the five criteria evaluated. Only the Environmental Index performed close to the excellent level of maturity, highlighting the need for further environmental sustainability measures.

A. R. Romeiro · L. F. Lima (✉) · M. D. A. Maciel · O. Sarcinelli
Institute of Economics, University of Campinas (UNICAMP), Campinas, Brazil

R. C. G. Maciel · O. F. Oliveira
Center for Applied Legal and Social Sciences, Federal University of Acre (UFAC), Rio Branco, Brazil
e-mail: raimundo.maciel@ufac.br; oleides.oliveira@ufac.br

J. A. C. Mangabeira · S. G. Tôsto
Ministry of Agrarian Development and Family Farming (MDAF), São Paulo, Brazil
e-mail: joao.mangabeira@mda.gov.br

L. C. Pereira
Brazilian Agricultural Research Corporation (EMBRAPA), Jaguariúna, Brazil
e-mail: lauro.pereira@embrapa.br

B. Jarboui et al. (eds.), *Data-Driven Waste Revolution*, Computational Intelligence Methods and Applications, https://doi.org/10.1007/978-981-96-5890-9_9

Keywords Performance indicator · Biodiversity · Amazon biome · Multicriteria decision analysis · Sustainable development

1 Extractive Reserves and Sustainable Development

The developmentalism implemented by the Brazilian military government after 1960 resulted in recent transformations in the Amazon, with the disarticulation of traditional extractivism, such as native rubber. This led to severe socioeconomic and environmental problems, such as the disappearance of traditional populations, the struggle for land ownership, and immense deforestation in the region [16]. From the 1970s onward, the rubber tapper social movement began fighting for land ownership, seeking to maintain the Seringals and forests that guaranteed their livelihood. Together with the national and international environmentalist movement, this made it possible to create Conservation Units (CUs) in the Amazon, such as RESEXs. The RESEXs are innovative proposals because they are simultaneously land and environmental policies: guaranteeing land ownership and the well-being of extractive families, as well as forest conservation (and its ecosystem services) [4, 15]. In recent years, assessing sustainability in protected areas in the Amazon has become more critical due to the increased deforestation pressure within these areas. Chico Mendes RESEX has suffered from invasions, fires, and deforestation, and in 2022, it was the protected area in the Amazon most pressured by deforestation [1]. This highlights the need to seek actions to guarantee the resilience of this ecosystem and the livelihoods of its population by promoting an effective sustainable development.

2 Sustainable Development Indicators

The impacts of anthropogenic activities are decisive for climate change worldwide, especially in the generation of so-called greenhouse gases (GHGs), which have raised the average temperature of planet Earth [27]. The United Nations Conference on Environment and Development, known as Eco-92, in Rio de Janeiro, was a milestone in raising global awareness of the implications of economic development, which notably has "dangerous anthropogenic interference in the climate system" concerning socio-environmental issues [6]. At the third Conference of the Parties (COP-3) to the United Nations Framework Convention on Climate Change, held in Kyoto, Japan, there was a consensus to adopt a protocol in which industrialized countries would commit to reducing GHG emissions—mainly carbon dioxide [6]. However, given the difficulties in implementing the Kyoto Protocol, a new, more comprehensive agreement called the Paris Agreement replaced it at COP-21. This agreement's primary objective, in article 2 (1. a), is "holding the increase in the global average temperature to well below 2°C above pre-industrial levels" [28, p. 3].

On the other hand, alternative policy approaches should be encouraged in the search for comprehensive and sustainable forest management. The benefits of these approaches go beyond the carbon associated with them [28]. In this way, the social carbon approach becomes essential in discussing effective sustainable development. The concept of social carbon is all carbon absorbed/reduced, taking into account actions that enable and improve the living conditions of the communities involved in emission reduction/climate change projects to ensure well-being and citizenship without degrading the resource base [24, p. 73]. The basic idea behind social carbon is to include sustainable development issues in carbon sequestration projects, seeking to transparently measure and value the community's social gains and adequately value the environmental services of the benefited region [24]. Thus, it is imperative to investigate the sustainability level of a livelihood at different scales, from the individual through the family and community to regions or countries. Therefore, "indicators of sustainable development need to be developed to provide solid bases for decision-making at all levels and to contribute to a self-regulating sustainability of integrated environment and development systems" [19, Ch. 40]. According to Veiga [29], the adoption of Agenda 2030 by the 193 Member States of the United Nations (UN) and the 17 Sustainable Development Goals (SDGs) condemned to obsolescence the search for better sustainability indicators of almost all the initiatives of the previous 40 years. A humanistic, constructivist, and participatory multicriteria methodology is necessary to assess the criteria surrounding sustainable development. In this research, the Multicriteria Decision Analysis (MCDA) will be used to assess and measure the sustainability of the Chico Mendes RESEX in the areas of Governance, Environmental, Social, Economic, and Agronomic.

3 Multicriteria Decision Analysis

The historical process of developing decision analysis goes back a long way and is evolutionary nature in terms of its content and name. In the eighteenth and nineteenth centuries, the foundations of decision analysis focused on the intertwined concepts of subjective probability and utility [12]. In the twentieth century, according to Lima [14], some decision problems were approached using statistical methods. Since then, there has been a wider dissemination of methods applied to mathematically well-structured problems involving uncertainties and possibilities of sampling or experimentation. These results formed a body of results known as Bayesian or statistical decision theory [21, 23, 25]. As this field of research matured, researchers isolated the main recurring problems. They used advanced mathematical techniques to obtain analytical solutions, and the analysis of complex and ill-structured problems became an advanced analysis of well-structured problems [22]. From this, the term decision analysis emerged in the literature [11, 13, 14] defines decision analysis as "philosophy, articulated by a set of logical axioms, and a methodology and collection of systematic procedures, based upon those axioms, for responsibly analyzing the complexities inherent in decision problems." However, according to Fantinatti [9], traditional problem-solving methods based on mathematical foundations are only suitable for clearly defined situations, such as production and inventory control. They are inappropriate for complex contexts, characterized by uncertainties, conflicts of values and distinct power relations, multiple dimensions of evaluation and information, etc. Nevertheless, no methodological tool considered the subjective part, i.e., the information from experts' judgments about uncertainties or the identification and structuring of multiple conflicting objectives [14, 20]. It was only in the last decades of the twentieth century that this problem was solved. According to Keeney [13], there are two ways of thinking about decisions: (1) Alternative-focused thinking, based on relevant characteristics of the actors and (2) Value-focused thinking, based on the intrinsic values of the actors. He also argues in favor of this last form of thinking since it requires focusing primarily on values, which are crucial in any decision situation, and considering alternatives as the next step to achieving them. In other words, a complex decision must consider the values, aspirations, and perceptions of all the decision-makers involved. In complex situations, subjective aspects are generally built in and need to be measured qualitatively and quantitatively. Alternatives must also be analyzed in terms of their impact on the established objectives [2, 3]. The MCDA, used in this work to instrumentalize the Social Carbon method, meets the requirement of thinking focused on the values of the residents of the Chico Mendes RESEX and serves as the basis for the creation of the Multicriteria Sustainability Index (MSI).

4 Materials and Methods

4.1 MSI for the Chico Mendes RESEX

The methodology for evaluating multicriteria sustainability, based on the concept of social carbon, is supported by EMBRAPA's methodological guide on the generation of performance indicators and a multicriteria sustainability index [17], which is being worked on in partnership with Institute of Economics of UNICAMP and the Federal University of Acre. The research field includes workshops and interviews with strategic players in the region under study who are responsible for developing and implementing production systems. Based on the meetings with the key players, it will be possible to determine the best technique for collecting data from the target audience of these systems, which could be through questionnaires. The object of the study is the families of the Chico Mendes Extractive Reserve (RESEX)—Law nº. 99.144 of March 12, 1990—which has an area of approximately 970.570 hectares (ha), managed by ICMBio, granted for rubber farmers sustainable use, covering the municipalities of Rio Branco, Xapuri, Epitaciolândia, Brasiléia, Assis Brasil, Sena Madureira, and Capixaba, in the state of Acre, Brazil [16], as shown in Fig. 1. There

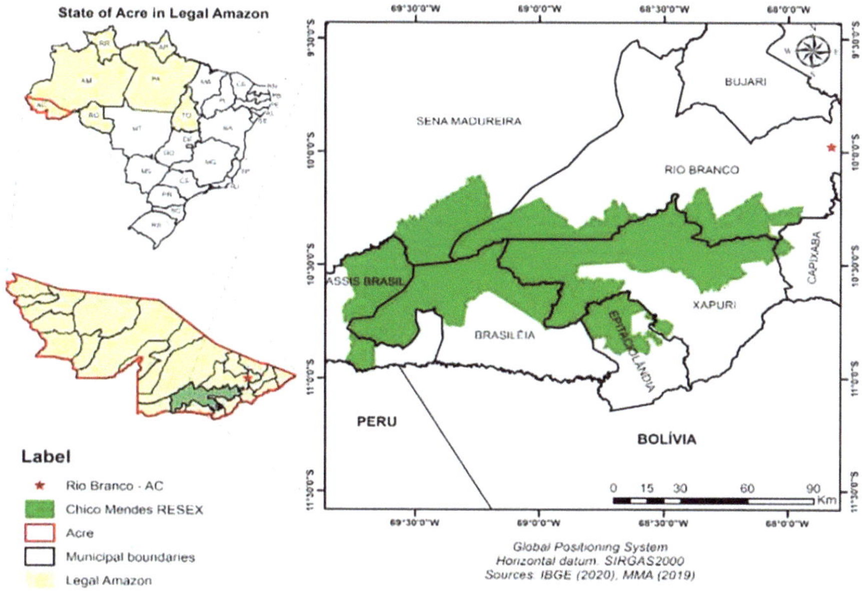

Fig. 1 Location of Chico Mendes RESEX, Acre, Brazil. Source: Silva et al. [26]

are currently around 2,000 family production units, known as "colocações" (family production units).

4.2 *Operational Plan*

According to Table 1, nine representatives and leaders of the Chico Mendes RESEX communities took part in the workshop to define the indicators/multicriteria sustainability index for this conservation unit from September 4 and 5, 2023, on the campus of the Federal University of Acre (UFAC), in Brasilé2a, Acre. This workshop sought to discuss and adjust the criteria and indicators/indices proposed in the methodology for an adequate assessment of the production systems and characteristics of the ecosystem to be studied.

The entities/communities represented at the workshop are described below:

1. AMOPREX—Association of Residents and Producers of the Chico Mendes Extractive Reserve (RESEX) in Xapuri, Acre
2. AMOPREAB—Association of Residents and Producers of the RESEX in Assis Brasil, Acre
3. ASPAFA—Association of Agro-Extractive Producers of Floresta and Adjacent Seringals, Xapuri, Acre
4. COOPERXAPURI—Cooperative of Agro-Extractivists of Xapuri, Acre

Table 1 Representatives and leaders of the communities that participated in the workshop to define the Multicriteria Sustainability Index for the Chico Mendes RESEX, Acre, 2023

Entities/communities	Municipality	Quantity
Amopreab e STR	Assis Brasil	2
STR e Coletivo Jovem Varadouro	Brasiléia	1
SOS Amazônia	Rio Branco	1
Amoprex, CooperXapuri, Aspafa e STR	Xapuri	5
Total		9

5. STR Xapuri—Rural Workers' Union in Xapuri, Acre
6. STR Brasiléia—Rural Workers' Union in Brasiléia, Acre
7. STR Assis Brasil—Rural Workers' Union in Assis Brasil, Acre
8. Fundação SOS Amazônia—a Brazilian nongovernmental organization dedicated to protecting and conserving the Amazon

The workshop was held in three stages: (1) presentation, discussion, and validation of the criteria/indicators, (2) prioritization and definition of the weights of the sustainability criteria, and (3) validation of the MSI with the application of questionnaires to community representatives; their "colocações" will serve as a reference (proxy) or approximation to estimate the MSI of the Chico Mendes RESEX.

4.2.1 Field Research in Porongaba Seringal

For field validation in the Chico Mendes Extractive Reserve's Seringais, the research used the study areas of the project approved in the Amazônia +10 call for proposals, in particular the production units of the Porongaba Seringal, located in the municipality of Epitaciolândia, Acre, under the responsibility of the Association of Producers and Residents of the Chico Mendes Extractive Reserve of Brasiléia and Epitaciolândia (Amoprebe), the Concessionary Association of the Conservation Unit. The Porongaba Seringal has seen several initiatives by government agencies and nongovernmental organizations to promote the sustainable use of forest resources, restore forests, improve the quality of life of the local community, and reduce deforestation. This Seringal was chosen as the study area precisely because it is facing strong pressure for further deforestation, given its geographical location, which has led some of the "colocações" to register environmental liabilities for exceeding the limits set for deforestation in the RESEX management plan, which is 10% of the total or up to 30 hectares of the production unit's area. Deforestation in this region has degraded permanent preservation areas (APP), such as riparian forests and springs. In an active search process among the allotments that had environmental liabilities, with degraded areas and agroforestry systems (SAFs) that

Table 2 Governance criteria and their respective performance indicators

Criterion/communities	Indicators
Governance	IG1: participation in a social organization
	IG2: exercising freedom of expression
	IG3: cooperation, trust, and coexistence
	IG4: transparency of the SPO's management
	IG5: equity in the distribution of benefits and losses of the SPO

Source: Adapted from [17]

Table 3 Environmental criteria and their respective performance indicators

Criterion/communities	Indicators
Environmental	IA1: availability of water for human consumption and agricultural production
	IA2: diversification of species cultivated by the farmer
	IA3: compliance with environmental legislation on rural property
	IA4: disposal of solid waste (domestic, pesticide packaging, effluents, etc.)
	IA5: protection of crop or pasture systems with exotic or native species (windbreaks)
	IA6: perception of wildlife biodiversity on the farm or in the surrounding area

Source: Adapted from [17]

were already in place as alternatives to deforestation, six families expressed an interest in taking part in the research to recover the APPs, as well as two others in other regions that already had mature SAFs that served the research objectives. Thus, given the region's lack of systematic evaluation and monitoring programs over time, the application of the MSI, validated for the Chico Mendes RESEX in the Porongaba Seringal, aimed to provide participatory evaluation and monitoring of the effectiveness of alternative initiatives to deforestation in the southwest of the Brazilian Amazon, which could subsidize effective public policies for sustainable development. To define the MSI, the criteria and performance indicators were discussed; the idea was to adapt them to the region and ecosystem studied, as shown in Tables 2, 3, 4, 5, and 6.

Governance Criteria

The governance criteria are related to corporations' collective management of production systems (associations and cooperatives). Basic corporate governance principles, such as transparency, fairness, and accountability, were considered.

Table 4 Social criteria and their respective performance indicators

Criterion/communities	Indicators
Social	IS1: access to the health system
	IS2: access to the education system and quality of teaching
	IS3: physical infrastructure for access to communities
	IS4: succession: continuity of productive activities by children/grandchildren
	IS5: access to communication means (internet/telephone)

Source: Adapted from [17]

Table 5 Economic criteria and their respective performance indicators

Criterion/communities	Indicators
Economic	IE1: income from self-consumption
	IE2: income from the product's sales from the production unit
	IE3: indebtedness of the agricultural production unit
	IE4: wear and tear on rural property and movable assets (machinery)
	IE5: extra farm income

Source: Adapted from [17]

Table 6 Agronomic criteria and their respective performance indicators

Criterion/communities	Indicators
Agronomic	IAG1: organic matter and soil color in the agricultural production unit or territory
	IAG2: soil fertility in the agricultural production unit or territory
	IAG3: soil cover (living/dead) on the agricultural production unit or territory
	IAG4: soil compaction on the agricultural production unit or territory
	IAG5: soil particle structure/aggregation on the agricultural production unit or territory
	IAG6: state of residues on top of the soil in the agricultural production unit or territory
	IAG7: incidence of diseases in the agricultural production unit or territory
	IAG8: incidence of pests on the agricultural production unit or territory
	IAG9: general appearance of foliage on the agricultural production unit or territory
	IAG10: plant growth/development on the agricultural production unit or territory

Source: Adapted from [17]

Environmental Criteria

High biological (vegetal and animal) biodiversity was considered an important environmental criterion for a sustainable production system. Respect for current environmental legislation, such as the Rural Environmental Registry (CAR) preparation and the adequate disposal of solid waste, was also considered.

Social Criteria

In family-based agricultural production units, social performance criterion is defined basically by access to the health system, good schooling, means of communication, and physical infrastructure. Another critical social indicator that represents a significant challenge for the maintenance and continuity of production systems is the generational succession of production units.

Economic Criteria

Some simplified indicators stand out for the composition of the economic criteria: Whether the production unit prioritizes production for the market or for self-consumption, the level of debt, the degree of depreciation of fixed capital, and wages outside the production units.

Agronomic Criteria

To determine agronomic performance, we considered adapting indicators that are easy for agroecological farmers to understand, precise and straightforward to interpret, sensitive to environmental changes, and impacts of soil management and cultivation practices. These indicators integrate the soil's physical, chemical, and biological properties underlying ecosystem processes.

The MSI is determined by aggregating the metrics assigned in the questionnaire to the performance indicators for each criterion (governance, environmental, social, economic, and agronomic), multiplied by their respective weights—validated at the workshop, according to Equation 1.

Equation 1: MSI in the Chico Mendes RESEX

$$MSI = (IAG * 0{,}22) + (IAA * 0{,}21) + (IAS * 0{,}20) + (IAE * 0{,}19) + (IAAG * 0{,}18)$$

where: MSI—Multicriteria Sustainability Index; IAG—Aggregate Governance Index; IAA—Aggregate Environmental Index; IAS—Aggregate Social Index; IAE—Aggregate Economic Index; IAAG—Aggregate Agronomic Index.

The MSI varies between 0 and 1. The closer it is to 1, the better the family's living conditions and the sustainability of the family-based production units (or

Table 7 Value scale degree (in ascending order)

Scale degree	Description
Unsustainable	Between 0.0 and less than 0.20
Low sustainability	Between 0.21 and less than 0.40
Moderate sustainability	Between 0.41 and less than 0.60
Sustainable	Between 0.61 and less than 0.80
High sustainability	Between 0.81 and 1.00

Source: Adapted from [17]

the territories mentioned). In other words, the MSI is based on a value scale, as described in Table 7.

The value scale degree reflects the level of the agricultural production unit sustainability maturity, individually or collectively (association or cooperative). A preliminary diagnosis using this tool classifies family production units according to their maturity level, aiming to identify the unit's stage of sustainable development. On an increasing scale, the first results of this qualitative assessment show the steps to be taken before the agricultural production unit or social organization reaches a level of excellence in its management in terms of multicriteria sustainability. Knowing the level of maturity is fundamental to taking safer steps toward sustainable development. Thus, classifying the production unit according to its level of maturity will serve as a basis for advice focused on improving its situation or solving the difficulties it faces. Therefore, this classification must make it easier for decision-makers to see the parameters presented in Table 10 in Appendix A.

5 Results and Discussions

The first important aspect of the validation of the MSI with the representatives of the Chico Mendes RESEX communities was precisely the positive impact the results had on them when they saw their areas and positions in the performance indexes and the MSI shown in Table 8 and Fig. 2.

The MSI for the Chico Mendes RESEX showed a value of 0.73, equivalent to the sustainable scale (Table 7). In terms of the performance of the sustainability criteria indexes, the Governance Aggregate Index (IAG) had the highest value, followed by the Agronomic Aggregate Index (IAAG), both of which correspond to the excellent level of sustainability scale. On the other hand, the Social Aggregate Index (IAS) and the Aggregate Economic Index (IAE) show the most significant attention points, as they are at the sustainable level, close to a moderately sustainable scale. The Governance Aggregate Index (IAG) performance lives up to the greater weight given to this criterion by the RESEX residents since it is essential for maintaining this conservation unit and is enshrined in the area's management plan. The governance of extractive areas has been sought/worked for since the rise of the rubber tapper movement in the struggle for ownership of their areas and has been consolidated in the RESEX. Figure 3 shows the excellent performance of

Table 8 Performance indexes with scores, weights, and MSI for Chico Mendes RESEX

Performance Indexes	Score	Weight	MSI Chico Mendes RESEX
IAG	8.8	0.22	0.73
IAA	6.9	0.21	
IAS	6.3	0.20	
IAE	6.4	0.19	
IAAG	8.1	0.18	

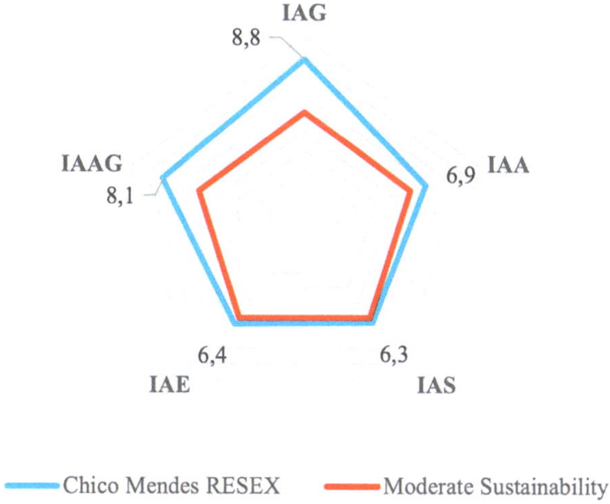

Fig. 2 Performance of the sustainability criteria indexes at Chico Mendes RESEX, Acre

Table 9 Performance indexes with scores, weights, and MSI for Porongaba Seringal

Performance Indexes	Score	Weight	MSI Porongaba Seringal
IAG	6.4	0.22	**0.66**
IAA	7.9	0.21	
IAS	6.5	0.20	
IAE	6.5	0.19	
IAAG	5.5	0.18	

the governance index (IAG), above 8.0, compared to the other indices, with few variations between other "colocações." On the other hand, the social (IAS) and economic (IAE) indexes show significant variations in sustainability between the production units. The IAS shows an inequality in sustainability of around 41% between the highest (8.5) and lowest (5.0—moderate) sustainability scores. The inequality of sustainability in the IAE is even more worrying since the difference between the extreme production units is around 60%, with the lowest-performing unit having an index of 3.6 (low sustainability) and the best-performing unit having 8.7 (high sustainability), according to Fig. 3.

In Fig. 4, according to the MSI, all the "colocações" were classified on a sustainable scale, with around 60% of them being closer to the moderate sustainable

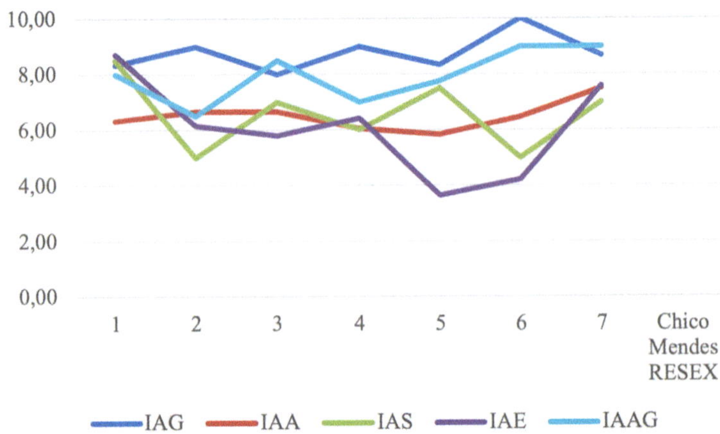

Fig. 3 Classification of the sustainability criteria indexes' performance in the "colocações" Chico Mendes RESEX, Acre, 2023

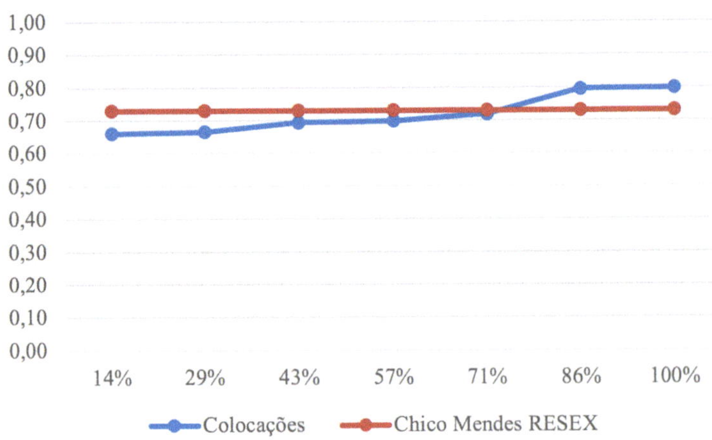

Fig. 4 Classification of the MSI in the "colocações" in Chico Mendes RESEX, Acre, 2023

scale—below 0.70—and only 14% being closer to the high sustainability level, i.e., above 0.80.

Thus, using the tested and validated MSI at the workshop as a reference (proxy), a field survey was carried out in the Porongaba Seringal, in the Chico Mendes RESEX, in Brasiléia, Acre. Because the questionnaires were used to calculate the sustainability indicators/indexes between the "colocações," the aim was to compare their results with those achieved in the research proxy.

5.1 The MSI of the "colocações" and Porongaba Seringal

The Porongaba Seringal was chosen as the MSI of the "colocações" and Porongaba Seringal a study area for the Amazon +10 Initiative, which is looking for alternatives to deforestation in the region. As mentioned before, this Seringal has been much pressured on its edges to deforest for extensive cattle breeding because of a federal highway (BR-317) and cattle ranches nearby. The region's cattle ranchers propose the extractivists share the cattle ranching in their areas, with all the (economic) costs paid by them, implying the deforestation within the Chico Mendes RESEX [10]. Of course, the extractivist families bear the socio-environmental costs. They generate environmental liabilities in some areas and are facing embargoes or have been embargoed by ICMBio, which is responsible for managing the conservation unit. These environmental liabilities call into question the sustainability of the allotments within the Chico Mendes RESEX. It is no coincidence that two allotments in Porongaba Seringal were defined as reference units for implementing Agroforestry Systems (AFS) in Permanent Preservation Areas (PPAs) as they were degraded by pastureland. The other allotments, which are part of the study's object, already had Agroforestry Systems (AFS) in place, but many have abandoned their subsistence crops in favor of cattle breeding. The questionnaires used to calculate the MSI were applied to these referenced production units to verify the performance of the indicators of the sustainability criteria and compare them to the results of the proxy generated and validated at the workshop. The results shown in Table 10 and Fig. 5 indicate that Porongaba's MSI of 0.66 is around 8% lower than that of Chico Mendes RESEX (Table 8), which is close to the moderate sustainability scale. This situation is corroborated by the performance of the sustainability criteria indices, of which five of the six indicators perform close to or within the moderately sustainable range. It is worth noting that only the Environmental Index (IAA) shows the situation on a sustainable scale, demonstrating that the forest is still resilient to deforestation in some production units.

Nevertheless, according to Table 10 and Fig. 5, it is essential to highlight the almost moderate sustainability of the governance index (IAG) of 0.64, which is still on the sustainable scale, showing that one of the pillars of the Chico Mendes RESEX, according to Table 10 and Fig. 2, may be being undermined by the pressures for more significant deforestation and the establishment of cattle ranching

Table 10 Performance of the sustainability criteria indices and the MSI for the Porongaba Seringal—Chico Mendes RESEX, Acre, 2023

Performance indexes	Score	Weight	MSI Chico Mendes RESEX
IAG	6.4	0.22	0.66
IAA	7.9	0.21	
IAS	6.5	0.20	
IAE	6.5	0.19	
IAAG	5.5	0.18	

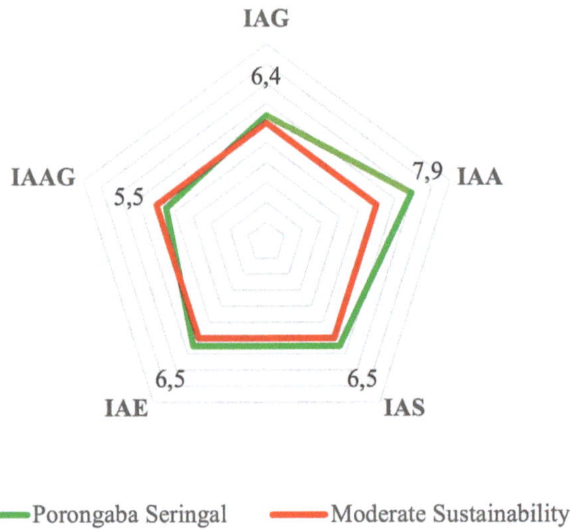

Fig. 5 Performance of the sustainability criteria indexes for Porongaba Seringal (Seringal)—Chico Mendes RESEX, Acre, 2023

in the region, which led some extractivists to violate both the legal norms and the informal rules the residents of the conservation unit are subjected to. According to Fig. 6, the environmental index (IAA) shows similar performance of most of the production units inside the sites surveyed in Porongaba, except one site with a moderately sustainable performance of 0.52. This site is precisely the one with the most significant environmental liabilities due to a level of deforestation above what is allowed in the Chico Mendes RESEX management plan, which is 10. On the other hand, the other sustainability indices show significant sustainability inequalities, with variations between the extremes of the best and worst performance in the IAAG, IAG, IAS, and IAE indices of 66However, the production unit with the worst performance in the agronomic index (IAAG), with 0.25 (low sustainability), is one of those that has degraded areas due to deforestation and pastures; but it is an exception, along with the other that has declared environmental liabilities, with 0.53 (moderate sustainability). Therefore, considering the other production units, all are considered sustainable. Thus, the preponderance of the discussion once again turns to governance performance (IAG) (Fig. 6).

When comparing the MSI between the "colocações" and the average value of the Porongaba Seringal with the Chico Mendes RESEX proxy (according to Fig. 7), Porangaba's current MSI of 0.66 (sustainable scale) is pulled toward moderate sustainability by 67% of the "colocações," none of which reached the MSI of the Chico Mendes RESEX proxy.

Based on the analysis of Porongaba's MSI relative to the Chico Mendes RESEX's MSI, it is possible to analyze which indicators have a positive (or negative) influence on Governance, Environmental, Social, Economic, and Agronomic issues.

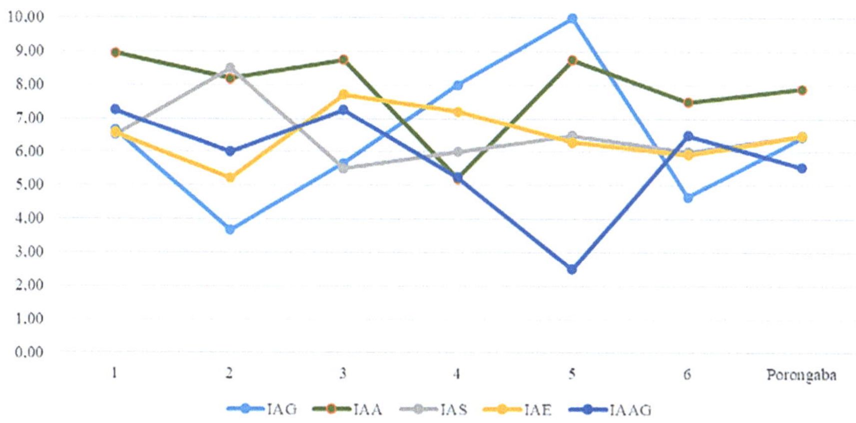

Fig. 6 Performance classification of the sustainability criteria indexes in the "colocações" of the Porongaba in the Chico Mendes RESEX, Acre, 2023

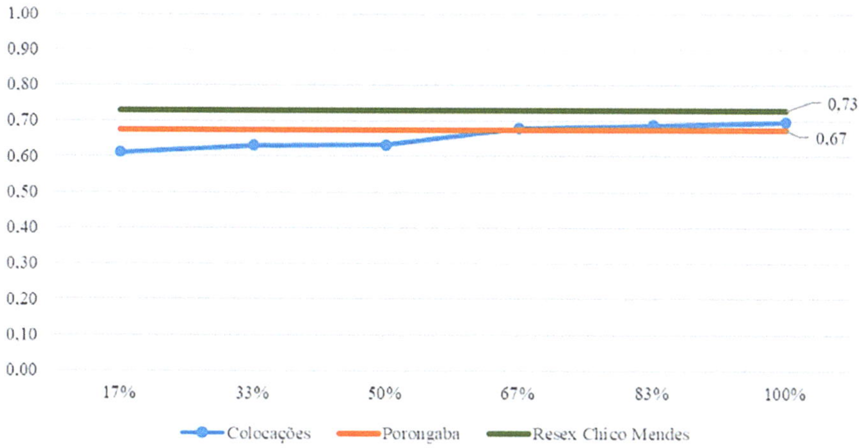

Fig. 7 Ranking of the MSI in the "colocações" and an average of Porongaba Seringal and Chico Mendes RESEX, Acre, 2023

5.2 *Highlights (Positive and Negative) of Porongaba's Indexes*

In the area of governance, the indicators that positively influenced the IMS of the residents of Porongaba Seringal were (1) the high rate of membership in the residents' association or cooperative of rural producers in the region and (2) the clarity in sharing information with members/cooperators. On the negative side, the residents highlighted (1) the low level of participation in the association's meetings, showing little active involvement in decision-making, and (2) the lack of distribution of economic gains from selling their products. Regarding the environment, the following aspects positively influenced the IMS: (1) the availability of quality drinking water and (2) the presence of birds, monkeys, and wild animals on the property, i.e., the high animal biodiversity. On the negative side, the following aspects were emphasized: (1) the inappropriate disposal of domestic sewage in the open or black cesspit and (2) the incorrect disposal of solid waste (recyclable and nonrecyclable), burned or buried by Seringal residents. On the social side, the following stand out positively: (1) the proximity to the city of Brasiléia, which facilitates trade and social interaction with city residents, and (2) the presence of children interested in continuing to work in the property. Negatively, there are the following conditions: (1) the access road quality to the community is always poor and (2) access to education for children and grandchildren is complicated. Concerning the economic criterion, the following stand out positively: (1) the absence of hunger due to the availability of nonmonetary income, i.e., locally, there is food production for self-consumption, such as eggs, meat, fruit, vegetables, and white crops such as rice, corn, and manioc, and (2) the absence of debt (financing or rural credit) linked to agricultural production or personal consumption. In negative terms, the following stand out: (1) the low monetary income, i.e., the low volume of production destined for sale—representing less than half of the family's income, and (2) the need to receive a pension or government aid to supplement the family's income. Finally, regarding the agronomic criterion, the following stand out positively: (1) the high content of organic matter and animals in the soil (earthworms, spiders, armadillos, and several types of ants), indicating good soil quality for agroforestry cultivation, and (2) soil cover, with plants or straw. On the other hand, the indicators that negatively influence this criterion are (1) soil compaction in deforested areas on the allotments and (2) diseases and pests on the plantations.

6 Conclusion

This study sought to contribute to creating a Multicriteria Methodology for assessing the level of sustainability for the forestry and farming practices of the extractivist families of the Chico Mendes RESEX, Acre, in the southwest of the Brazilian Amazon, called the "Multicriteria Sustainability Index (MSI)." The basic idea behind the research is that, after the initial studies, the IMS can be appropriated by the communities in the region and becomes an effective tool for strengthening families' livelihoods. The assessment of the sustainability of the Chico Mendes RESEX, using the MCDA and Social Carbon methodologies, demonstrated the importance of understanding the RESEX's maturity degree and sustainability levels, using criteria and indicators validated by the community involved. It allowed us to identify the real needs of the residents, which, on the one hand, can indicate the dimensions or criteria that need more detailed studies and, on the other, should be forwarded to public policy managers to ensure effective sustainable development in the region. The Chico Mendes RESEX's MSI, validated in this study with the conservation unit's leaders/residents, was 0.73. Regarding the sustainability criteria, the highlights were the governance and agronomic indexes, which showed excellent levels of sustainability maturity. The social and economic indexes showed that sustainability performance was close to moderate. The implementation of the MSI was validated by the residents/productive units of the Porongaba Seringal, which make up the Chico Mendes RESEX. In this region, the Seringal's MSI was 0.66, below the RESEX proxy (0.73), and four of the five evaluated criteria represented it on the moderate sustainability scale. Only the environmental index performed close to the excellent level of maturity. Thus, according to these results, strengthening governance, curbing deforestation, providing income from forest products (notably rubber and Brazil nuts), and eliminating environmental liabilities in the region should be priority activities for public policymakers and managers, as well as research centers, to achieve sustainability in the five dimensions suggested in this study. Alongside the results of this study, the integration that has taken place between the scientific community, residents, and public managers is fundamental to the development of activities that add economic value to the standing forest and agroforestry products, which can promote forest recovery, ensuring the conservation of biodiversity and ecosystem functions, responsible for the well-being of the residents of the Chico Mendes RESEX.

Acknowledgments The authors would like to thank the São Paulo Research Foundation (grant no. 2O22/10403-4), the Institute of Economics of the University of Campinas (UNICAMP), the Federal University of Acre (UFAC), the Ministry of Agrarian Development and Family Agriculture (MDA), and the Brazilian Agricultural Research Corporation (EMBRAPA) for their financial and institutional support.

Competing Interests The authors have no conflicts of interest to declare that are relevant to the content of this chapter.
All authors equally contributed to the conceptualization, methodology, and drafting of the initial and final versions of the text.
This study does not involve human subjects and clinical trials that the Institutional Ethics Committee should approve.

References

1. Amorim L, Santos B, Ferreira R, Brandão I, Dias M, Ribeiro J, Souza Junior C (2022) Ameaça e Pressão de Desmatamento em Áreas Protegidas: SAD de Julho a Setembro de 2022. Belém: 2022. https://imazon.org.br/publicacoes/43951-2/>. Accessed on Nov. 10, 2023
2. Bana e Costa CA (1993a) Processo de apoio à decisão: actores e acções; estruturação e avaliação. Publicação CESUR, S.l 618:31
3. Bana e Costa CA (1993b) Três convicções fundamentais na prática do apoio à decisão. Pesquisa Operacional S.l 13(1):9–20
4. Barbosa De Almeida MW, Allegretti MH, Postigo A (2018) O legado de Chico Mendes: êxitos e entraves das Reservas Extrativistas. Desenvolvimento e Meio Ambiente, vol 48. Edição especial: 30 Anos do Legado de Chico Mendes, pp 25–55
5. Bernoulli D (1738) Specimen theoriae novae de mensura sortis. Papers Imp Acad Sci St Petersburg 5:175–192

6. Brasil (1997) Protocolo de Quioto à Convenção sobre a Mudança do Clima: o Brasil e a Convenção Quadro das Nações Unidas. Ministério da Ciência e Tecnologia. 1997. http://mudancasclimaticas.cptec.inpe.br/~rmclima/pdfs/Protocolo-Quioto.pdf. Accessed on Nov. 10, 2023
7. CMMAD (1991) Nosso Futuro Comum. 2. ed. Rio de Janeiro: Editora da Fundação Getúlio Vargas, 430 p
8. De Finetti B (1937) La prévision: ses lois logiques, ses sources subjectives. In Annales de l'institut Henri Poincaré. pp 1–68
9. Fantinatti PAP (2011) Abordagem MCDA como ferramenta de mudança de paradigma no planejamento dos recursos hídricos. Tese de Doutorado. [sn]. Universidade Estadual de Campinas, Campinas, Brasil
10. Fittipaldy MCPM, Castelo CEF (2021) Há boi pastando, há desmatamento e outras coisas mais: o retrato da Resex Chico Mendes. CONEXÕES, Belém **9**(2):35–68
11. Howard RA (1966) Decision analysis: applied decision theory. In Proceedings of the Fourth International Conference on Operational Research. WileyInterscience, New York, pp 55–71
12. Keeney RL (1982) Decision analysis: an overview. Oper. Res. 30(5):803–838
13. Keeney RL (1992) Value-focused thinking: a path to creative decision making. Harvard University Press, Cambridge
14. Lima LF (2012) Processo sociotécnico MACBETH de apoio multicritério à decisão e a organização de comunidades tradicionais: o caso da comunidade do Marujá no Vale do Ribeira-SP. Dissertação de Mestrado. Universidade Estadual de Campinas, Campinas, Brasil
15. Maciel RCG, Cavalcanti FCS. Souza EF, Oliveira OF, Cavalcante Filho PG (2018) The "Chico Mendes" extractive reserve and land governance in the Amazon: some lessons from the two last decades. J Environ Manag 223:403–408
16. Maciel RCG, Mangabeira JAC, Kassai JR (2021) Reserva Extrativista ¨Chico Mendes': a socioeconomia 25 anos depois. 1ª. ed. Triunfo: Editora Omnis Scientia, vol 1, 164 p
17. Mangabeira JAC, Pinto DM, Scarazatti B (2021). Guia Metodológico: geração de indicadores de desempenho e índice multicritério de sustentabilidade para agricultura familiar no bioma Amazônia
18. Millennium Ecosystem Assessment (2005) Ecosystems and Human Well-being: Synthesis. Island Press, Washington, DC. https://www.millenniumassessment.org/documents/document.356.aspx.pdf. Accessed on Nov. 10, 2023
19. ONU (Organização das Nações Unidas) (1992) Conferência das Nações Unidas Sobre Meio Ambiente e Desenvolvimento—Agenda 21. Rio de Janeiro
20. Paiva Sobrinho R (2014) Apoio à decisão em sistemas socioecológicos complexos: uma proposta metodológica aplicada na avaliação ex-ante de políticas públicas utilizando moeda complementar. Tese de Doutorado. [sn]. Universidade Estadual de Campinas, Campinas
21. Pratt JW, Raiffa H, Schlaifer R (1965) Introduction to statistical decision theory. McGraw-Hill, New York
22. Raiffa H (2007) Decision analysis: a personal account of how it got started and evolved. In Edwards W, Miles JRF, Winterfeldt DV (eds) Advances in decision analysis. Cambridge University Press, New York, pp 57–70
23. Raiffa H, Schlaifer R (1961) Applied statistical decision theory. Harvard Business School. Division of Research, Boston
24. Rezende D, Merlin S (2003) Carbono social—agregando valores ao desenvolvimento sustentável. São Paulo: Editora Peirópolis. https://fdocumentos.tips/document/carbono-2012-12-5-carbono-social-a-agregando-valores-ao-desenvolvimento-sustentvel.html. Accessed on: Nov. 10, 2023
25. Schlaifer R (1959) Probability and statistics for business decisions. McGraw-Hill, New York
26. Silva AG, Silva FC, Yamada T (2019) Social reproduction of traditional populations and livestock in the Chico Mendes extractive reserve: reflections from the life projects of young extractivists. Desenvolvimento e Meio Ambiente 52:235–260

27. UNEP (2024) United Nations Environment Programme. Global Resources Outlook 2024—Bend the Trend: Pathways to a liveable planet as resource use spikes. International Resource Panel, Nairobi
28. United Nations (2015) Paris agreement. https://unfccc.int/sites/default/files/english-paris-agreement.pdf. Accessed on: Nov. 10, 2023
29. Veiga JE (2019) Deplorável inércia. CIÊNCIA E CULTURA, vol 71, pp 27–33
30. Wald A (1950) Statistical Decision Functions. Wiley, New York

Plastics Recycling Dilemma: A Value Chain Perspective on Integrating Mechanical and Chemical Recycling for Closing the Plastic Loop

Paolo Trucco, Carol Maione, and Yulia Lapko

Abstract The current deployment of recycling technologies is not sufficient to close the loop of plastics, and plastics pollution remains a drastic environmental problem. While there is a good understanding of a broad range of factors influencing the development of plastic recycling capacity at the supply chain or sectoral levels, only a limited number of studies highlight the need for examining how these factors manifest at different value chain levels in order to ensure an aligned deployment of both mechanical (MR) and chemical recycling (CR). This study aims to address this gap in the context of the Lombardy region, Italy, by surveying 115 stakeholders of the plastic value chain. Results show that the adoption of CR could mitigate economic, technical, and environmental barriers of MR, thus enabling the circular economy (CE) of plastic. Based on findings, this study suggests four areas of interventions to advance recycling through systemic integration of MR and CR and better alignment of plastics value chain functions. Finally, it provides implications for practitioners and policy-makers for the development of a CE of plastic and suggests prospects for further research.

Keywords Plastic recycling · Barriers and drivers · Plastic value chain · Lombardy · Circular economy

1 Introduction

Global plastic demand has increased exponentially in the last decade, reaching 390.7 million tons of plastics produced in 2021 [44]. This exceptional growth inevitably results in increased waste generation, which is posing growing pressure on the waste management sector to find alternatives for handling plastic waste and minimize its environmental impact [20, 33].

P. Trucco · C. Maione (✉) · Y. Lapko
Politecnico di Milano, Milan, Italy
e-mail: carol.maione@polimi.it; carol.maione@polimi.it; yulia.lapko@polimi.it

B. Jarboui et al. (eds.), *Data-Driven Waste Revolution*, Computational Intelligence Methods and Applications, https://doi.org/10.1007/978-981-96-5890-9_10

Scholars and institutions urge to accelerate the transition to a circular economy for plastic for sustainably managing plastic materials and waste. This implies fostering plastics recycling through the combination of actions across the entire value chain [5, 17, 23, 49], which eventually lead to decoupling plastic demand from fossil fuels extraction [30].

Recycling is typically divided into mechanical and chemical recycling (see [25] for a detailed review). Each technological solution presents advantages and disadvantages related to the characteristics of recyclables and the process that waste materials undergo. Mechanical recycling technologies (such as injection molding, blow molding, and screw extrusion) constitute more than 98% of recycling operations worldwide [43]. From the 1970s, mechanical recycling has been employed to treat single-polymer plastics [1], thanks to its economic viability [57], low energy demand, and limited environmental impact [27, 50]. However, the efficacy of mechanical recycling heavily depends on the quality of recyclables due to low tolerance to mixed and contaminated waste streams [1, 16, 53].

Chemical recycling includes more complex, energy-intensive technologies (such as chemolysis, thermal cracking, gasification, and pyrolysis) [1, 26, 35]. Despite its initial stage of development, chemical recycling has already demonstrated remarkable success in treating contaminated and mixed plastic waste that cannot be mechanically recycled [1, 47, 53, 55]. Hence, the drawbacks of mechanical recycling lead to a growing interest in chemical recycling from both industrial and institutional stakeholders [50].

Despite increasing interest in technological solutions that integrate both mechanical and chemical recycling in plastic's end-of-life management [13, 53], the present deployment of recycling technologies results ineffective and non-optimized to close the loop of plastics [29], due to barriers of diverse nature (e.g., economic, technical, regulatory, etc.) that can manifest at different stages of the plastic life cycle [39, 52]. One part of the problem relates to the composition of plastics products that can affect their ability to be recovered through recycling and then be reintegrated back in the production [52]. The second part of the problem lays on the limited efficiency of current waste collection and sorting practices [39], the readiness level of available recycling technologies [47, 53], as well as the economics of their deployment at industrial scale [53]. This highlights the need for aligned efforts across the entire value chain if the plastics loop is to be closed.

While there is a good understanding of a broad range of factors that can influence the development of plastic recycling [39, 41, 52], only a limited number of studies examined how these factors affect different value chain actors and several open issues with these studies persist. First, authors (e.g., [4, 42]) generally present the implications of barriers on plastic circularity, without addressing the dual nature of factors (e.g., drivers and opportunities along the value chain) or acknowledging that factors can manifest differently for different value chain players [37]. Typically, existing studies only address particular value chain positions, such as actors of the reverse supply chain, neglecting representation of the entire industrial system. Furthermore, previous research on factors is predominantly focused on mechanical recycling or does not explicitly distinguish implications for mechanical and chemical recycling, hence failing to capture specific aspects that could determine the adoption of chemical recycling and its integration with mechanical recycling to achieve a higher level of plastic's circularity.

Recognizing the importance of filling these gaps, this study specifically focuses on the examination of the potential for improving recycling capacity via scale-up of mechanical recycling and adoption of chemical recycling to advance the circular economy for plastic. The investigation was carried out in the context of the plastic value chain in Lombardy, Italy. This context is particularly relevant for this study because plastic recycling is well established and developed in Lombardy—a leading European region with more than 10 million inhabitants and about €370 billion GDP—where 54.9% of collected plastic waste is recycled (compared to 32% recycled at national level and 38% at EU level) (Regione Lombardia internal report). Moreover, stakeholders of Lombardy's plastic industry demonstrate growing interest in the issues that can hinder the transition to a circular economy for plastic and are therefore inclined to make progress toward a circular economy. Finally, there is a favorable ground for implementing circular economy policies at the regional level, including availability of public investments to support circularity interventions.

2 Barriers and Drivers of Plastic Recycling and Open Gaps

Several studies report multiple barriers and drivers that can inhibit or help the development of a circular economy for plastic, with implications at supply chain level, for specific economic sectors, or in relation to specific plastic materials. These studies generally aggregate factors into six macro-categories, namely economic, technical, environmental, regulatory, informational, and sociocultural. Each factor could act as either a barrier or driver based on its negative or positive impact on the development of plastic recycling. Commonly acknowledged barriers include the higher cost of recycled plastics that makes them less economically attractive than fossil raw materials for both plastic manufacturers and recyclers, especially when factoring in the limited marginal profits and small production volumes [42, 52]. Technical barriers are often associated with limited technological capacity and efficiency (e.g., of sorting and recycling processes) that can significantly affect the final quality of recyclates and, consequently, their ability to compete with virgin plastics [39]. Major concerns are also attributed to the presence of chemical substances and additives that can affect the ability of plastics to be recycled and bear environmental impacts when plastic waste is disposed of [23]. A fourth obstacle concerns regulatory factors, such as a lack of incentives from the use of recycled materials, or a lack of recycling standards for both upstream and downstream companies [4]. On the informational side, limited traceability of recyclable volumes can cause a mismatch between demand and supply for recycled plastics [39]. Finally, at the sociocultural level, consumer behaviors or poor communication of recycling information to end users can further reduce the demand for recycled plastics [49].

Despite studies have increasingly explored circularity barriers and drivers, the following gaps remain. First, only a few studies acknowledge that the same factor can manifest differently (in nature, type, and impact) for different value chain actors. For example, [37] observed that the properties of plastic materials (technical factor) acted as a driver of circular design for upstream industrial actors, while they could become a barrier for downstream actors during waste management activities, as they could affect the ability of plastic products to be recycled. Similarly, [52] indicated that consumer awareness (sociocultural factor) was generally indicated as a driver of circular economy by most industrial actors, but brand owners raised concerns about the significant challenges of translating this awareness into actual purchase decisions. Consequently, studying the specific implications of these factors along the value chain can bring deeper insights on the role of different actors in the circular economy transition.

Second, most studies tend to discuss barriers and drivers in isolation, neglecting to acknowledge that value chain actors may experience multiple factors, both simultaneously and over time. A few notable exceptions started to analyze how, for example, regulatory or technical barriers interact with the other barriers (e.g., [4]), with implications for their cumulative effects. In these studies, analyzing the interconnections between different factors depicts a better representation of existing challenges and opportunities along the value chain, thus enabling to remove barriers and leverage drivers of circularity.

Third, studies generally identify engagement of and collaboration between all value chain actors as a main driver of successful recycling [34], thanks to improved exchange of information along the value chain [14], standardized use of materials [39], and implementation of circular product design [34]. While this argument is often highlighted, only a few studies have explored it in detail. Besides a few notable exceptions [21, 52], existing studies claim the need for cross-value chain coordination in general, without addressing specific actions at different value chain stages and obstacles that need to be removed to foster system-level collaboration.

Finally, studies examining the impacts of influential factors on the plastic recycling value chain generally discuss them in relation to mechanical recycling or without distinguishing specific implications for mechanical and chemical recycling (e.g., [15, 39]). Assessing barriers and drivers in relation to mechanical and chemical recycling, respectively, would yield a more comprehensive understanding of the specific areas that require improvement for successfully integrating in both plastic's end-of-life management.

In this chapter, we employ the list of influential factors proposed by Maione et al. [37] and examine their role in the circular economy transition in the context of the Lombardy's plastic value chain. Specifically, this study intends to address the above-mentioned limitations by surveying the perceptions of specific companies along the plastic value chain concerning the main barriers and drivers toward the scale-up of mechanical recycling and the adoption of chemical recycling at industrial scale.

3 Empirical Context

3.1 Legislative Landscape of Plastic Waste Management

The transition to a circular economy for plastic is high on the socioeconomic and political agendas. In line with the objectives of the EU Circular Economy Action Plan to set new recycling targets (e.g., using 100% reusable and/or recyclable plastic packaging by 2030; recycling at least 50% of plastics waste by 2030) and scale-up recycling capacity (e.g., an increase of 5.8 million tons/year by 2020; foster collaboration for adoption of chemical recycling), Italy adopted a nationwide action plan in 2020 (D.Lgs. 2020/116). The plan focuses on three main directives: (i) Directive 2018/850/EC on the landfill of waste; (ii) Directive 2018/851/EC on waste; and (iii) Directive 2018/852/EC on packaging and packaging waste. Concerning the use of plastic materials, in 2021, Italy adopted a new regulation to reduce the use and incidence of single-use plastics on the environment that limited the trade, use, and import of a number of single-use plastics (D.Lgs. 2021/196). Additionally, a plastic tax was introduced to reduce the generation of nonrecyclable plastic packaging waste. The plastic tax is based on the "polluter pays principle" and imposes a fee on every extra kg of nonrecycled plastic packaging (€0.80/kg). However, due to a great disagreement within the plastic industry, the enforcement of the plastic tax has been temporarily postponed.

In addition to alignment with national directives, the management of plastic waste in Lombardy is regulated by the Regional Waste Management Plan (Piano Regionale di Gestione dei Rifiuti, PRGR). The plan seeks to: (i) improve the management of plastic waste at all stages of the supply chain; (ii) encourage the reduction of the use of plastic packaging in favor of alternative materials; (iii) promote awareness and acceptance of recycled products in the economic system; (iv) combat plastic pollution and losses to the marine environment; and (v) promote the use of structural funds to support the development of new technological solutions [48].

Furthermore, in 2019, Lombardy adopted a detailed roadmap (EU Circular Economy Action Plan) to accelerate the transition to a circular economy by 2027, hence supplementing the lack of specific indications on the achievement of circularity targets in EU frameworks. Lombardy's roadmap focuses on seven priority areas including wider adoption and industrial maturity of innovative closed-loop recycling technologies, with the double aim of obtaining high-quality and abundant volumes of secondary raw materials and increasing profitability of recycling business models [48]. Therefore, for example, complex waste streams (such as multilayer packaging, black plastics, and plastics waste from electrical and electronic equipment) can be recovered via integrated use of advanced mechanical and chemical technologies [48]. Projections estimate that this technological innovation will enhance the capacity of current recycling systems by unlocking potential for implementing circular plastic value chains at the regional level [48].

3.2 *Plastic's End-of-Life Management*

The total plastic waste managed through separate waste collection in Italy in 2021 was equal to 1,335,181 tons/year, which is roughly 90.5% of all collected plastics from domestic and industrial sources [8]. Over 70% of all plastics consumed came from packaging applications, with a large majority of polyethylene (PE), polyethylene terephthalate (PET), and polypropylene (PP) [8]. In 2021, 663,082 tons of plastic waste (approximately 32% of the total collected waste) were recycled using mechanical recycling technologies [18]. An additional 20,594 tons of plastic waste were transformed via secondary reducing agent (SRA) treatments, and only 939 tons (less than 0.1%) were treated via chemical recycling, for an overall total of 684,615 tons of plastic waste recycled (recording a 9.5% increase in plastic recycling compared to 2020 and a 15.9% increase compared to 2019) [18].

Postconsumer plastic recycling rates are generally high thanks to (i) capillary waste collection and sorting infrastructure (33 waste sorting plants on the national territory); (ii) a high number of recycling plants (65 recycling plants on the national territory); and (iii) efficient governance models for plastic and plastic packaging waste, including the presence of packaging consortia like CONAI (National Consortium of Packaging) and COREPLA (National Consortium for the Collection and Recycling of Plastic Packages) [18, 36].

The Lombardy region currently holds the highest recycling rate nationwide, with 54.9% of collected plastic waste sent to recycling facilities (Regione Lombardia internal report). According to the region's projections, the recycling rate will reach between 62% and 67% by 2027, well above the national objective of 55% recycled plastics by 2025 and 65% by 2035 (Regione Lombardia internal report). A main reason for Lombardy's leading role in recycling is that the region presents the largest plastic industry and solid waste management sector of the country, with 3,618 operating firms, which is about one fourth of the total in Italy.

Despite these promising recycling rates, the potential of plasmix is currently uncaptured within the plastic's material system. At the national level, in 2021, of the total 407,777 tons of plasmix, 50.7% were sent to cement factories (including 16 national plants and 27 plants located across Europe), 14.2% were sent to waste-to-energy plants (including 12 plants located in Italy and 1 in Europe) [8, 18], and 35.1% were sent to landfill [8, 18].

4 Study Design and Methodology

The study uses a semiquantitative approach based on the development of a structured survey to characterize the plastic value chain in Lombardy and assess the perception of different value chain functions on barriers and drivers of recycling. This approach is well suited for determining influential factors in the circular economy transition and capturing the needs and concerns of individual companies of the plastic industry.

4.1 Survey Design and Data Collection

The study's initial population consisted of firms operating in the plastic industry (NACE code C22; N=2,492) and solid waste management sector (NACE code 38; N=1,126) in Lombardy, selected using the AIDA database. For sample preparation, the following exclusion criteria were used: firm in liquidation, firm without website, firm's website does not list contacts, invalid contacts, firm has the registered office in Lombardy but operates in other regions, firm does not treat plastic materials. After cleaning, a total sample of 1,718 companies were obtained. Given the relatively small population, the entire pool of firms was sampled to ensure that the results obtained were representative of the entire population. Of the total, 318 companies (18.5%) responded to the questionnaire over two consecutive waves conducted between June 1 and June 15, 2021 and June 21 and June 30, 2021. This timeframe is consistent with the works of the Plastics Working Group of Lombardy region. Following the approach of De Lucia and Pazienza (2019), a 95% confidence level with a marginal error smaller than 10% was considered to determine the minimum acceptable sample size (111 observations). After data cleaning, we obtained 115 usable responses (marginal error 8.86%). We only included in the analysis responses that answered all sections of the questionnaire (in relation to at least one recycling technology) to avoid nonresponse biases.

Figure 1 shows the diversity of the pool of respondents. The analysis is based on the assumption that the survey was filled in by the most competent/knowledgeable informant within each firm to fully capture the company's point of view on the subject matter.

The survey was administered in Italian via email. The questionnaire was composed of 18 questions including: (i) socio-professional demographics to gain general knowledge of the respondents; (ii) value chain attributes; (iii) company's experience in plastic recycling, inputs and outputs of recycling processes, and

(a) Respondent's role in the company (Q1)

ID	Role in the company	Frequency
1	CEO	34 (29.6%)
2	Director	21 (18.3%)
3	Sales & business manager	14 (12.2%)
4	Technical manager	9 (7.8%)
5	Project manager	6 (5.2%)
6	Administrative manager	5 (4.3%)
7	Environment/sustainability manager	5 (4.3%)
8	Quality manager	5 (4.3%)
9	Legal representative	4 (3.5%)
10	Operation manager	3 (2.6%)
11	Associate	3 (2.6%)
12	CFO	2 (1.7%)
13	Human resources manager	1 (0.9%)
14	No answer	3 (2.6%)

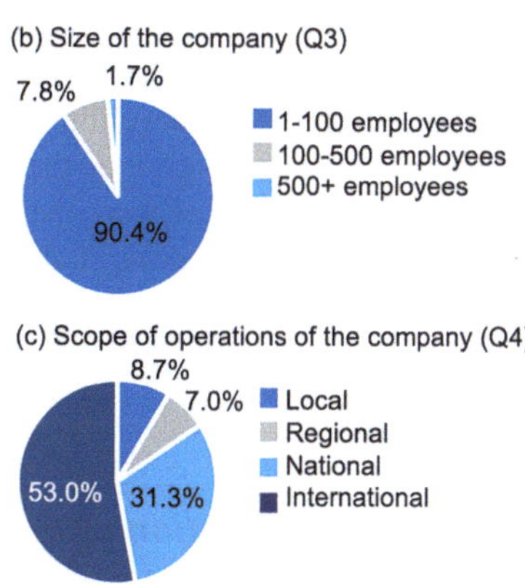

Fig. 1 Company characteristics and socio-professional demographics of respondents (N=115)

recycling technologies in use; and (iv) influential factors in plastic recycling, categorized as follows: economic, technical, environmental, regulatory, informational, and sociocultural.

The analysis of influential factors was conducted specifically to assess respondents' perceptions of barriers and drivers in the scale-up of mechanical recycling and adoption of chemical recycling. For the purpose of this study, we employed the framework of influential factors proposed by Maione et al. [37]. This choice is driven by the comprehensiveness of the framework, as it was derived from a systematic review of recent literature on barriers and drivers toward the circular economy for plastic in the context of the Italian plastic industry. Survey respondents could use a 5-point Likert-like scale ranging from very negative (1) to very positive (2) for ranking each of the 18 influential factors as barrier (defined as a factor with a negative impact on plastic recycling) or driver (defined as a factor with a positive impact).

To enrich the quantitative assessment, a qualitative investigation was carried out to collect insights and additional feedback on barriers and drivers of recycling, as prescribed by Paletta et al. [42]. A complementary, nonobligatory open-ended question was introduced at the end of the survey to help the authors contextualize the respondents' perceptions of circularity factors, further characterize barriers and drivers, and avoid wrong expectations. Through this analysis, we collected 52 additional comments pertaining to economic (N=7), technological (N=18), environmental (N=1), regulatory (N=15), informational (N=2), and sociocultural (N=9) aspects.

4.2 Data Analysis

Data analysis was conducted using SPSS. First, descriptive statistics (frequencies, distributions) were conducted for all answers to map Lombardy's plastic industry. Second, descriptive statistics (crosstabs) were employed to map the 18 factors, their influence (barriers/drivers) toward the scale-up of mechanical recycling and adoption of chemical recycling, respectively, and their mean values, with respect to the value chain position of the respondents. Finally, for open-ended questions, respondents' text entries were then manually coded according to the framework of factors identified by Maione et al. [37].

5 Results

5.1 Characterization of the Plastic Value Chain in Lombardy

In Lombardy, the upstream tier of the plastic's value chain is globalized, including small- and medium-sized enterprises, as well as larger firms with operations ranging from regional to international level, while the downstream tier appears to be more localized and structured under municipal- or regional-level systems. Out of the 115 respondents to the survey, 98 companies (85.2%) reported primary functions in the upstream tier (oil and gas companies, raw material suppliers, polymers producers, plastic converters, retailers, and distributors), and 17 (14.8%) indicated primary functions in the downstream tier (waste collectors, waste sorting companies, and recyclers). Such proportion reflects the allocation of upstream and downstream industrial functions found in the initial population of firms in Lombardy, where more than two thirds of the selected firms belonged to the pool of plastic converters, and approximately one third was represented by waste management companies. Of the total, 26 companies covered two or more value chain functions. Table 1 presents an overview of the typical value chain distinction in terms of functions, including the primary and secondary functions of each respondent, the main material streams handled by the surveyed companies, the use of recycled plastics in the manufacturing of new products, and engagement in recycling activities.

Results show that the majority of companies handled plastics and at least another material (recyclables, solid waste, or others), and only two companies (1.7%) handled plastics exclusively. When asked about their role in plastic recycling, 82 respondents (71.3%) stated that they employed recycled plastics and, of these, 75 (91.5%) indicated primary functions in the upstream tier. Out of the total, 32 companies (27.8%) reported that they performed recycling activities. Notably, 27 of these companies (84.4%) had core functions in the upstream tier, such as polymer production, plastic conversion, and distribution.

The typical thermoplastic polymers used on the Lombardy market are PE (including low-density polyethylene (LDPE) and high-density polyethylene (HDPE)), PET, PP, polystyrene (PS), and polyvinyl chloride (PVC), showing similarities in terms of plastic composition with other Italian regions [19] and European countries (e.g., [7, 17, 50]). More specifically, PP was the most common type of recycled plastic used by 45.1% of companies employing secondary raw materials, followed by LDPE (40.2%), HDPE (32.9%), rPET (19.5%), PS (17.1%), and PVC (11.0%). Other recycled plastics, such as acrylonitrile butadiene styrene (ABS), polyamide (PA), polycarbonate (PC), polymethyl methacrylate (PMMA), polyurethane (PUR), and ultrahigh molecular weight (UHMW) PE, were used by less than 10% of respondents using recycled materials.

Concerning inputs for recycling, HDPE and PP were the main plastic types reported by 43.8% of companies performing recycling activities, followed by LDPE (37.5%), PET and ABS (25.0%), PVC (12.5%). Other plastics such as PS, plasmix, PMMA, multilayer materials, PC, PA, expanded PS (EPS), and PUR were treated by less than 10% of respondents performing recycling. However, recycled EPS and plasmix were not used on Lombardy's market, suggesting that the potential of these recyclable materials remains non- or underutilized and currently escapes the plastic material loop.

Our results showed that mechanical recycling was the preferred route (98.3%), and sampled companies performed mechanical techniques such as grinding, regranulation, extrusion, and molding at industrial scale for the conversion of plastics into flakes, pellets, and granulate. These results resonate with those of other studies

Table 1 Overview of respondents (N=115) and their engagement in plastic recycling

Value chain positioning (Q5)									Material streams (Q6)	Use of recycled plastics (Q8)	Performing recycling activities (Q10)	Total
Raw material production		Primary plastic production		Use		Waste management						
Oil and gas companies	Raw material suppliers	Polymer producers	Plastic converters	Retailers	Distributors	Waste collectors	Waste sorting compa-nies	Recyclers				
o		x	x					x	Plastics, other materials	-Yes: 2 (66.7%) -No: 1 (33.3%)	-Yes: 2 (66.7%) -No: 1 (33.3%)	3
	o								Plastics, other materials	-Yes: 1 (33.3%) -No: 2 (66.7%)	-No: 3 (100.0%)	3
x		o	x					x	Plastics, recy-clables	-Yes: 4 (66.7%) -No: 2 (33.3%)	-Yes: 3 (50.0%) -No: 3 (50.0%)	6
	x		o	x	x	x	x	x	Plastics, recy-clables, other materi-als, munici-pal solid waste	-Yes: 64 (78.0%) -No: 18 (22.0%)	-Yes: 21 (25.6%) -No: 61 (74.4%)	82

				o					Plastics, recyclables	-Yes: 2 (100.0%)	-No: 2 (100.0%)	2
					o	x	x	x	Municipal solid waste, other materials	-Yes: 2 (100.0%)	-Yes: 1 (50.0%) -No: 1 (50.0%)	2
x						o	x	x	Recyclables, other materials, municipal solid waste	-Yes: 7 (53.8%) -No: 6 (46.2%)	-Yes: 2 (15.4%) -No: 11 (84.4%)	13
x						x	o	x	Plastics, municipal solid waste	-No: 2 (100.0%)	-Yes: 1 (50.0%) -No: 1 (50.0%)	2
								o	Plastics	-No: 2 (100.0%)	-Yes: 2 (100.0%)	2

Note: o=primary value chain function, x=secondary value chain function

investigating plastic waste in Europe, where mechanical recycling covers about 98–99% of all plastic recycling operations [43]. One respondent indicated that their company performs plastic micronization at the laboratory scale. In our sample, only two companies performed chemical recycling, including glycolysis at the laboratory scale, with conversion of plasmix into polymers, and gasification at industrial scale, with conversion of plasmix into gas.

5.2 Barriers and Drivers of Plastic Recycling

In the second part of the survey, the experts were consulted about their perceptions of the main drivers and barriers to the integration of mechanical and chemical recycling in the plastic value chain. Main constraints to scaling up recycling were identified in economic, technical, and environmental aspects.

It should be noted that some factors presented values near neutrality (when the value is equal to or near 3.0), which could indicate shared caution or ambiguity of the perceived role of these factors. However, in some cases, values near neutrality can indicate internal disagreement when respondents report different perceived values for the same factor. This variability is notable within and across the sample subsets. For example, in Table 2, energy consumption and environmental impact factors were perceived as barriers by raw material producers and waste managers, while manufacturers and users indicated them as drivers, resulting in values near neutrality (3.03) at the sample level. On the one hand, this suggests the need to appreciate the specific differences in the perception of these factors from the perspective of different value chain actors. However, it indicates that more effort is required at the industry level in order to develop a circular economy for plastic, considering the specific and sometimes contrasting needs and priorities along the value chain.

Considering the marginal error of 8.86%, the values near neutrality may provide controversial implications for the entire population. For example, in Table 2, process cost assumes the value 2.89 for the entire sample, with the confidence interval ranging from 2.63 (barrier) to 3.15 (driver). Hence, while the factor was perceived as a barrier by our respondents (115 firms), it could likely act as a barrier or a driver for the entire population (1,718 sampled firms). In this perspective, it should be noted that more barriers toward recycling could arise at the industry level (entire population).

A detailed analysis of barriers and drivers of mechanical and chemical recycling, respectively, is presented below. Notably, a higher number of responses was recorded for influential factors in mechanical (115 responses) than chemical (103) recycling, which may suggest a more limited knowledge of the challenges and needs associated with chemical recycling.

5.2.1 Barriers and Drivers of Mechanical Recycling

Table 2 presents the aggregated responses for barriers and drivers toward increasing capacity of mechanical recycling. Based on the empirical analysis of barriers and drivers, all respondents identified three core issues pertaining to (i) the high costs associated with recycling and employment of recycled materials, (ii) inefficiencies in the waste management infrastructure, and (iii) constraints related to the presence of contaminants and toxic additives in plastic products and waste streams. An in-depth analysis of the most significant barriers and drivers gleaned from multiple firms is discussed below.

Generally, economic barriers pose the strongest limitation to scaling up mechanical recycling. The causes are rooted in different parts of the recycling value chain. For example, sampled companies in the raw material production tier and primary plastic and product manufacturing tier emphasized the high costs of secondary raw materials limiting the employment of recycled plastics in production processes. Raw material producers indicated the insufficient consumer demand for recycled plastics and that fierce price competition makes recycled materials unattractive to most plastic users (see also [42, 52]. According to respondents in the waste management tier, the scale-up of mechanical recycling is mainly constrained by the high process costs associated with pre-recycling and recycling operations, combined with an increase in operational expenditures (e.g., the cost of energy consumption, general expenses, direct production costs, and other fixed costs), which altogether result in generally low recycling margins, as it was also observed in similar studies [3, 4, 31]. Moreover, downstream companies also complained about difficulties in finding recyclable materials, and one recycler, in particular, raised concerns regarding the plastic tax, as it may further reduce the volumes of available recycling inputs. On the technical side, raw material producers and waste management companies reported that urban infrastructure for municipal solid waste collection and separation is still unable to capture and recover all plastic waste streams, and avoid waste contamination at all stages of the waste management process, thereby restraining the use of mechanical recycling in favor of other recovery routes (e.g., landfilling, energy recovery, or chemical recycling). These results are expected, as they reflect a broader infrastructural problem that has been observed at both national [37, 42] and European [4, 22, 23] levels, bearing repercussions at the regional level as addressed in the following. According to surveyed raw material producers, inefficiencies in the urban infrastructure can also be coupled with inefficiencies in the industrial infrastructure and technology maturity. To address this conundrum, one respondent emphasized that, in a circular value chain, building new recycling plants with an increased capacity to process recyclables could bear a positive impact on the entire plastic industry, including "*a consequent increase in sustainability and resilience of the entire Italian production chain of plastic products.*" Another challenge is represented by the presence of toxic additives and substances in plastic products and waste streams. Material contamination is a core barrier because it can occur at various stages of the plastic's life cycle, from product design to all stages of waste management, and it significantly affects the quality of recyclates. To reduce material

Table 2 Barriers and drivers of mechanical recycling by value chain position (N=115)

		Total (N=115)		Raw material production (N=6)		Primary plastic and product manufacturing (N=88)		Use (N=4)		Waste management (N=17)	
Macro-area	Factor	Avg	CI	Avg	CI	Avg	CI	Avg	CI	Avg	CI
Economic	Process cost	**2.89***	0.25	2.67	0.23	2.94*	0.25	3.00*	0.26	2.65	0.23
	Consumer demand	3.80	0.33	2.67	0.23	3.84	0.33	3.75	0.32	4.00	0.34
	Market shift	3.42	0.29	3.17*	0.27	3.43	0.29	3.00*	0.26	3.53	0.30
	Competition vir-gin/recycled plastics	3.25	0.28	3.00*	0.26	3.28	0.28	2.75	0.24	3.29	0.28
Technical	Material properties	3.10*	0.27	3.00*	0.26	3.09*	0.27	3.25*	0.28	3.18*	0.27
	Technology readiness	3.18*	0.27	2.67	0.23	3.16*	0.27	3.75	0.32	3.35	0.29
	Industrial infrastructure development	3.13*	0.27	2.83*	0.24	3.09*	0.27	3.50	0.30	3.35	0.29
	Urban infrastructure development	**2.98***	0.26	2.33	0.20	3.03*	0.26	3.00*	0.26	2.94*	0.25

Environmental	Toxic additives and substances	**2.57**	0.22	2.67	0.23	2.63	0.23	2.25	0.19	2.29	0.20
	Energy consumption	3.03*	0.26	2.00	0.17	3.10*	0.27	3.50	0.30	2.94*	0.25
	Environmental impact	3.03*	0.26	2.00	0.17	3.10*	0.27	3.50	0.30	2.94*	0.25
Regulatory	Circular economy and sustainable development directives	3.61	0.31	2.67	0.23	3.67	0.32	3.50	0.30	3.65	0.31
	Incentive/tax schemes	3.39	0.29	4.00	0.34	3.38	0.29	3.25*	0.28	3.29	0.28
	Market regulation	3.31	0.28	3.50	0.30	3.30	0.28	3.00*	0.26	3.41	0.29
	Engagement in extended producer responsibility	3.13*	0.27	3.33	0.29	3.03*	0.26	3.25*	0.28	3.53	0.30
Informational	Data transparency and traceability along the value chain	3.33	0.29	2.83*	0.24	3.36	0.29	3.50	0.30	3.06*	0.26
Sociocultural	Consumer awareness	3.64	0.31	3.00*	0.26	3.61	0.31	4.00	0.34	3.94	0.34
	Consumer behavior	3.54	0.30	3.00*	0.26	3.55	0.31	3.50	0.30	3.71	0.32

Note: Values below 3 indicate a negative impact (barrier) of the factor, and values above 3 represent a positive impact (driver).
* Impact of the factor could change (8.86% margin of error).

contamination in plastic production, upstream firms stressed the need to introduce end-of-life considerations during the design and development stages to facilitate the disassembly and recyclability of plastic products and packaging after disposal. Waste contamination was appointed by all respondents as constraining mechanical recycling, due to the inability of this technology to treat contaminated plastics. Like previous studies [4, 13, 52], respondents stressed that the quality of recycling outputs is generally difficult to upgrade up to that of virgin materials. In particular, a recycler commented that waste contamination affects "*the mechanical performance of recycled plastics, in their multiple applications, and we still need new plastics to face the plastic demand*." Additionally, shifting to raw material producers and waste management companies, a new set of environmental barriers arises for increasing the capacity of mechanical recycling, including higher energy requirements and environmental impact. In extant literature, these barriers are often interrelated with waste contamination and, altogether, reduce the ability of recycled plastics to compete with their virgin counterparts, thereby resulting in weaker demand for recycled materials [4]. Notably, respondents in the raw material production tier indicated that circular economy and sustainable development directives, and data transparency and traceability along the value chain can constitute additional limiters to scaling up mechanical recycling. Our respondents in the primary plastic and product manufacturing segment showcase, in particular, how the current lack of incentives to employ recycled materials can hinder their transition to sustainable plastics. Some plastic converters expressed the need to introduce financial mechanisms such as "*tax reduction*," "*financial benefits*," or "*reduction of/exemption from environmental contributions*" for industries utilizing recycled plastics. All respondents agreed on the key role of government and consumers to increase mechanical recycling. Concerning the first point, government can create economic opportunities in the green market to foster utilization of recycled plastics, regulate market dynamics to shift competition, or increase societal pressures for sustainable plastics, among others [4, 15, 21, 39, 52]. While there is consensus that the scale-up of mechanical recycling can only happen with interventions on the institutional level, sampled companies stressed the need to modify and adapt existing directives to align EU standards for recycling. For example, a plastic converter condemned that "*little attention has been paid to the regulatory issue. The Italian legislation on the subject is still in 1973 and this is an obstacle when compared to the European regulations to which other countries refer to directly*." Concerning the second point, consumers are key to scaling up mechanical recycling, as they have the power to steer the demand for sustainable plastics and cooperate in plastic recycling [21]. Among sampled upstream companies, the majority indicated consumer awareness and behavior as mainstream drivers of mechanical recycling, and one plastic converter, in particular, reported a "*growing awareness of the consumer (end user) about their key role in the recovery-recycling chain*." To this end, several respondents indicated that it is necessary to strengthen transparency and communication with the consumers on the impacts of plastic materials and their correct management throughout the entire plastic's life cycle to foster consumer engagement. This is especially evident in the following quote from a plastic converter: "*Plastic in general is a material that can really play a leading role in a circular economy, but you need to know it, and you need to know the differences. The impact that good information can have is that the areas in which plastic is truly an added value and a precious commodity could be preserved and the areas in which it is used inappropriately and therefore demonized could be changed.*"

5.2.2 Barriers and Drivers of Chemical Recycling

Table 3 presents the aggregated responses for barriers and drivers toward adopting chemical recycling technologies. Based on the empirical analysis of barriers and drivers, all respondents identified four core issues pertaining to (i) unattractiveness of chemically recycled feedstock from an economic point of view, (ii) limited technological and infrastructural maturity, (iii) environmental concerns for scaling up this technology, and (iv) ambiguity and unclarity of the regulatory framework. An in-depth analysis of the most significant barriers and drivers gleaned from multiple firms is discussed below.

Unsurprisingly, also the adoption of chemical recycling is significantly restrained by economic challenges. Designing recycling routes for chemical recycling that are economically feasible and can guarantee profit margins in the initial utilization phase remains a challenge [30]. Moreover, raw material producers indicated the lock-in with the existing production model, centered around fossil resources (see also [52]). On one hand, the higher price of chemically recycled feedstock and the price fluctuation of fossil feedstock negatively affect the competitiveness of chemical recycling, further reducing consumer demand. On the other hand, the configuration of the existing supply chain, which is optimized for fossil raw materials, and the lack of a strong market pull for plastic-to-plastic conversion discourage the market shift toward chemically recycled materials.

Shifting to the downstream industrial tier, technical feasibility of chemical recycling depends on the degree of technological and infrastructural maturity. At present, only one third of available technologies have reached commercialization (TRL=9), namely conventional pyrolysis (thermal cracking), catalytic cracking, and conventional gasification; and even fewer technological options enable plastic-to-plastic conversion [47, 53]. In addition to recycling technologies, a recycler expressed the need to "*develop sorting technologies in order to consolidate and improve the yields of mechanical recycling and allow the development of chemical recycling, understood as depolymerization—for example, for the production of Styrene oil—and physical recycling through solvolysis—for example, dissolution of EPS waste from the demolition of buildings*" for the production of new plastic materials.

Another issue undermining the adoption of chemical recycling is the presence of environmental barriers, as indicated by all respondents and acknowledged by previous studies on plastic recycling (e.g. [47]). A first concern is that, despite the ability of chemical recycling to treat a wider range of materials and mixed plastic waste streams [1], the presence of additives and toxic substances in plastic waste is

Table 3 Barriers and drivers of chemical recycling by value chain position (N=103)

Macro-area	Factor	Total (N=103)		Raw material production (N=4)		Primary plastic and product manufacturing (N=78)		Use (N=4)		Waste management (N=17)	
		Avg	CI	Avg	CI	Avg	CI	Avg	CI	Avg	CI
Economic	Process cost	**2.58**	0.22	2.75	0.24	2.55	0.22	3.00*	0.26	2.59	0.22
	Consumer demand	3.23*	0.28	2.75	0.24	3.23*	0.28	3.50	0.30	3.29	0.28
	Market shift	3.13*	0.27	2.75	0.24	3.17*	0.27	3.25*	0.28	3.00*	0.26
	Competition virgin/recycled plastics	3.24*	0.28	3.25*	0.28	3.22*	0.28	3.75	0.32	3.24*	0.28
Technical	Material properties	3.27*	0.28	3.25*	0.28	3.28	0.28	3.25*	0.28	3.24*	0.28
	Technology readiness	3.16*	0.27	3.50	0.30	3.19*	0.27	3.25*	0.28	2.88*	0.25
	Industrial infrastructure development	3.10*	0.27	3.00*	0.26	3.10*	0.27	3.50	0.30	3.00*	0.26
	Urban infrastructure development	**2.99***	0.26	3.00*	0.26	3.05*	0.26	3.25*	0.28	2.65	0.23
Environmental	Toxic additives and substances	**2.67**	0.23	3.00*	0.26	2.67	0.23	2.50	0.22	2.65	0.23
	Energy consumption	**2.90***	0.25	2.75	0.24	2.99*	0.26	2.50	0.22	2.65	0.23
	Environmental impact	**2.90***	0.25	2.75	0.24	2.99*	0.26	2.50	0.22	2.65	0.23

Regulatory	Circular economy and sustainable development directives	3.18*	0.27	3.25*	0.28	3.18*	0.27	3.25*	0.28	3.18*	0.27
	Incentive/tax schemes	3.25*	0.28	4.00	0.34	3.26*	0.28	3.50	0.30	3.00*	0.26
	Market regulation	3.25*	0.28	3.25*	0.28	3.19*	0.27	3.00*	0.26	3.06*	0.26
	Engagement in extended producer responsibility	**2.96***	0.25	3.00*	0.26	2.92*	0.25	3.25*	0.28	3.06*	0.26
Informational	Data transparency and traceability along the value chain	3.20*	0.28	3.50	0.30	3.22*	0.28	3.25*	0.28	3.06*	0.26
Sociocultural	Consumer awareness	3.29	0.28	3.75	0.32	3.26*	0.28	3.50	0.30	3.29	0.28
	Consumer behavior	3.31	0.28	4.00	0.34	3.29	0.28	3.50	0.30	3.18*	0.27

Note: Values below 3 indicate a negative impact (barrier) of the factor and values above 3 represent a positive impact (driver).
* Impact of the factor could change (8.86% margin of error).

still a major obstacle to its scale-up, as decontamination and removal of impurities are difficult, costly, and less understood processes [4, 50]. Regarding this matter, a recycler pointed out that "*the quality of the chemically recycled material can be identical to that of the virgin products, which can therefore be used for any application, e.g., contact with food, high tech, etc., but [decontaminated] high-quality recycling outputs are economically uncompetitive when compared to virgin materials.*" Other environmental concerns about this technology include higher energy intake (with critical values for depolymerization and re-polymerization processes) [13] and emissions that make chemical recycling less environmentally friendly than conventional mechanical alternatives [6].

From a regulatory perspective, chemical recycling emerged as a contested solution due to the diverse nature of its outputs (see also [11]). Today, one of the main end markets for chemical recycling outputs from gasification and pyrolysis is crude diesel, making plastic-to-fuel a preferred route [58]. A recycler raised questions concerning whether this process can be considered recycling at all, as the principal recycling outputs are not products or materials to be used for their original purposes (e.g., gas from gasification process), thereby undermining the principles of circular economy and decarbonization agendas (e.g., EC, 2015, 2018a). Italian regulation on recycling and waste recovery (D.Lgs. 152/06, art. 181) defines "recycling" as every plastic-to-plastic or plastic-to-chemicals process resulting in the production of monomers to be employed in the production of new polymers, or chemical substances (such as hydrogen) for the production of chemicals other than polymers and fuels (Parlamento Italiano, 2006). In other words, plastic-to-fuel and plastic-to-energy processes cannot be considered recycling and accounted for toward the achievement of national recycling targets. That said, another recycler contested that plastic-to-fuel outputs should still be considered for the achievement of circularity targets as "*this process is circular in the way that it creates a substitute to new fossil fuels and addresses the present market demand for sustainable fuels.*" To further complicate this situation, at the European level, there is no such differentiation between mechanical and chemical recycling in the waste treatment hierarchy (Quicker et al., 2022; Zwe2019); therefore, the role of plastic-to-fuel treatments in the circular economy transition remains unclear. In our analysis, a waste sorting company raised the point that "*it is necessary to regulate that, in the treatment hierarchy, priority is given to mechanical over chemical recycling.*"

Additionally, primary plastic and product manufacturers indicated the lack of extended producer responsibility (EPR) schemes as a further constraint to the adoption of chemical recycling, which, in literature, is often regarded as a relevant obstacle to supplying continuous and abundant flows of recyclables and is strictly interrelated with economic and technical barriers [3, 9, 10].

In conclusion, many of the respondents see chemical recycling as an attractive option to increase plastic circularity over ideally infinite cycles, transforming plastic waste into raw materials that can be reintroduced at different stages of the plastic production process. This is particularly evident in this quote from a recycler: "*chemical recycling complements mechanical recycling in terms of both technology and to loop back to more performing materials, similar to virgin ones, and it can extend the useful life of plastics thanks to the high performance of its recycling outputs.*" In this perspective, similar to our findings for mechanical recycling, the results of this analysis show a fair alignment among different value chain functions on the leading role of government and consumers in the adoption of chemical recycling.

6 Discussion

6.1 A Value Chain Perspective on Mechanical-Chemical Recycling Integration

This research enriches extant knowledge on the barriers and drivers of recycling from the perspective of individual actors of the plastic's value chain in Lombardy, Italy. Our results emphasize the need to remove economic barriers, infrastructural inefficiencies, and environmental concerns linked to recycling processes to increase recycling capacity, which is the most critical bottleneck to enhancing plastic circularity. Although respondents across the value chain seem to align on the main barriers and drivers of recycling, when analyzing different perspectives, our analysis yielded tier-specific implications. For example, respondents in the upstream tier stressed the central role of consumers in overcoming the current economic lock-ins by steering demand for sustainable plastics and pushing the creation of new end markets for recycled materials. Downstream firms, instead, highlighted the centrality of consumers in fostering plastic waste separation post-consumption with the aim of guaranteeing high-quality recyclables. This complexity supports the need for exploring the diversified impacts of externalities and the related implications for different value chain actors to undermine existing barriers and leverage drivers in an efficient way.

Similarly, survey respondents were aligned on the importance of developing integrated mechanical-chemical solutions. These results also align with those found at the national level [37], denoting an apparent consensus among the Italian plastic industry. The results of this analysis show that the integration of mechanical and chemical recycling in plastic's end-of-life management would enable structural changes in both the forward and reverse supply chain, addressing some of the existing challenges to plastic recycling. Respondents emphasized that adoption of chemical recycling in the current recycling value chain could mitigate economic, technical, and environmental barriers associated with both recycling options.

First, integrating chemical recycling in the current recycling system would increase the economic competitiveness of employing mixed and contaminated plastic wastes as feedstock for recycling rather than for energy recovery, land filling, or cross-sectoral recovery. Second, as chemical recycling enables receiving more diverse and contaminated waste streams than mechanical recycling, respondents in the downstream industrial segment highlighted that its adoption would increase the volume of plastic recycled, hence supplying the current paucity of high-quality recyclables on one hand, and expanding and diversifying existing end markets for secondary raw materials on the other hand. Third, raising the competitiveness of chemical recycling would motivate the development and adoption of chemical technologies at the commercial scale to enable treating larger volumes of plastic wastes (in line with the findings of [38, 39]). Consequently, a fourth advantage indicated by our respondents is that chemical recycling could supplement existing barriers of technological nature (e.g., difficulty to treat plastic waste based on their material properties) and environmental nature (e.g., difficulty to treat contaminated waste streams) that are typically associated with mechanical recycling. Finally, respondents in the upstream segment of the value chain emphasized that a balanced integration of mechanical and chemical technologies supports different pathways for both recyclables and recycled materials toward closing the loop of plastic. In fact, it can improve the quality of recycled materials, offer a valuable option to virgin materials, and increase the use of recycled materials in plastic manufacturing toward closing the loop of plastics.

6.2 Future Pathways for Plastic Recycling

While some of the problems highlighted in the previous subsection have already been stated in previous studies and some solutions have been advanced, a roadmap to enhance the circular economy at the regional level is still missing. In fact, we need to understand problems and possible alternatives altogether to find synergies and co-solutions for both recycling technologies.

To this end, based on the highlighted drivers and comments from our survey's respondents, in this subsection, we frame four groups of interventions required for higher plastic circularity pertaining to (1) leveraging economies of scale for sustainable plastics, (2) regulation of contaminants and toxic additives in production and recycling systems, (3) deployment of best available recycling technologies and development of emerging ones at industrial scale, and (4) improved waste collection and separation at all stages of waste management, including diversification of input flows for mechanical and chemical recycling. As shown in Table 4, each area addresses conundrums identified by our sampled companies (value chain functions column) that manifest in the production and/or recycling system (reference system column) and can limit mechanical and chemical (technology column), due to interrelated economic, technical, environmental, or regulatory challenges (barriers and problems columns). Following, the authors explored alignment with the Lombardy Region circular economy roadmap and interventions present in the literature or existing policies to advance solutions aimed at removing existing barriers.

The dual value of this approach lies in the definition of areas that require priority intervention based on our survey results, and the identification of possible actions and measures to address such conundrums in the existing regulatory framework

Table 4 Future pathways for plastic recycling with related problems and opportunities

Value chain functions	Reference system		Tech		Barriers	Problems	Lombardy Region circular economy roadmap (2021–2027)	Examples of interventions
	PS	RS	MR	CR				
(1) Leveraging economies of scale for sustainable plastics								
A, B	x		x	x	- Process cost - Consumer demand - Market shift - Market regulation	- High cost of secondary raw materials - Misaligned supply-demand dynamics - Difficulty of finding applications for recyclables	- Incentives to create end markets for recyclates in the short/medium period - Structural funds to support employment of recycled plastics into production cycles (PR FESR 2021–2027)	- All plastic packaging to be reusable, recyclable, or recycled in a cost-effective manner by 2030 (EU Circular Economy Action Plan) - New market dynamics: shift from price competition to competition based on sustainability benefits (competitive sustainable plastics, new product categories) (Siltaloppi and Jähi, 2021)

(continued)

Table 4 (continued)

Value chain functions	Reference system		Tech		Barriers	Problems	Lombardy Region circular economy roadmap (2021–2027)	Examples of interventions
	PS	RS	MR	CR				
(1) Leveraging economies of scale for sustainable plastics								
A, B, C, D		x	x	x	- Process cost - Energy consumption - Incentive/tax schemes	- High waste management costs - Low revenues from recycling - High operating expenses - Increased cost of energy (up to 30% of total revenues) - Lack of incentives for recycling - High gate fee for recycling	- Subsidizing waste management systems focused on more efficient collection, separation, reuse, and recycling of plastics (PR FESR 2021–2027) - Selective disassembly planning to save production energy and costs	- Increased supply of plastic waste to recyclers leading to economies of scale for improved recycling outputs (Milios et al., 2018) - Fourfold increase in sorting and recycling capacity in the period 2015–2030 (EU Circular Economy Action Plan)

A, B, C, D	x		x		- Toxic additives and substances - Environmental impact - Competition virgin/recycled plastics	- Lack of ecodesign, design for recycling/disassembly - Complex multilayer/multi-material packaging - Lower quality of recyclates		- Design for recycling for higher separation and recycling efficiency (Roosen et al., 2020) - Abolish or reduce the use of multi-material multilayer packaging (mono-materials) (De Mello Soares et al., 2022) - REACH regulation (1907/2006/EC) on chemicals used in plastic production - Ecodesign Directive (2009/125/EC) on improving environmental performance of products via implementation of ecodesign principles, extended producer responsibility for sustainability, measures on raw materials and production processes

(continued)

Table 4 (continued)

Value chain functions	Reference system		Tech		Barriers	Problems	Lombardy Region circular economy roadmap (2021–2027)	Examples of interventions
	PS	RS	MR	CR				
(1) Leveraging economies of scale for sustainable plastics								
A, B, C, D		x	x	x	- Toxic additives and substances - Environmental impact	- Difficult decontamination - Higher energy requirements from decontamination process - Higher environmental impacts from decontamination process	- Guaranteeing high sorting quality and deployment of robotics sorting systems for fast identification of materials and components (in-line material identification through optical systems)	- Developing reverse logistics, extended producer responsibility, and innovative business models to reduce waste contamination (EU Circular Economy Action Plan) - Cost-efficient, simplified pretreatment (e.g., modular unit operations) (Qureshi et al., 2020) - Research and development in deinking processes and other decontamination technologies (Horodytska et al., 2018)

(3) Deployment of best available recycling technologies and development of emerging ones at the commercial scale								
A, D		x	x	x	-Recycling technology readiness	- Limited capacity of recycling plants - Limited plastic-to-plastic applications for chemical recycling - Difficult recycling due to poor waste separation (lower-quality recyclables) - Lack of preselection of recyclables (materials, cleanliness, quality)	- Improving existing technologies or development of emerging ones at the commercial scale - Innovative integration of advanced mechanical and chemical recycling to increase recovery rate - Developing pilot applications to reduce recycling costs at the plant level (techno-economic feasibility) and validate high-risk investments at the industrial level	- Innovative solutions for advanced sorting, chemical recycling, and improved polymer design (EU Circular Economy Action Plan) - Collaboration between the chemical industry and plastic recyclers for wider and higher value applications (chemical feedstock) (EU Circular Economy Action Plan) - Integration of chemical recycling in the recycling system (Siltaloppi and Jähi, 2021)

(continued)

Table 4 (continued)

Value chain functions	Reference system		Tech		Barriers	Problems	Lombardy Region circular economy roadmap (2021–2027)	Examples of interventions
	PS	RS	MR	CR				
(4) Improved waste collection and separation at all stages of waste management, including diversification of input flows for MR and CR								
A, B, C, D		x	x	x	- Urban infrastructure development - Industrial infrastructure development - Incentive/tax schemes	- Limited waste separation at collection stage - Difficult waste separation post-consumption - Difficulty to find recycling facilities - Prioritization of technology route based on waste hierarchy - Policy misalignment at regional, national, and international level - Lack of incentives for waste collection.	- Structural funds to support innovative collection schemes and recycling technologies (PR FESR 2021–2027) - Improving plastic waste management at all stages of the value chain - Improved infrastructure for inspecting and tracking input material flows (traceability, information sharing)	- Improving quality of separate waste collection and sorting (EU Circular Economy Action Plan) - Long-term binding agreements between waste suppliers and recyclers to guarantee quality and quantity of recyclables (Qureshi et al., 2020) - Systematic development of plastic recycling infrastructure via multi-actor collaboration (Siltaloppi and Jähi, 2021) - Harmonizing national measures and management of plastic packaging waste, and defining recycling standards - Increase of 5.8 million tons/year in waste recycling capacity (€5.5 billion investment) by 2020 (EU Circular Economy Action Plan) - Introduction of diversified collection/separation targets for different fractions of plastic waste to increase recyclability (Schuch et al., 2023)

Note: A=raw material production, B=primary plastic and product manufacturing, C=use, D=waste management; PS=production system, RS=recycling system; MR=mechanical recycling, CR=chemical recycling

or found in literature, which can offer a starting point to accelerate the circular economy transition.

6.2.1 Leveraging Economies of Scale for Securing Recycled Plastics Supply

As shown above, the interrelation of economic with other barriers constitutes a core conundrum of the circular economy transition. To break down this gap further, one part of the problem originates in the production system, where price-based competition makes it difficult for companies to incorporate socio-environmental performance and sustainability benefits at competitive prices in the short run, creating misalignment between supply and demand for sustainable products [52]. Furthermore, the fragmentation of the plastic industry complicates the collaborative development of end markets for recycled plastics [28], and, consequently, there is an unused supply of recyclables. The second part of the problem originates in the recycling system, where economic, environmental, and regulatory aspects intertwine, undermining the efficiency of the recycling infrastructure.

The findings of our survey accentuate the need to upscale the supply of plastic waste to recyclers, with consequent improvement of recycling outputs, thus leading to economies of scale (see also [39]). As recent research suggests, product manufacturers and users cover a central role in steering the demand for recycled plastics that are functionally and economically competitive with virgin materials [52]. This means that they can set new competitive priorities and benchmarking standards based on environmental performance, systemic impacts, and footprint of new plastic products. At the same time, collaboration is required across the value chain to monitor market dynamics, guarantee adequate uses for recycled plastics, and secure quality and quantity of recycling inputs. Furthermore, availability of upgraded recycling outputs and chemically recycled feedstock for plastic-to-plastic applications can bolster plastic circularity, raising the market value of plastic waste on one hand, and addressing the scarcity of sustainable chemical feedstock on the other [52, 56].

Our survey respondents stressed the centrality of the government in regulating these new supply-demand dynamics and incentivizing engagement in recycling. The incentive is there: Lombardy's PR FESR 2021–2027 (Regional Program for the European Regional Development Fund) concentrates on identifying possible forms of support for innovative collection schemes, recycling technologies, and employment of recycled plastics into production cycles, through the introduction of structural funds to support regional supply chains toward the development of new solutions and technologies [48]. In line with EU regulation, this program can help identify more effective policies to subsidize waste management systems focused on more efficient collection, separation, reuse, and recycling of plastics, including incentives to create end markets for recyclates in the short and medium period.

6.2.2 Regulation of Contaminants and Toxic Additives in Production and Recycling Systems

At the recycling stage, material contamination arises from collection (e.g., organic and inorganic residues), mis-sorting, and difficulties in the decontamination process, particularly bounding mechanical recycling [50]. At the production stage, the presence of unwanted substances (e.g., additives, adhesives, labels) results from a lack of standardized production requirements and ecodesign principles [23, 54].

At the EU level, several policy frameworks have been implemented to improve material efficiency and reduce contaminants, such as the REACH regulation (1907/2006/EC), the Ecodesign Directive (2009/125/EC), and more in general the European Strategy for Plastics in a Circular Economy. However, in reality, the chemical composition of plastic products makes polymer contamination unavoidable, even in the presence of optimized sorting systems. In this perspective, chemical recycling offers an appealing alternative as it can open up more business opportunities and upcycle plastic waste into valuable chemicals [16, 56].

6.2.3 Deployment of Best Available Recycling Technologies and Development of Emerging Ones at the Commercial Scale

The level of maturity of recycling technologies has been appointed as a significant restraint to both upscaling mechanical recycling capacity and adopting chemical recycling at the commercial scale. Existing policy frameworks call for advancing mechanical recycling and integrating chemical recycling, with particular attention to plastic-to-plastic conversion, in the plastic value chain. However, there is no indication of specific solutions and actions to achieve these goals.

To optimize plastic waste recovery, it is imperative to pursue the preferential end-of-life option based on the quality of the recyclates and their ability to substitute fossil feedstock. In reality, based on the trade-offs associated with existing technologies, the most desirable option is given by a combination of recycling techniques deploying the best available technologies [39]. Lombardy Region is investing in pilot processes and demonstration facilities to foster innovation, reduce recycling costs at the level of a single plant (techno-economic feasibility), and to validate high-risk investments before emerging technologies reach the industrial scale within a regional model of circular economy [48]. This requires new collaborations and the development of competences along the value chain for wider and higher value applications for recycled plastics.

6.2.4 Improved Waste Collection and Separation at All Stages of Waste Management, Including Diversification of Input Flows for Mechanical and Chemical Recycling

Without significant improvement of the waste management infrastructure at the urban and industrial levels, recycling capacity is unlikely to raise [23]. Against this drawback, our respondents emphasized the importance of rethinking plastic waste separation at all stages of waste management, including proper recovery of compostable, biodegradable, and bio-based materials, creation of separated collection channels for industrial and domestic plastic waste, and dedicated recovery systems for contaminated materials.

To function properly, the waste management infrastructure necessitates interventions on different fronts. At the value chain level, cooperation is required to secure provision of improved recycling inputs, including binding agreements between waste suppliers and recyclers [46], multi-actor collaboration [52], and joint ventures.

On the regulatory side, there is a need to better define collection, sorting, and recycling targets, including diversified targets for specific plastic types [51], and harmonize them with national and EU standards. As it has also been observed in the literature [32, 39], there is a growing impetus to promote regulations aimed at increasing waste recovery toward a circular economy for plastic. Concerning this matter, one of the objectives of the Lombardy Region circular economy roadmap reads: "*improving management of plastic waste at all stages of the value chain, from prevention at source, to collection and handling operations, to waste treatments*" [48]. Consequently, the region allocated structural funds to support the adoption of innovative waste collection and recovery schemes.

Finally, monitoring material flows across the entire plastic's life cycle is critical to ensure that no plastic waste flows are left uncaptured and to detect potential loopholes along the value chain. From a technological perspective, peculiar monitoring of plastic waste streams allows for acquisition of distinct material characteristics (e.g., contamination, decomposition, downgrading) that can act as a discriminant to diversify input waste flows for mechanical or chemical recycling. Hence, industrial stakeholders and policy-makers should recognize the value of chemical recycling as a complementary technique rather than a direct competitor of mechanical recycling, to obtain value-added chemicals and upcycled plastics from heterogenous and contaminated waste streams [56]. Additionally, effective mechanical-chemical integration requires the development of the entire plastic value chain. Overall improvements are needed in the waste collection and sorting infrastructure to make optimal use of different technological options based on recycling inputs, as well as structural changes to the production system to handle growing flows of recycled materials and reintroduce them in the value chain [52]. This development calls for collaborative efforts across all value chain tiers, alongside governmental support for the plastic industry to orchestrate industrial interventions, cope with existing necessities, and guarantee system-level benefits.

7 Conclusions

Our study explored the potential for scaling up plastic recycling to accelerate the transition to the circular economy for plastic. It deployed a semiquantitative approach to characterize the plastic value chain in Lombardy and identify different barriers and drivers toward the scale-up of mechanical recycling and the adoption of chemical recycling at the regional level. Based on our results, the following contributions to research and practice can be identified. First, the present study demonstrates how the same factor can play different roles (barrier or driver) when perceived by different value chain functions. This suggests the need to develop and implement diversified policy instruments to eliminate existing obstacles and leverage opportunities along the value chain toward plastic circularity. Second, the study demonstrates that although mechanical and chemical recycling share similar barriers and drivers, there are important differences to be carefully addressed for a successful integration of both mechanical and chemical recycling in plastic's end-of-life management. The commonality of problems indicated by the study's results suggests that addressing economic, technical, and environmental barriers would potentially foster both recycling technologies, hence retaining plastic materials longer in the loop. Third, we propose a number of areas of intervention that could support such development and set the successful path toward the transition to a circular economy for plastic based on a concurrent and well-balanced set of actions in the plastic production and recycling systems, aimed at removing existing economic, technical, environmental, and regulatory barriers, while leveraging the favorable policy ground. Fourth, to the best of our knowledge, this is the first study to provide empirical evidence for improving plastic recycling in the Lombardy region, Italy. Fifth, although the results are based on this empirical context, the study provides managerial implications that can be generalized and extended beyond the region's boundaries, hence supplementing the lack of studies on firm-specific perspectives on plastic recycling. Implications for industrial actors include the interdependence between actors of the forward and reverse supply chain in the development of a circular economy for plastic, and the need for cross-value chain collaboration to implement the circular economy. Implications for policy-makers are relative to the development of policy instruments that consider the heterogeneity of perceived barriers and drivers by actors across the plastic value chain, and the need for adopting a system perspective.

Some limitations of this analysis are related to the sample selection and uneven representation of different value chain functions in our pool of respondents. Hence, increasing diversity of responses could enrich our analysis of individual perspectives, supporting firm-specific recommendations to foster the circularity transition. Another suggestion for further research includes the use of a mixed method approach, for example, quantitative research could be supported by the qualitative examination of priorities, strategies, and tactical decisions taken by different value chain actors. Finally, the proposed methodology could be extended to the examination of cross-sector studies, e.g., to differentiate barriers and drivers across different economic sectors, or cross-country studies, as value chains are global.

References

1. Al-Salem SM, Lettieri P, Baeyens J (2009) Recycling and recovery routes of plastic solid waste (PSW): A review. Waste Manag 29(10):2625–2643. https://doi.org/10.1016/j.wasman.2009.06.004
2. Al-Salem SM, Lettieri P, Baeyens J (2010) The valorization of plastic solid waste (PSW) by primary to quaternary routes: from re-use to energy and chemicals. Prog Energy Combust Sci 36(1):103–129. https://doi.org/10.1016/j.pecs.2009.09.001
3. Bashirgonbadi A, Lase IS, Delva L, Van Geem KM, De Meester S, Ragaert K (2022) Quality evaluation and economic assessment of an improved mechanical recycling process for post-consumer flexible plastics. Waste Manag 153:41–51. https://doi.org/10.1016/j.wasman.2022.08.018
4. Bening CR, Pruess JT, Blum NU (2021) Towards a circular plastics economy: interacting barriers and contested solutions for flexible packaging recycling. J Clean Prod 302:126966. https://doi.org/10.1016/j.jclepro.2021.126966
5. Bharadwaaj SK, Jaudan M, Kushwaha P, Saxena A, Saha B (2024). Exploring cutting-edge approaches in plastic recycling for a greener future. Results Eng 102704. https://doi.org/10.1016/j.rineng.2024.102704
6. Brivio L, Tollini F (2022) Chapter Six—PET recycling: review of the current available technologies and industrial perspectives. Adv Chem Eng 60(1):215–267. https://doi.org/10.1016/bs.ache.2022.09.003
7. Brouwer M, Picuno C, van Velzen EUT, Kuchta K, De Meester S, Ragaert K (2019) The impact of collection portfolio expansion on key performance indicators of the Dutch recycling system for Post-Consumer Plastic Packaging Waste, a comparison between 2014 and 2017. Waste Manag 100:112–121. https://doi.org/10.1016/j.wasman.2019.09.012
8. COREPLA (2022) Relazione sulla gestione 2021. https://www.corepla.it/sites/default/files/documenti/corepla_relazione_sulla_gestione_2021_1.pdf
9. Da Cruz NF, Simões P, Marques RC (2012) Economic cost recovery in the recycling of packaging waste: the case of Portugal. J Clean Prod 37:8–18. https://doi.org/10.1016/j.jclepro.2012.05.043
10. Da Cruz NF, Ferreira S, Cabral M, Simões P, Marques RC (2014). Packaging waste recycling in Europe: is the industry paying for it?. Waste Manag 34(2):298–308. https://doi.org/10.1016/j.wasman.2013.10.035
11. Das S, Liang C, Dunn JB (2022) Plastics to fuel or plastics: life cycle assessment-based evaluation of different options for pyrolysis at end-of-life. Waste Manag 153:81–88. https://doi.org/10.1016/j.wasman.2022.08.015
12. De Lucia C, Pazienza P (2019) Market-based tools for a plastic waste reduction policy in agriculture: A case study in the south of Italy. J Environ Manag 250:109468. https://doi.org/10.1016/j.jenvman.2019.109468
13. De Mello Soares CT, Ek M, Östmark E, Gällstedt M, Karlsson S (2022) Recycling of multi-material multilayer plastic packaging: current trends and future scenarios. Resour Conserv Recycl 176:105905. https://doi.org/10.1016/j.resconrec.2021.105905
14. De Paula IC, De Campos EAR, Pagani RN, Guarnieri P, Kaviani MA (2020) Are collaboration and trust sources for innovation in the reverse logistics? Insights from a systematic literature review. Supply Chain Manag Int J 25(2):176–222. https://doi.org/10.1108/SCM-03-2018-0129
15. Dijkstra H, van Beukering P, Brouwer R (2020) Business models and sustainable plastic management: a systematic review of the literature. J Clean Prod 258:120967. https://doi.org/10.1016/j.jclepro.2020.120967
16. Dogu O, Pelucchi M, Van de Vijver R, Van Steenberge PH, D'hooge DR, Cuoci A, Van Geem KM (2021) The chemistry of chemical recycling of solid plastic waste via pyrolysis and gasification: State-of-the-art, challenges, and future directions. Prog Energy Combust Sci 84:100901. https://doi.org/10.1016/j.pecs.2020.100901

17. Eriksen MK, Astrup TF (2019) Characterisation of source-separated, rigid plastic waste and evaluation of recycling initiatives: effects of product design and source-separation system. Waste Manag 87:161–172. https://doi.org/10.1016/j.wasman.2019.02.006
18. Fondazione per lo Sviluppo Sostenibile (2021) L'Italia del Riciclo 2021. http://www.fondazionesviluppoostenibile.org/wp-content/uploads/ITALIADELRICICLO2021web.pdf
19. Foschi E, D'Addato F, Bonoli A (2021) Plastic waste management: a comprehensive analysis of the current status to set up an after-use plastic strategy in Emilia-Romagna Region (Italy). Environ Sci Pollut Res 28(19):24328–24341. https://doi.org/10.1007/s11356-020-08155-y
20. Geyer R, Jambeck JR, Law KL (2017) Production, use, and fate of all plastics ever made. Sci Adv 3(7):e1700782. https://doi.org/10.1126/sciadv.1700782
21. Gong Y, Putnam E, You W, Zhao C (2020) Investigation into circular economy of plastics: the case of the UK fast moving consumer goods industry. J Clean Prod 244:118941. https://doi.org/10.1016/j.jclepro.2019.118941
22. Hahladakis JN, Iacovidou E (2018) Closing the loop on plastic packaging materials: what is quality and how does it affect their circularity? Sci Total Environ 630:1394–1400. https://doi.org/10.1016/j.scitotenv.2018.02.330
23. Hahladakis JN, Iacovidou E (2019) An overview of the challenges and trade-offs in closing the loop of post-consumer plastic waste (PCPW): focus on recycling. J Hazard Mater 380:120887. https://doi.org/10.1016/j.jhazmat.2019.120887
24. Horodytska O, Valdés FJ, Fullana A (2018) Plastic flexible films waste management–A state of art review. Waste Manag 77:413–425. https://doi.org/10.1016/j.wasman.2018.04.023
25. Hossain MU, Ng ST, Dong Y, Amor B (2021) Strategies for mitigating plastic wastes management problem: a lifecycle assessment study in Hong Kong. Waste Manag 131:412–422. https://doi.org/10.1016/j.wasman.2021.06.030
26. Hu B, Wang S, Yan J, Zhang H, Qiu L, Liu W, Guo Y, Shen J, Chen B, Shi C, Ge X (2024) Review of waste plastics treatment and utilization: efficient conversion and high value utilization. In Process Safety and Environmental Protection. https://doi.org/10.1016/j.psep.2024.01.024
27. Jeswani H, Krüger C, Russ M, Horlacher M, Antony F, Hann S, Azapagic A (2021) Life cycle environmental impacts of chemical recycling via pyrolysis of mixed plastic waste in comparison with mechanical recycling and energy recovery. Sci Total Environ 769:144483. https://doi.org/10.1016/j.scitotenv.2020.144483
28. Kahlert S, Bening CR (2022) Why pledges alone will not get plastics recycled: comparing recyclate production and anticipated demand. Resour Conserv Recycl 181:106279. https://doi.org/10.1016/j.resconrec.2022.106279
29. Kubiczek J, Derej W, Hadasik B, Matuszewska A (2023) Chemical recycling of plastic waste as a mean to implement the circular economy model in the European Union. J Clean Prod 406:136951. https://doi.org/10.1016/j.jclepro.2023.136951
30. Larrain M, Van Passel S, Thomassen G, Kresovic U, Alderweireldt N, Moerman E, Billen P (2020) Economic performance of pyrolysis of mixed plastic waste: open-loop versus closed-loop recycling. J Clean Prod 270:122442. https://doi.org/10.1016/j.jclepro.2020.122442
31. Larrain M, Van Passel S, Thomassen G, Van Gorp B, Nhu TT, Huysveld S, Billen P (2021) Techno-economic assessment of mechanical recycling of challenging post-consumer plastic packaging waste. Resour Conserv Recycl 170:105607. https://doi.org/10.1016/j.resconrec.2021.105607
32. Larrain M, Billen P, Van Passel S (2022) The effect of plastic packaging recycling policy interventions as a complement to extended producer responsibility schemes: a partial equilibrium model. Waste Manag 153:355–366. https://doi.org/10.1016/j.wasman.2022.09.012
33. Lebreton L, Andrady A (2019) Future scenarios of global plastic waste generation and disposal. Palgrave Commun 5(1):1–11. https://doi.org/10.1057/s41599-018-0212-7
34. Lisiecki M, Damgaard A, Ragaert K, Astrup TF (2023). Circular economy initiatives are no guarantee for increased plastic circularity: a framework for the systematic comparison of initiatives. Resour Conserv Recycl 197:107072. https://doi.org/10.1016/j.resconrec.2023.107072

35. Liu Q, Martinez-Villarreal S, Wang S, Tien NNT, Kammoun M, De Roover Q, Len C, Richel A (2024) The role of plastic chemical recycling processes in a circular economy context. Chem Eng J 155227. https://doi.org/10.1016/j.cej.2024.155227
36. Lombardi M, Rana R, Fellner J (2021) Material flow analysis and sustainability of the Italian plastic packaging management. J Clean Prod 287:125573. https://doi.org/10.1016/j.jclepro.2020.125573
37. Maione C, Lapko Y, Trucco P (2022) Towards a circular economy for the plastic packaging sector: insights from the Italian case. Sustainable Prod Consumption 34:78–89. https://doi.org/10.1016/j.spc.2022.09.002
38. Maisels A, Hiller A, Simon FG (2022) Chemical recycling for plastic waste: status and perspectives. ChemBioEng Rev 9(6):541–555. https://doi.org/10.1002/cben.202200024
39. Milios L (2018) Advancing to a Circular Economy: three essential ingredients for a comprehensive policy mix. Sustain Sci 13(3):861–878. https://doi.org/10.1007/s11625-017-0502-9
40. Milios L, Christensen LH, McKinnon D, Christensen C, Rasch MK, Eriksen MH (2018) Plastic recycling in the Nordics: a value chain market analysis. Waste Manag 76:180–189. https://doi.org/10.1016/j.wasman.2018.03.034
41. Mohammadi M, Younes GR, Yousefzade O, Ajji A (2024) Challenges in Polymer Recycling. In Reference Module in Materials Science and Materials Engineering. https://doi.org/10.1016/B978-0-323-95486-0.00048-X
42. Paletta A, Leal Filho W, Balogun AL, Foschi E, Bonoli A (2019) Barriers and challenges to plastics valorisation in the context of a circular economy: case studies from Italy. J Clean Prod 241:118149. https://doi.org/10.1016/j.jclepro.2019.118149
43. PlasticsEurope (2018) Plastic the facts—2018: an analysis of European plastics production, demand and waste data. https://plasticseurope.org/knowledge-hub/plastics-the-facts-2018/
44. PlasticsEurope (2022) Plastics—the facts 2022. https://plasticseurope.org/knowledge-hub/plastics-the-facts-2022/
45. Quicker P, Seitz M, Vogel J (2022) Chemical recycling: a critical assessment of potential process approaches. Waste Manag Res 40(10):1494–1504. https://doi.org/10.1177/0734242X221084044
46. Qureshi MS, Oasmaa A, Pihkola H, Deviatkin I, Tenhunen A, Mannila J, Laine-Ylijoki J (2020). Pyrolysis of plastic waste: opportunities and challenges. J Anal Appl Pyrolysis 152:104804. https://doi.org/10.1016/j.jaap.2020.104804
47. Ragaert K, Delva L, Van Geem K (2017) Mechanical and chemical recycling of solid plastic waste. Waste Manag 69:24–58. https://doi.org/10.1016/j.wasman.2017.07.044
48. Regione Lombardia (2020) Lombardy Roadmap for Research and Innovation on Circular Economy, RI priorities enhancing Circular Economy in Lombardy. http://clustercollaboration.eu/sites/default/files/eu_initiatives/roadmapforrioncirculareconomy.pdf
49. Reijonen H, Bellman S, Murphy J, Kokkonen H (2021) Factors related to recycling plastic packaging in Finland's new waste management scheme. Waste Manag 131:88–97. https://doi.org/10.1016/j.wasman.2021.05.034
50. Roosen M, Mys N, Kusenberg M, Billen P, Dumoulin A, Dewulf J, De Meester S (2020) Detailed analysis of the composition of selected plastic packaging waste products and its implications for mechanical and thermochemical recycling. Environ. Sci. Technol. 54(20):13282–13293. https://doi.org/10.1021/acs.est.0c03371
51. Schuch D, Lederer J, Fellner J, Scharff C (2023) Separate collection rates for plastic packaging in Austria–A regional analysis taking collection systems and urbanization into account. Waste Manag 155:211–219. https://doi.org/10.1016/j.wasman.2022.09.023
52. Siltaloppi J, Jähi M (2021) Toward a sustainable plastics value chain: Core conundrums and emerging solution mechanisms for a systemic transition. J Clean Prod 315:128113. https://doi.org/10.1016/j.jclepro.2021.128113
53. Solis M, Silveira S (2020) Technologies for chemical recycling of household plastics–A technical review and TRL assessment. Waste Manag 105:128–138. https://doi.org/10.1016/j.wasman.2020.01.038

54. Van Doorsselaer K (2022) Chapter 12—The role of ecodesign in the circular economy. Circ Econ Sustain 1(2022):189–205. https://doi.org/10.1016/B978-0-12-819817-9.00018-1
55. Volk R, Stallkamp C, Steins JJ, Yogish SP, Müller RC, Stapf D, Schultmann F (2021) Techno-economic assessment and comparison of different plastic recycling pathways: a German case study. J Ind Ecol 25(5):1318-1337. https://doi.org/10.1111/jiec.13145
56. Vollmer I, Jenks MJ, Roelands MC, White RJ, van Harmelen T, de Wild P, Weckhuysen BM (2020) Beyond mechanical recycling: giving new life to plastic waste. Angew Chem Int Ed 59(36):15402–15423. https://doi.org/10.1002/anie.201915651
57. Zia KM, Bhatti HN, Bhatti IA (2007) Methods for polyurethane and polyurethane composites, recycling and recovery: a review. React Funct Polym 67(8):675–692. https://doi.org/10.1016/j.reactfunctpolym.2007.05.004
58. ZWE (Zero Waste Europe) (2019) El Dorado of Chemical Recycling. State of play and policy challenges. https://zerowasteeurope.eu/wp-content/uploads/2019/08/zerowasteeuropestudychemicalrecyclingupdateden.pdf

BCpbP: Blue Circular Post-branding Project, a Local Portuguese Case Study for a Blue Circular Economy

Carlos A. M. Duarte, Isabel Farinha, Rui Miguel, Gabriel Pestana, Nuno Sá Leal, Sónia Seixas, and José Manuel Robalo

Abstract Portugal has committed itself to the 2030 Agenda for Sustainable Development, adopted by the United Nations in September 2015. Since its creation, Portugal has prioritised the conservation and sustainable use of the ocean. Promoting these goals is a key strategic priority for Portugal.

According to BCpbP data, around 2400 tons of fishing gear are destroyed or abandoned every year in Portugal, based on fishing activities in the fishing ports of the Cascais Captaincy. This case study in Portugal aimed to (1) create value from marine waste, (2) promote sustainable design alternatives to current consumption patterns, and (3) raise awareness and literacy about ocean sustainability, contributing to the Blue Economy.

An R&D methodology adapted to the concept of Living Labs was adopted, in which innovations such as services, products, or application improvements are created and validated in a collaborative metacontext environment.

C. A. M. Duarte (✉) · I. Farinha
Universidade Europeia, UNIDCOM/IADE, Lisbon, Portugal
e-mail: carlos.duarte@universidadeeuropeia.pt; isabel.farinha@universidadeeuropeia.pt

R. Miguel
Universidade da Beira Interior, Covilhã, Portugal
e-mail: rmiguel@ubi.pt

G. Pestana
Escola Superior de Tecnologia - Instituto Politécnico de Setúbal, Setúbal, Portugal
e-mail: gabriel.pestana@estsetubal.ips.pt

N. Sá Leal
Universidade de Trás-os-Montes e Alto Douro, Vila Real, Portugal
e-mail: nunosaleal@utad.pt

S. Seixas
Universidade Aberta, Lisbon, Portugal
e-mail: sonia.seixas@uab.pt

J. M. Robalo
Guarda, HR Consultors, Guarda, Portugal
e-mail: jmrobalo@hrconsultores.pt

B. Jarboui et al. (eds.), *Data-Driven Waste Revolution*, Computational Intelligence Methods and Applications, https://doi.org/10.1007/978-981-96-5890-9_11

As a result, a macro-process was established as a chain of custody for the circular economy called the Blue Big Bag.

From this, it can be said that BCpbP can be an example of a blue circular economy approach. Its main purpose is to combat excessive consumption of goods and reduce unnecessary waste, while at the same time sensitising our partners to environmental education. On the other hand, it also aims to contribute to the removal of tons of end-of-life materials and products from the ocean, namely fishing nets and gear (DFG) and other marine litter, [5] in order to create new and sustainable products.

Keywords Blue circular economy · Social awareness · Service design · Sustainable design

1 Introduction

The main objective of this case study was to address the possible positive impact on fishing communities in Portugal, with the project promoted by Docapesca—Portos e Lotas, SA with the support of APLM—Associação Portuguesa do Lixo Marinho (Portuguese Marine Litter Association), called "Pescar por um mar sem lixo" [9] (Fishing for a litter-free sea) (Fig. 1), focusing on improving waste management on board fishing vessels and in fishing ports by raising awareness among fishermen of the importance of adopting or maintaining good environmental practices. By promoting the selective collection of waste generated on board and caught in fishing gear and by providing adequate infrastructure for its reception on land, this case study has made it possible to draw up a set of guidelines for involving local fishing communities and the organisations responsible for managing fishing ports, in order to improve the environmental conditions of the Portuguese coastal zone and preserve marine ecosystems.

In pursuit of these objectives, a sustainable value-added business model was tested and implemented with the aim of setting up a public-private collaboration network that would involve the different players in the process, so that together it would be possible to add value to this waste taken from the sea or from fishing

Fig. 1 Corporate image of the project "A pesca por um mar sem lixo/ Fishing for a sea without litter"

activities, thus promoting the creation of a market based on the principles of social and environmental sustainability.

In pursuit of this goal, an initiative called the Blue Circular Post-branding Project (BCpbP) was created, https://bluecircular.org/https://bluecircular.org/—supported by A2S—Associação para o Desenvolvimento Sustentável da Região Saloia (Association for the Sustainable Development of the Saloia Region), with co-funding from the Mar2020 Programme: It is a clear example of a blue circular economy approach. On the one hand, the aim was to combat the excessive consumption of goods and reduce the production of unnecessary waste, while at the same time promoting awareness and raising awareness of environmental education among our partners. On the other hand, the aim was to remove tons of end-of-life materials and products from the ocean, namely fishing nets and marine litter, in order to promote new and sustainable products.

The BCpbP's main objective was to promote a "recycling journey", transforming existing materials and products into new items that could be reused. In the first instance, it aimed to add value to marine litter, recognising it as a valuable resource and not just as a pollutant. To this end, a "network" was created for the interaction of various players, including the communities that play a key role in collecting and disposing of these materials, local and solid waste collection organisations, shredding and sorting facilities, distributors, and the processing industry. The ultimate aim of setting up this network is to make it possible to remove tons of end-of-life materials and products from the oceans, in particular fishing nets and plastic.

The authors advocate creating partnerships with entities that use sustainable methods and materials and incorporate best practices throughout their supply chain, emphasising social responsibility as a fundamental asset. They consider this initiative to be pioneering, as no other known project in Portugal has established such a robust supply chain, involving a wide range of players—from fishermen and local communities to waste removal and treatment companies, research units, NGOs, and industry. This collaboration has succeeded in recycling and reincorporating marine waste into the manufacturing industry, producing items such as fishing nets, footwear, accessories, textiles, and decoration.

The project complies with the European Strategy for Plastics in the Circular Economy, [15] which aims to transform the way plastic products are designed, produced, used, and recycled in the EU, as well as with Portuguese legislation, following the Decree-Law No. 78/2021, of 24 September, which transposes EU directive 2019/904, on reducing the impact of certain plastic products on the environment.

Knowing that freeing the oceans from plastic is a huge task, and not just for the maritime sector: It must also involve other sectors such as design, new materials, waste management, and behavioural psychology [18].

For this purpose, the authors present a set of general guidelines in the form of a business plan, as the final result of the case study, aimed at structuring a platform for managing end-of-life fishing gear and plastic materials, to be collected by fishing communities in their coastal areas and consequently reused from a blue economy perspective, as a collaborative process and value creation—service design.

In summary, the BCpbP aimed to help establish a robust supply chain capable of creating value from marine litter, promoting sustainable design alternatives to current consumption patterns and raising awareness of ocean sustainability, contributing to the Blue Economy. The BCpbP has established the general principles for setting up a networking platform for marine litter suppliers and distributors as a nonprofit organisation. The initiative aimed to build bridges and links between government, NGOs, consumers and communities, with the conviction that this model can be replicated in other areas to improve ocean health and community well-being. The principles adopted for the development of a platform of this nature essentially aim to unite various efforts to promote the health of the marine ecosystem, to be supported by a web-based platform and a mobile application, to be initially supported by sponsorship.

2 Business Opportunity (Problem)

2.1 Marine Plastic Pollution and the Sustainable Developmental Goals

The impacts of marine plastic are broad and include environmental effects and impacts on marine industries and the livelihoods of communities that rely on them. Marine plastic pollution (MPP) is an emerging environmental and socioeconomic problem that originates mainly from terrestrial sources and severely impacts the marine ecosystem. It undermines the livelihood of biodiversity [4]. Plastics permeate the entire width and depth of seas and oceans, both close to well-developed coastal areas and in the most remote locations, decreasing the economic and social values of the oceans, especially in terms of fishing productivity and tourism [12]. Marine plastic pollution can be defined as "any persistent solid material, manufactured or processed that has been discarded, abandoned and eventually reaches the marine or coastal environment" (UN [29, p.VII]). Global plastic production in 2017 was about 335 million tons (Plastics Europe [22]), and estimates predict a doubling global plastic production in 20 years [10]. Marine plastic pollution impacts the environment, biodiversity, and industries such as tourism, shipping, and fishing, representing a potential risk to food security and human health [4].

At the European Union (EU) level, the EU Water Framework Directive (60/2000/EC) and the EU Marine Strategy Framework Directive (2008/56/EC) include provisions on reducing pollution and marine litter, respectively. With the launch of the Circular Economy Action Plan, the European Commission committed to "adopt a strategy for plastics in the circular economy, addressing issues such as recyclability, biodegradability, the presence of hazardous substances in certain plastics and marine waste" (COM/2015/0614) [6]. The international community has recognised marine plastic pollution as an essential impediment to sustainable development (UNEP [28]) since at least 8 million tons of plastic are dumped into the oceans annually—which is equivalent to dumping the contents of a garbage truck into the ocean every minute [10]. If no action is taken, these numbers are expected to double by 2030 and quadruple by 2050 [12].

At the United Nations (UN) General Assembly (2015, September), Heads of State and Governments agreed on 17 SDGs as a framework for the 2030 Agenda for Sustainable Development [30]. The 2030 Agenda for Sustainable Development galvanised the unanimous global commitment to address the unsustainable use of plastic and MPP. Sustainable Development Goal (SDG) 14, "Life below water", perseveres to conserve and sustainably use the oceans, seas, and marine resources for sustainable development. Some of the guidelines included in SDG 14 are preventing and reducing marine litter, limiting the impact of fishing on the marine environment, and adapting fishing to protect species. It also promotes the protection, restoration, and sustainable management of marine and coastal ecosystems and marine biodiversity, fosters the local development of coastal communities, increases scientific knowledge, and develops research and marine technology capacities, among others. Goal 14.1 deals specifically with marine litter by 2025: preventing and significantly reducing marine pollution of all types from terrestrial activities, including marine debris and pollution of nutrients.

Regarding objective number 12, it aims at "Responsible consumption and production". It comprises numerous orientations, among which the development of the circular economy with a focus on dematerialisation, collaborative economy and sustainable consumption, product design, efficient use and enhancement of resources; the change in production and consumption models; the increase in global and sectorial collection, recycling and recovery rates for the different materials; the promotion of public, ecological, and sustainable purchasing practices, among others.

However, there are quite a few risks associated with the success of such a venture, such as technological risks associated with the recycling of these materials and their high processing costs at the expense of the value of virgin materials.

While DFG is a more complicated and mixed material than end-of-life fishing gear, a general waste stream for fishing gear is the first step toward material recycling. Implementing waste fishing gear collection in fishing ports is one of the recently drafted claims of the European Plastics Directive [25]. The MARELITT Baltic recycling study provides the first baseline for the recycling options for derelict and end-of-life fishing gear. Given the effort required to recycle DFG, two requirements are identified:

- Retrieved DFG needs to be incorporated into the existing waste management infrastructure in fishing ports.
- Retrieval and preprocessing efforts by fishers, divers, or other professionals need to be financially supported by municipalities or national authorities to establish a DFG value chain.

2.2 Platform Background: Blue Circular Economy

A circular economy can be defined as an economic model that aims to efficiently use resources by minimising waste, striving for long-term value retention, reducing the need for primary resources, and producing products within environmental protection limits.

As presented in (Fig. 2), the circular economy seeks to rebuild capital, be it financial, manufactured, human, social, or natural, ensuring improved flows of goods and services. The diagram illustrates the continuous flow of technical and biological materials around the "circle of value". We may conclude that a circular economy model has the potential to lead to sustainable development while decoupling economic growth from the negative consequences of resource depletion and environmental degradation [20].

A circular economy is increasingly seen as a possible solution to address sustainable development [14]. An economic system that minimises resource input introduction and waste, emission, and energy leakage out of the system is hoped to mitigate adverse impacts without jeopardising growth and prosperity [14]. Combining the challenges of putting circular economy into reality and the practice-oriented approach of business model innovation leads to circular business models (CBMs). This concept describes adequate business models for the circular economy by incorporating elements into the organisation where waste and emission leakage

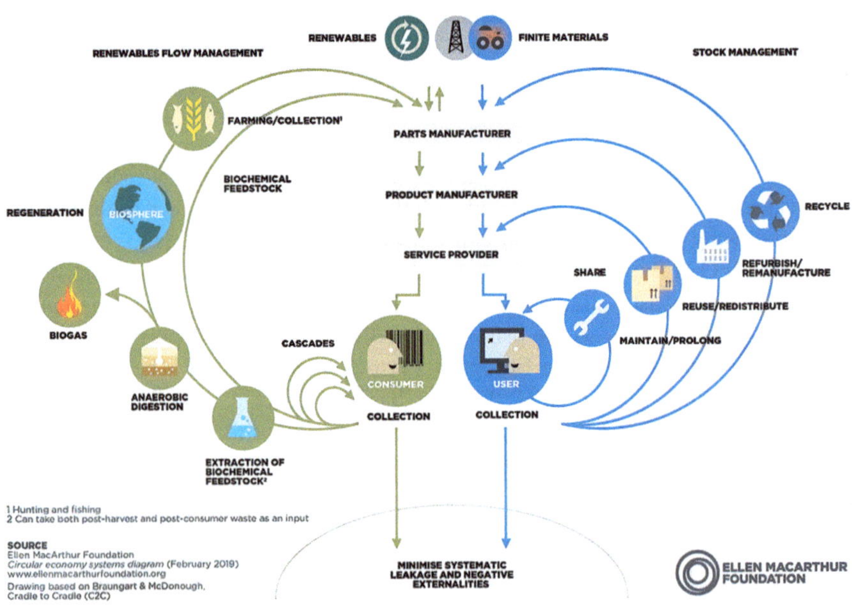

Fig. 2 System diagram through the "circle of value" (Ellen MacArthur Foundation [11])

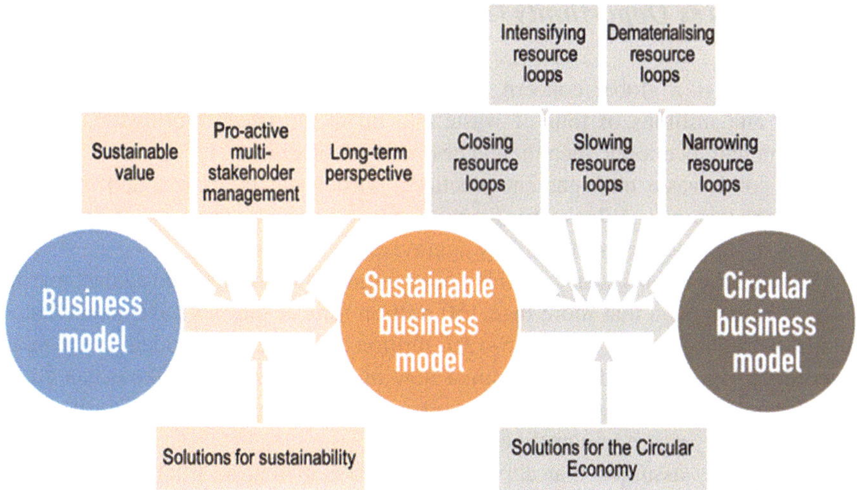

Fig. 3 Comparison of traditional, sustainable, and circular business models (Geissdoerfer et al. [14])

out of the system is minimised (Fig. 3) [14]. The Blue Economy's basic principles propose that the local environment, with its specific ecological features, is the basis for sustainable solutions. In total, the Blue Economy is defined by the following six principles [13]; quoting Blue Economy, 2016:

1. It must be local so that companies use what they have close to them.
2. It should be efficient, where organisations should "substitute something with nothing".
3. The surrounding nature is mimicked with a systemic approach.
4. It should aim at profitable solutions through the optimisation and generation of "multiple cash flows".
5. It should satisfy "all basic needs".
6. It calls for an innovative culture to create change [2].

It is also essential that social criteria, such as employment creation, are desirable outcomes of an economic transition [13]. With this in mind, we believe that the BCpbP aligns with these principles, as we have created a value chain that powers up the interrelation between fishermen who collect marine litter locally and Portuguese brands that take this debris and transform it into new products. Basically, we can say that it combines two strands: the blue circular economy and regenerative design, with the purpose of reintroducing materials and waste into the system or their metamorphosis into new valuable resources at the end of a product's useful life [13].

2.3 Business Opportunity (Problem): BCpbP Platform

Marine litter is a global concern, affecting all the world's oceans. Every year, millions and millions of tons of waste end up in the ocean worldwide, causing environmental, economic, health, and aesthetic problems.

Poor solid waste management practices, wastewater collection and treatment (including stormwater), lack of infrastructure, and public awareness of the consequences of their actions substantially aggravate the situation.

Cleaning the oceans is an option but not the most effective method against marine litter. It is estimated that more than 150 million tons of plastics have accumulated in the world's oceans, while 4.6–12.7 million tons [16] are added every year. According to recent studies, the annual flow of plastic waste into the ocean could almost triple by 2040 to 29 million metric tons per year, equivalent to 50 kg of plastic for every metre of coastline worldwide.

It is broadly assumed that approximately 80% of marine litter is land based, with regional fluctuations (e.g. in the Northeast Atlantic, shipping and fishing are significant litter sources); at the same time, sea-based sources receive increasing attention, both because of the quantities of lost or abandoned fishing gear and because of the damage to marine life and the negative impact on fishing and economy. Marine litter can cause severe economic damage: losses for coastal communities, tourism, shipping, and fishing. Considering its accumulation and dissemination, marine litter may be one of the fastest-growing threats to the health of the world's oceans.

Pollution of the seas from plastics and microplastics is one of the three major areas of the strategy for plastics, adopted by the European Commission in 2018; most of the proposed actions are directly or indirectly related to marine litter, including its international dimension [21].

The directive on single-use plastics and fishing gear (DFG) introduces a series of ambitious measures:

- A ban on selected single-use plastic products for which there are alternatives on the market: cotton swabs, cutlery, plates, straws, stirrers, balloon sticks, as well as expanded polystyrene cups, food and beverage containers, and so on, all products made of oxodegradable plastic
- Measures to reduce consumption of plastic food containers and beverage cups and product-specific marking and labelling
- Extended farmer responsibility schemes covering the cost of cleaning up waste, applied to products such as tobacco filters and fishing gear
- A 90% separate collection target for plastic bottles by 2029 (77% by 2025) and the introduction of design requirements for connecting caps to bottles and a target to incorporate 25% recycled plastic into PET bottles from 2025 and 30% on all plastic bottles from 2030

In Portugal, the European Directive was recently transposed to Decree-Law no. 78/2021, of 24 September; the placing on the market of single-use plastic products and products made from oxodegradable plastic is now prohibited, including measures to raise awareness of the use of reuse systems and waste management options for plastic products single-use and fishing gears containing plastic.

Portugal has one of the largest maritime areas in Europe. According to Law No. 17/2014, related to maritime spatial planning and management, the national marine space extends from the baselines (as defined by UNCLOS) to the outer limit of the continental shelf beyond 200 nautical miles. However, despite all efforts, information on the distribution of marine litter along the Portuguese coast remains scarce.

The Portuguese coast is exposed to the accumulation of marine litter from different terrains and marine sources due to population concentration (more than half of the population) living close to the coast, intense tourism, high fishing, and because it is a vital marine transport route.

According to the https://ec.europa.eu/oceans-and-fisheries/system/files/ (Annual Report Portuguese Fishing Fleet—2019), the national fishing fleet comprises Madeira (RAM) and Azores (RAA) continental and peripheral regions. It carries out its fishing activity according to the areas of activity and the arts assigned, which can be divided into the following groups: gill nets and trammel nets (DFN); dredgers (DRB); trawl nets (DTS); traps (FPO); lines and hooks (HOK); xavega (beach fencing net) (MGO); seine nets (PS); beam trawls (TBB) and multipurpose vessels (MGP, PGP, and PMP). The Portuguese fishing fleet consisted of 7768 vessels with a total gross tonnage of 87,290 GT.

However, we also know that, in the international context, the Portuguese fishing fleet is not very expressive, namely, in comparison with Spain, in terms of the number of vessels and gross tonnage.

But according to data extrapolated by BCpbP, based on the fishing activity carried out in the fishing ports of the Cascais Captaincy, it is estimated that in Portugal alone, around 2400 tons of fishing gear—DFG are destroyed or abandoned per year.

This is the reason that leads us to believe, as important as it is, that the existence of a management platform that does not yet exist in the country and that we know nothing similar exists yet in Europe comes to allow the identification of these materials, their transport, and the identification of needs by the processing industry so that they can be reused in terms of circular economy.

3 Solution—Product/Service Concept

3.1 Platform Background: BCpbP—As a Local Pilot Project

The BCpbP is an example of a blue circular economy approach as a service design. On the one hand, its main objective is to combat excessive consumption of goods and reduce unnecessary waste, while raising awareness of environmental education among its partners. On the other hand, it also wants to be able to contribute to the removal of tons of end-of-life materials and products from the ocean, namely fishing nets, fishing gear (DFG), and other marine litter, in order to boost the creation of new and sustainable products.

As a value proposition, the BCpbP aimed to create a robust supply chain through a wide variety of actors, from fishermen and their communities to companies, research units and NGOs, who in cooperation can (i) create value from waste, seeing the ocean as an opportunity, and (ii) promote sustainable design alternatives to current consumption patterns, while raising awareness of ocean sustainability. To this end, this initiative was intended to be aligned and in line with the "European Strategy for Plastics in a Circular Economy", which aims to transform the way plastic products are designed, produced, used, and recycled in the EU.

By acknowledging the accumulation of plastics in the ocean as a big problem, as the plastics can even turn into microplastics and affect not only marine animals but even human health, [24] the BCpbP sought to establish the principles for the creation, design and management of a platform for the management of marine litter, plastic materials, fishing nets and other artefacts (DFG), as well as promoting the production of durable articles with an environmental, social, visual, and conscious consumer impact, from marine litter, as well as fostering a transparent and integrated circular economy process, boosting socioeconomic development based on the principle of "buying what is ours" and improving sustainability, based on an ethical and creative design process focused on a high standard of quality and environmental awareness, based on conveying the idea that, in a transparent process, there is life in the plastic collected from the oceans, (re)transforming it into new products. The innovative and unique character of the product concept proposed by BCpbP platform lies in identifying the waste as a valuable opportunity for the socioeconomic development of fishing communities and an eco solution to the huge issue of marine life pollution. Inclusively, in 2019, when International Coastal Cleanup Day was marked for the first time, the Oceano Azul Foundation (https://www.oceanoazulfoundation.org/pt-pt/#m150pagehttps://www.oceanoazulfoundation.org/pt-pt/#m150page) created a "marine litter counter" whose numbers result from the actions organised that year. Reportedly, 138,801 kg of garbage was collected by 14,672 volunteers in 868 cleanups from across the country. The BCpbP, within a blue circular economy approach, which involves academia, the fishing community, civil society, the political and business sphere, [7] has as its goal the metamorphosis of daily discarded garbage into something new and with value capable of promoting conscious and globally sustainable consumption. This project started in 2020 as a pilot project in Ericeira and Cascais (Capitania de Cascais). For the collection process, the pilot project has as partners:

- The "Fishing for a sea without litter" project, from the national authority on port infrastructures (Docapesca, Portos & Lotas) and the nonprofit/nongovernmental organisation Portuguese Association of Marine Debris (APLM) started in 2016. Docapesca provides bins to fishermen for their boats. The materials are collected and brought to land. Our project has also separated the sorting channels by making differentiated containers for fishing nets in collaboration with Docapesca.

Fig. 4 BCpbP—blue circular economy process

- The fishermen community that brings and collects marine litter to land and the discarded fishing nets; for analysis of the collected material and storage, we have Tratolixo—Tratamento Resíduos Sólidos—Emp as a partner. Intermunicipal, S.A., a certified intermunicipal company 100% owned by AMTRES—Association of Municipalities of Cascais, Mafra, Oeiras, and Sintra for solid waste treatment.
- Due to their social responsibility program, we are counting on a big supply chain for the transportation process.
- Ambiberica Waste Management, an industrial partner, does the dismantling process in the north of Portugal.
- All the phases of this study, its conception, development, and implementation were centred on the Ericeira Business Factory in Ericeira, managed by the Mafra City Council, which aims to support companies and other entities linked to the Sea Economy, Tourism, and Technology, under the banner of sustainability.

The project's objectives are based on transmitting the idea that, under a transparent process, there is life in the plastic collected from the oceans, (re)transforming it into new products (Fig. 4).

In short, to follow a blue circular economy model, the most varied organisations and brands must be willing to collaborate so that all parties are on the same page towards a future with fewer plastics and more solutions and innovative products made from end-of-life marine plastics. Indeed, sustainable development aims at satisfying current needs without harming future generations' ability to satisfy their needs while considering limitations in the Earth's resources in the face of human development, as well as synergies and trade-offs between economic, environmental, and social goals [14].

The BCpBP platform is thus a unique agglutinating element of all these actions, as a repository of data and respective monitoring, and as a facilitator and mobiliser in cooperation with all the entities involved and the communities.

3.2 Solution—Product/Service Concept: BCpbP Platform

Today, digital transformation forces companies to take productivity and digital evolution to a new level. Digital transformation requires the adoption of more agile business processes and the development of new customer-oriented digital services. It means creating scale through reusable services and enabling self-service consumption of those services. Business processes and transactions can be automated through a combination of microservices. Indeed, the composability principle allows microservices to provide value to the business from multiple perspectives. For data governance in a blue circular economy, interoperability between various systems is foreseen as a standardisation requirement. The guiding principle is to promote the need to adapt to the rapidly changing needs of the business and outline the importance of business process automation for the organisation to acquire know-how in a "just-in-time diachronic dialogue" with all its stakeholders.

Knowledge is considered the most valued intangible asset of the digital economy. On the other hand, knowledge management and innovation seem to be an axial paradigm for the survival of business models interested in analysing and understanding the market's perception of specific products or brands in the digital world. Providing decision-makers with a situational awareness of consumer opinion and dynamics (i.e. suggestions, criticisms, ideation) has emerged as a critical concept. The transformation of raw data into information, then into knowledge, and finally into "embedded wisdom" (e.g. know-how) is a differentiation factor for the competitive advantage of a successful organisation. The synchronised management of subjective (tacit) and objective (explicit) knowledge generates embedded knowledge in a diachronic, just-in-time management flux.

This transformative flux of data into know-how—allowed by a constant incremental technological loop and stakeholders empowerment—raises the fundamental question: Can organisations gather, measure, learn, and incorporate knowledge, embedding it into their organisational cultural and operational modus operandi, thus developing exclusive differentiating know-how?

The BCpbP adopts a model supported by the Matrioskas sequence framework, [23] a service design project empirically conceived—as an organisational ontology—to test and answer this fundamental question. Funded on three conceptual pillars: processes, technology, and people and skills, the experimental development of the Matrioskas Sequence for Knowledge and Innovation (MATSKI) empirically revealed three sub-modelisations that might be used in a living lab for analysing the blue circular economy paradigm for economic growth.

This holistic framework establishes a collaborative just-in-time omnichannel, empowering operational management to achieve (a) performance control (input and output) and continuous improvement, (b) an innovation co-creative process, and (c) value chain logistics and efficiency—through a synchronised AI-based metric and alert system, for the transformation of knowledge (e.g. data semantics, context awareness, and visual data analytics) into performance efficiency monitoring, valued insights, and added know-how to the operation.

The MATSKI co-creative process combined with a service design approach will collect, analyse, and transform data into information, generate knowledge, control the value chain and logistics, and continuously improve performance. Support decision-makers in fostering innovative approaches to a blue circular economy.

The outline of the proposed solution was designed in conjunction with end users, addressing and enabling the visual analysis of data, the creation of a metadata catalogue and the configuration of interactive dashboards with indicators presented in the form of advanced or predictive scorecards. This visualisation strategy will complement the adoption of formal business process specifications (using BPMN) to track and monitor customer opinions, perceptions, and attitudes (i.e. behaviour changes) about the usefulness and added value of the blue circular economy. Customer satisfaction and advocacy are other aspects involved in the deployment of the Living Lab; therefore, in addition to all other stakeholders involved in the recycling chain, gamification methods are also planned to attract and motivate the active participation of users (e.g. the fisherman's results of the sea garbage retrieval).

In this context, the BCpbP, as a nonprofit initiative to be set up, as mentioned above, intended to contribute a set of principles for structuring a platform for managing marine waste, namely plastic materials, nets, and other fishing artefacts (DFG), as a means of interaction between all the partners involved in the whole process—as customer relationship.

4 R&D, IP, and Pipeline

4.1 Research Methodology

The BCpbP, in the view of its authors, is a concrete example of a blue circular economy approach. On the one hand, we aim to combat excessive consumption of goods and reduce unnecessary waste, while promoting environmental education awareness with our partners. On the other hand, we aim to contribute to removing tons of materials and end-of-life products from the ocean, namely fishing nets and gear (DFG), and other marine litter, to create new and sustainable products.

The following methodology was adopted:

- The research plan is an empiric service design case study executed by a mixed-method (triangulation) approach, organised according to a Living Lab methodology.

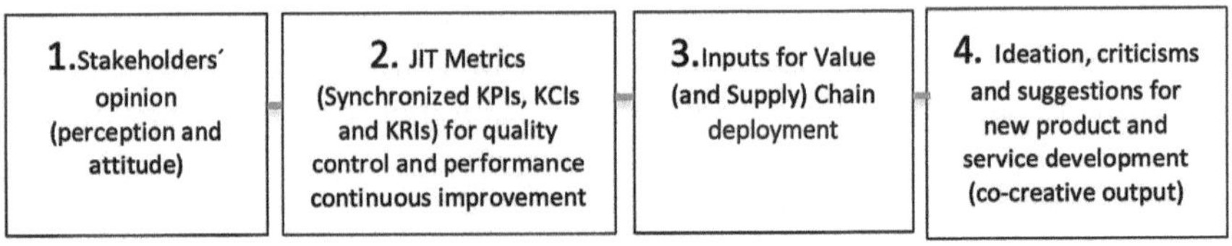

Fig. 5 Matski Living Lab outputs for governance

- The Living Labs Concept refers to an R&D methodology where innovations, such as services, products, or application enhancements, are created and validated in a collaborative metacontext environment. It is a user-centric research methodology for sensing, prototyping, validating, and refining complex solutions in multiple evolving, real-life contexts. According to Mitchell [19], a Living Lab aim is to bring laboratory experimentation to real-life environments with the belief that this will provide improved insights into solution validity and product usefulness while at the same time surfacing new and unexpected patterns of use and user groups.
- The Matski proposed Living Lab will provide the system tools for creating a governance workflow in a real-life context. This will support its decision-makers (e.g. the body of governance) with (1) stakeholders' opinions for a constant measure of the public perception and attitude of the project's impact, acceptability, and efficiency; (2) just-in-time metrics for performing quality control and continuous improvement; (3) logistics inputs for the value (and supply) chain deployment; (4) ideas and suggestions (e.g. the gold nuggets) for new product and service development based on a co-creative flow, within the blue circular economy. In short, the following Matski Living Lab outputs for governance will be produced (see Fig. 5).

A key characteristic of experimentation in a real-life context is identifying new applications and users' needs. Living Labs require more trials and subjects, [26, 27]. In contrast to traditional experimental sciences, Living Labs situate experimentation in multiple and context-rich environments, trying to achieve a high degree of observation [3]. Therefore, the objective is not to understand causal relationships, refute hypotheses, or validate theoretical propositions.

Instead, the aim is more exploratory and explanatory: to understand how a product or service is adopted and used and how its meaning is socially constructed in different contexts. Furthermore, von Hippel's [31] emphasis on the ability of lead users, rather than manufacturers, to create (mainly ICT) innovations must be considered.

In short, the Living Labs method represents a pragmatic approach to innovation (of ICTs and other artefacts), characterised by experimentation in users' real-life and active involvement. We strongly believe that this initiative is based on building the bridge and connecting the government, NGOs, consumers, and communities; it is extremely crucial and possible to be replicable and adopted in many other areas to improve the oceans and our own well-being, that is why we decided to apply the described method.

5 Operations

5.1 Macro-Process Representation: Circular Economy Chain of Custody: Blue Big Bag

As a nonprofit initiative managing a network platform of marine litter suppliers and distributors, the general lines designed and proposed for the BCpbP platform aim to achieve its objectives using a chain of custody for:

- Tracking marine litter, ensuring that it comes from the Portuguese coast or the European area.
- Promote cooperation between various stakeholders as a key factor for success: associations of fishing boat owners and fishermen, environmental organisations (a collection of marine plastic wastes, including old and unwanted nets), from beaches or fishing ports (NGOs), for deposit in containers and other facilities provided for this purpose.
- Establish partnerships with local authorities and waste disposal, treatment and dismantling companies, transport and storage, and transport of the processed material to the final industry.

The following diagrams (see Figs. 6 and 7) map macro-processes—to get an overview of the information flow that integrates each of the two stages of the Blue Big-Bag monitoring chain of custody.

The Notation Semantics

Standard process represents a standard process with no need for further informational detail. As a rule, it will correspond to an existing system or functionality(s) capable of being integrated with solutions held by the entity. In this case, it is essential to identify the business rules that must be verified to ensure compliance with the specified information flow.

Subprocess: It represents a subprocess, therefore, an abstraction of the level of detail. The objective is to help in the macro-understanding of the flow without

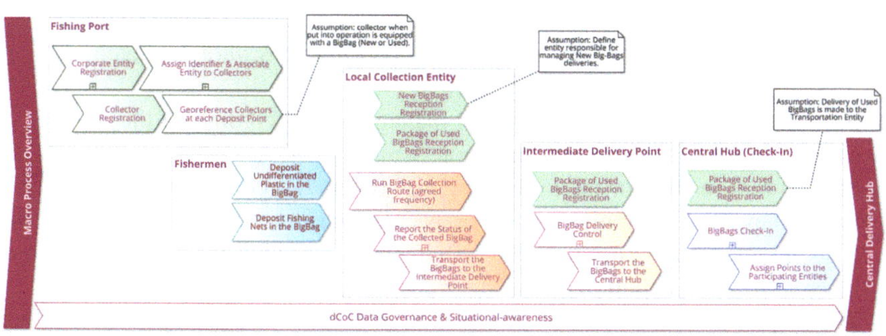

Fig. 6 Phase 1 of the chain of custody: start of the process

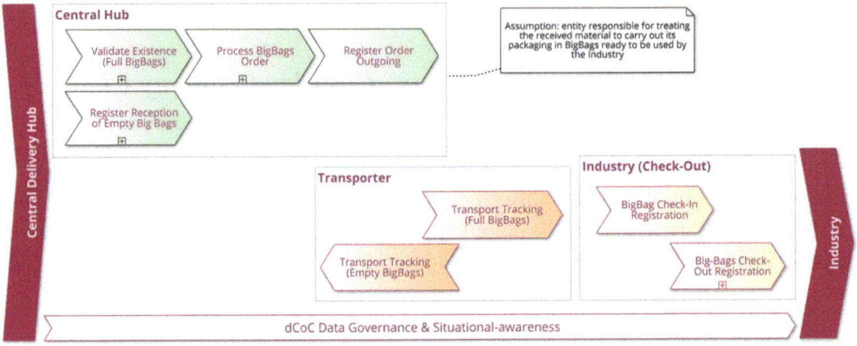

Fig. 7 Phase 2 of the chain of custody: central hub

dispersing the reading by aspects of operational detail; it presents a planning view of the information flow that it is important to model. To access the details of the information flow mapped by the subprocess, the user clicks on the symbol (+) to navigate to a BPMN diagram, where he will have a detailed description of the activities that are encapsulated by the subprocess.

5.2 *Critical Analysis of the Process Design/Specification of the Operational Scenario*

The segmentation of the chain of custody information flow is structured into two phases:

Phase 1—Selective collection feedstock:

- The BCpbP platform is the management entity of the process—a procedure that should be as simple and automated.
- Fishing port, the entity responsible for managing the participants (e.g. Shipowners or Fishermen)—registering the collective entities and georeferencing the collectors.
- Shipowners associations—registration by permillage reward management (e.g. vessel typology).
- Fishermen associations—registration management according to the number of members of the community/association.
- Entities responsible for the local plastics and nets (DFG) collection circuit (e.g. local council)—The entity should be accountable for managing new/used Big Bags. In addition to providing the containers, it should also assume the charges inherent to the installation process).
- The BCpbP is responsible for the registration of the new Big-Bag batch to which it assigns an identifier: the QR code of the batch.

- The BCpbP platform monitors the Big-Bag collection and replacement procedure, as well as registering additional data:
  - Collection volume: empty, medium, or full → input that would allow an analysis of the performance of the collectivity(ies) affecting the collector
  - collection status: new, used, damaged → input that would allow an analysis of the quality of the Big Bag (e.g. metadata about material durability, the number of reuses, eventually tracking of vandalism situations)
  - collector, analyse the interest/impact in monitoring situations of vandalism or damage that compromise the collector's operability
- The entity responsible for collecting Big Bags in the containers (those entities depending on agreements established with each fishing port) shall also be responsible for replacing an empty Big Bag in the container.
- Entities responsible for preprocessing and sorting; shredding; density separation; washing; grinding.
- Entity responsible for the intermediate delivery point is listed following our partner Docapesca and the "Fishing for a sea without litter" project.

Phase 2—Raw material treatment (central hub and conveyor):

- The entity serving as the central hub oversees the reuse process of used Big Bags, which is a vital aspect of the circular economy. Additionally, they are in charge of maintaining records of the delivery of these used Big Bags to the transporter.
- Carrier, responsible for transporting orders to the destination (end customer).
- Manufacturing Industry (end customer): In coordination with the central hub, the BCpbP platform manages the rewards. This includes registering prizes or rewards to be awarded to participating entities, which should be processed when the manufacturing industry places an order.

5.3 Key Resources

The key resources for the business model are based on the establishment of partnerships, with a particular focus on those entities that, under contract with Docapesca, are responsible for managing the removal of marine litter, including DFG, from fishing ports and for obtaining income by delivering the product under the conditions requested by the processing industry, and earn revenue:

- To establish relationships with all partners involved
- To manage revenue streams to reward the fishing community

Types of Key Resources

- Physical: IT infrastructure, specific collection sites at fishing ports; storage for sorting, shredding, density separation, washing, grinding, and logistics—transport to the processing industry. Development and establishment of sound

waste management systems in the landing of fishing ports (due to the fishing for litter Big Bags, the waste streams can be sorted out more quickly).
- Intellectual Property: resources such as trademarks, proprietary knowledge, copyrights, partnerships, and partner databases (to promote research to identify policy drivers of marine debris and encourage development and implementation of innovative solutions; patent the model created for the blue circular economy platform).
- Human: Raising awareness among the fishing community is laborious and time-consuming.
- Financial: As a nonprofit entity, it needs patronage support. To establish a more humanitarian society and improve the environment in which it operates, it will depend on the support and resources to be made available by partners and the tax incentives it may obtain (funding is an ongoing challenge, money is needed for storage, transport, and processing of the waste).

5.4 Channels

The BCpbP platform adopts owned direct channel. It allows to expand its reach and benefit from partner strengths. It aims to achieve the objectives through a chain of custody:

- To provide the screening of marine litter, ensuring that it comes from the Portuguese coast
- To establish cooperation between various stakeholders as a key success factor: fishing vessel owners' and fishermen's associations, environmental organisations (collecting marine plastic waste, including old and unwanted nets), from fishing ports or beaches (NGOs), for deposit in containers and other facilities provided for this purpose; partnerships with local authorities and waste disposal, treatment and dismantling companies, transportation and storage; and also transportation of the processed material to the final industry

6 Industry and Competition

The fishing activity where the waste of nets has been recovered is coastal fishing, in which polyamide nets predominate. In deep-sea fishing, the most used nets are made of polyethene and polypropylene. The differences are due to the different fish capture processes and the corresponding requirements regarding the mechanical behaviour of the nets, which is conferred on them, in part, by their chemical composition. Thus, the waste from the fishermen's nets for recycling within this project's scope is essentially made of polyamide. As is well known, one of the significant problems of the circular economy is the recycling of material mixtures. In the case of this project, at the end of the useful life of the products (fishing nets) and in their collection for recycling, it is necessary to proceed with the identification of the chemical composition, separation, and washing. These processes, similar to what happens in other recycling fields, have costs that need to be absorbed by the gain in value that the market can attribute to products that incorporate all or a percentage of recycled raw materials.

The gain in value in recycling is even more significant as the products are transformed into raw materials with characteristics as close as possible to the initial ones. This is not easy in mechanical recycling, and the resulting raw materials clearly lose quality. The best results are obtained in recycling by chemical processes, but these processes cannot always be used.

In the case of fishing nets, produced from synthetic yarns, mono or multifilaments, recycling with added value goes through the chemical way of transforming waste and extrusion of new filaments, either for circularity, giving rise to new nets, or for applications in other products. However, recycling processes, including the extrusion of filaments, are not universal; they are specific for each type of raw material and are very demanding concerning the unwanted presence of contaminating materials, such as other materials of different chemical compositions, salinity and sand.

In Portugal, several companies can recycle and extrude polyethylene and polypropylene waste, but not polyamide waste. Thus, polyamide, polyethene, and polypropylene waste have specific recycling and extrusion lines and require rigorous selection to avoid contaminants and an effective washing to eliminate salt and sand. Therefore, under current circumstances, Portugal has no conditions to chemically recycle and give the best value to the waste from fishing nets collected under this project.

Three possible solutions can be analysed: (1) transport waste from fishing nets to other countries (e.g. Spain) that meet these technical conditions, albeit with an increase in transport costs and CO2 emissions; (2) investment in Portugal in a polyamide extrusion line, with a broader market, enabling the recycling of waste; and (3) study other solutions for recycling polyamide waste, different from chemical processes and extrusion, potentially with lower added value.

6.1 The Recycling Market Value of DFG

Given that the BCpbP platform has no competition within its purpose, the major problem, as previously identified, is the difficulty in finding partners that are willing to cooperate with it and, particularly, to identify industries that are willing to incorporate these materials in their production, given the diversity of plastic materials and their incompatibility in mixing, for most primary sectors linked to the production of fishing nets and gear (DFG). In their genesis, they incorporate different types of plastics that, in the recycling process, are not likely to be mixed for reuse, given their various natures.

So, the different types of plastics are: PET—polyethylene terephthalate (resin code 1); HDPE—high-density polyethylene (resin code 2); PVC—polyvinyl chloride (resin code 3); LDPE—low-density polyethylene (resin code 4); PP—polypropylene (resin code 5); PS—polystyrene (resin code 6); EPS—expanded polystyrene; PE—polyethylene; PMMA—Poly (methyl methacrylate); PUR—polyurethane; PP&A—polyester, polyamide, and acrylic.

Fishing nets—DFG are composed of synthetic materials such as PP, PE, and in particular, PA (nylon), a high-value polymer on the recycling market. For new granulates, the market value of PA6 is 1.7–2.0 Euro/kg twice as high as that of most other common polymers (PP, LDPE, HDPE, PS, PET).

At the same time, PA monofilament fibres are remarkably pristine as additives are not needed in the production process. An efficient separation and cleaning process would allow lost and discarded fishing nets to return to the production cycle as a valuable raw material. Nets are predominantly composed of PA6 fibres; ROPs are in PET.

More information about the raw materials marketplace: www.plasticker.de www.plasticker.de and Global Plastic Recycling Market Report 2021: Rise in the Demand for Plastics in Diverse Industry Practices.

7 Market Analysis

7.1 Customer Segments—B2B: Building a Network of Suppliers and Distributors

B2B—building a network of suppliers and distributors—platform partners all use sustainable methods and materials and tend to incorporate better practices throughout their supply chain.

The BCpbP platform supports companies with corporate social responsibility above their business objective. They do not compromise or engage with stakeholders who do not have an ethical approach to business. What value do we deliver to the customer? Package of products and services that the platform could help to create value:

- Suppliers for the commercial fishing industry: fishing nets; mooring and towing ropes; cordon; twisted ropes
- Sports nets
- Safety nets
- Technical weavings, cordon, bird nets, and twisted ropes for agriculture
- A range of twines and ropes, synthetic oriented towards markets such as packaging and decoration
- Footwear industry
- Production of thermoplastic and cross-linked compounds

All the stakeholders are committed to practices within the circular economy and literacy, which involves the fishing community and volunteers in the collection of marine litter (plastic and derelict fishing gear—DFG), distribution logistics systems and the manufacturing industry that incorporates plastic as a raw material:

- Direct impact on the fishing and net industries (e.g. Sicor https://www.sicor.pt/enhttps://www.sicor.pt/en; Cadilhe & Santos, Lda. www.cadilhe-santos.pt/En/www.cadilhe-santos.pt/En/; or COTESI https://www.cotesi.com/https://www.cotesi .com/
- Or in thermoplastics, cross-linked compounds, and the footwear industry (e.g. Cabopol https://cabopol.comhttps://cabopol.com; Zouri https://www.zouri-shoes.com/https://www.zouri-shoes.com/)
- Or even in the manufacture of plastic bags and films, products with a transversal consumption in all sectors of the economy: food industry, agriculture, restaurants, hotels, hygiene and cleaning, construction, and commerce (e.g. Plasoeste http://www.plasoeste.pt/http://www.plasoeste.pt/

Platform partners are all those who use sustainable methods and materials and tend to incorporate better practices throughout their supply chain. They prefer and support companies where corporate social responsibility is above their business objective. They do not compromise or engage with stakeholders who do not have an ethical approach to business.

7.2 *Key Partners*

The network of suppliers and partners makes the business model work—optimisation and economies of scale. This is a collaborative effort between researchers, fishers' unions, harbour/local fish landing facilities, and local and national authorities:

- Docapesca—Portos e Lotas, SA, is a state-owned company. It provides public service on mainland Portugal by organising the first sale of fish and supports the fishing and fishing ports sector (http://www.docapesca.pt/en/ http://www.docapesca.pt/en/)—partnership under the project "A Pesca por um Mar Sem Lixo" (http://www.marsemlixo.com/http://www.marsemlixo.com/).
- Local waste collection and management entities (e.g. local authorities; waste collection and treatment companies (e.g. TratoLixo, https://www.tratolixo.pt/ and transport and storage companies)—the building of own spaces for marine litter (plastics and nets) deposit, collection, sorting, and shredding.
- Fishing vessel owners' and fishermen's associations.
- Insurance company—Mútua dos Pescadores or others develops its activity within a framework of social responsibility, sustainably intervening in communities, not only developing its economic mission as an insurance company, but also promoting and supporting environmental, social and cultural projects and ini-

tiatives (https://www.mutuapescadores.pt/https://www.mutuapescadores.pt/)—engagement fishing communities and managing the added value created by the industry in favour of those communities.

BCpbP promotes collaboration between:

- Waste processors (by improving recycling possibilities)
- Fishing ports (by establishing waste facilities/fishing port reception and recycling logistics)
- Gear producers
- NGOs (ALP, Movimento Ajude a Limpar a Praia; APLM; Associação Portuguesa do Lixo Marinho)
- Researchers (research data about which materials can be found in the waste stream)
- Policymakers and national authorities (coordinate rules on national, local, international, and EU levels)
- Fishing industry (by creating awareness/spreading the good stories and practices)
- The user of the recycled material

7.3 Key Activities: Business Model Supported by a Platform/Network of Partners

Besides assuming the purpose of acting following the principles of social responsibility in organisations, the BCpbP also sought to contribute to Portugal's environmental objectives (SGD 14.1 prevents and significantly reduces marine pollution of all kinds, including marine litter) and to the ecological obligations of the European Union.

Portugal, excluding the autonomous regions of the Azores and Madeira, currently has 22 fishing ports with fish auctions and another 28 with sales points, all managed by Docapesca—Portos e Lotas, SA.

According to the extrapolation made by the BCpbP, 2400 tons of fishing nets are expected to be collected yearly if the current Portuguese fishing fleet is maintained. Thus, the main objective of the BCpbP initiative is to finance the fishing communities through the realisation of capital gains that can be achieved from the collection and separation, either at sea or on land, of nets and other fishing gear at the end of their useful life, which, once processed and placed in different industries adopting a circular economy strategy, may provide revenue that, in turn, may be distributed proportionally by the fishing communities according to the collections made.

In this sense, the BCpbP business plan will reflect all actions with all associated costs and the final profit, which will be attached.

Segmentation of the BCpbP platform market as a supplier:

- Commercial fishing industry (fishing nets; mooring and towing cables; ropes; twisted ropes)
- Service and sports support industry (sports nets; safety nets)
- Agro-livestock support industry (technical weaves, cords, bird nets, and twisted ropes for agriculture
- Synthetic rope industry is oriented to markets such as packaging and decoration
- Footwear and accessories industry
- Production of thermoplastic and cross-linked compounds
- Others not yet identified

Discuss entry, namely entry barriers:

- Dependence by the accession of entities contracted by Docapesca that hold rights over the materials (DFG) abandoned in fishing ports
- Dependence on entities for processing (selection, grinding, washing)
- Dependence on logistical entities (transport)

8 Business Model and Strategy

These are some of the most important actions that must be taken for a successful operationalisation of the process of creating and delivering our value proposition.

The key actions for structuring the business model and strategy of the BCpbP platform are subdivided into several points of interaction as supply chain management (Fig. 8):

1. As the fishermen are the first and most important link in this circular economy project, we believe that the primary key action is to carry out awareness-raising actions, making known the benefits of our purpose, actions to be carried out in the leading 22 fishing ports of the continent, with the logistical support of Docapesca.
2. Ocean literacy—promotion of best practices and adoption of environmental awareness courses for fishing communities.
3. In parallel, to activate contacts for establishing a network with the current entities responsible for removing marine litter in the 22 fishing ports of the continent, managed by Docapesca. Its purpose is to explain the objectives of the platform, which, being a nonprofit entity, intends, on the contrary, to boost the placement of these materials through a stock exchange of entities capable of absorbing these materials for later selection, grinding, and washing, increasing the value of these materials and placing them later in the transformation industry.
4. Considering that the local authorities where these 22 fishing ports are located have the responsibility to promote good circular practices among the population, it would be expected, as a mission for the reduction of marine litter, to establish agreements with these municipalities for the provision of BIG BAGS, for the subsequent collection and transportation of nets and other fishing gear, for subsequent selection and treatment.

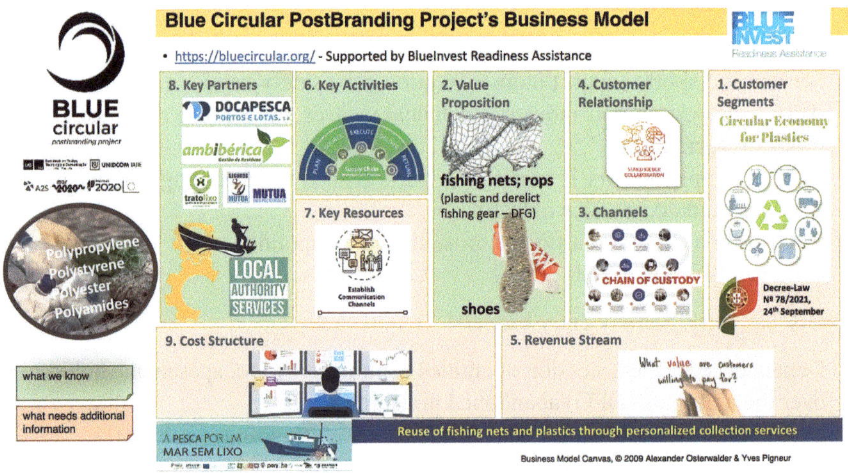

Fig. 8 The BCpbP business model

5. With the support of a logistical partner, the transport of materials, either before processing or already processed for subsequent insertion in granulate production lines and subsequent incorporation in the production of new products for a strategy of greater circularity by the processing industry, should be monitored by the platform [17].
6. Given the current globality of markets, consider the possibility of identifying other industries capable of absorbing these materials, in a circular economy perspective, either in Portugal or Spain or even in other EU countries.
7. Promotion of studies and research in terms of technological advances that can mitigate marine litter (plastic and abandoned fishing gear—DFG), knowing that new smart technologies can play a vital role in solving the plastic pollution crisis in our waters.
8. Stimulate revenue generation and reward involvement of the fishing community and provide resources to NGOs and volunteers involved in marine plastic litter collection.

Finally, the BCpbP initiative is also responsible for presenting a set of recommendations for the creation of a certification system in partnership with the Portuguese Pact for Plastics, https://www.pactoplasticos.pt/https://www.pactoplasticos.pt/, which includes, among other aspects, traceable labels for the processing industry. These labels would provide the final consumer with helpful information, such as the composition of the product and its percentage of incorporation of plastics taken from the ocean, the geographical origin of this material, and also the date of removal from the maritime environment, in addition to mapping the processes by which the product was subjected.

As the BCpbP purpose is to set itself up as a nonprofit entity, it is expected that it can be set up as a nongovernmental organisation to promote solidarity actions in the field of public policies. It will always be dependent on support and sponsorship from organisations, entities, or companies, in addition to possible community support, so that it can comply with the planned actions and thus contribute to the pursuit of environmental goals, benefiting through this economic motivation the fishing communities, the main link in the chain and the target of this initiative.

BCpbP platform's mission is to be an integrator of all partners to be involved, managing in each phase the fulfilment of agreements signed with entities and organisations, and once the income from this management has been obtained, to be able to distribute it equitably among the fishing communities involved.

Today, consumers are highly dependent on social media platforms such as Instagram, Facebook, LinkedIn, and Snapchat. The purpose of the BCpbP platform is to use digital marketing actions, namely, to experience future strategies and the consequent decision-making.

In this purpose and an initial phase, the BCpbP platform intends to create digital marketing strategies, knowing that in an immediate scenario, it will not aim to generate revenue, but to create conditions to attract investors that assume the values of the blue economy and to gain traction to the project namely increasing the engagement of the current partners and enlarging its number.

9 Milestones and Roadmap

Figure 9 provides a Gantt with a global overview of the project execution plan, considering 13 months until cruising control of the BCpbP platform. The Gantt is structured into four main work activities, planned as follows:

AS 1—Project Management and Coordination (duration: 52 weeks), this section aims to coordinate the successful delivery of the project, manage the progress of the project, and maintain efficient coordination and communication channels between project members and with the Project Officer. Perform a risk assessment to ensure the project outcomes will be resilient to any problems that may arise. Manage interactions with advisory boards and between other management structures. Ensure that the project goals are met within the given budget and time restrictions through carefully planned management activities.

It will list and manage indicators to monitor the flow of information and gains from recycling activity, following multiple perspectives (e.g. financial, envi-

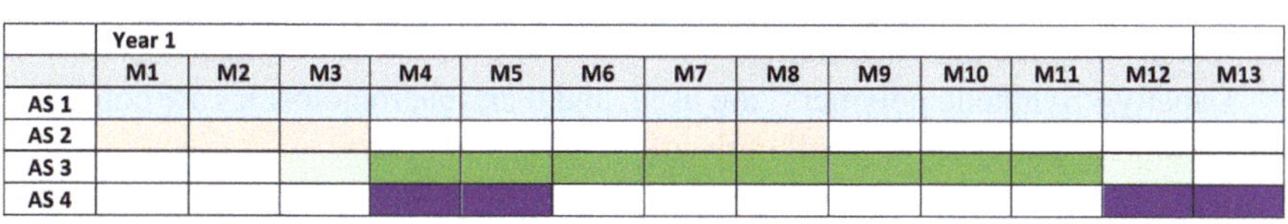

Fig. 9 The platform development roadmap activity sections (AS)

ronmental, social responsibility, compliance with stakeholder's requirements and expectations).

AS 2—Stakeholder requirements and technical specifications (duration: 16 weeks), this section aims to compile the knowledge and experience of the stakeholders (i.e. business users) to obtain a precise definition of their concerns, expectations, and motivations. Develop a legal and ethical framework to address users' confidentiality and data ownership concerns. Establish sustainable governance for the data infrastructure. Translate suitable scenarios, user cases, and structural requirements into technical specifications. Ensure that the collected information is sufficient for designing the system architecture.

AS 3—Pilot's setup and execution (duration: 32 weeks), this section aims to develop the solution and ensure the smooth deployment of the pilot, including testing the solution with the end users. Assure the proper functionality of MATSKI workflow solution in a real-life environment. Implement the data infrastructure according to the specifications previously defined, including data confidentiality and platform access requirements (e.g. authentication and authorisation). Adopt agile implementation to assure continuous improvement based on identified gaps or operational requirements. The pilot will act as a Living Lab, providing a natural context for testing the platform from multiple perspectives. These tests will be performed in close collaboration with the end-user group.

AS 4—Dissemination and exploitation (duration: 12 weeks), this section aims to set up an innovation management strategy and monitor relevant emerging opportunities for project results exploitation, guiding dissemination, and exploitation efforts. Through targeted events and actions, raise awareness of the project's activities, giving high visibility of the project outcomes to the scientific community and the organisational and social communities. Communicate the project's findings and results to all relevant stakeholders.

Actively engage in clustering initiatives with other projects to align through joint dissemination activities and identification of synergies. Build an exploitation roadmap and related business models [8] to adopt the project's results rapidly.

10 Risk Analysis

As previously highlighted, there are technological risks associated with recycling these materials and their high processing costs at the expense of the value of virgin materials. When discussing fishing nets, their characteristics are defined by the type of "catch" they are intended for.

Usually, "Synthetic polymers" are used, and their macromolecules are composed by linking several units of small molecules called monomers.

PVC (polyvinyl chloride) will perhaps be used most frequently, but polyethylene, polypropylenes, and others are equally suitable. Acrylonitrile-butadiene-styrene copolymer (ABS) is also used to make the deep buoys, and PUR (rigid polyurethane) is used to make other types of buoys used with the nets. Like Styrofoam, polystyrene foam is also widely used as surface buoys.

Nylon and fluorocarbon are commonly used to make the monofilaments and multifilament commonly used in “line fishing”.

This diversity must be considered when recycling is carried out, together with eliminating hooks, sand, and other foreign elements that the network may carry. It must also be degreased. Fish remains cling to the lines that make up the net, making it difficult for them to recover.

Factories may have their own systems to reduce the separated and cleaned plastic into small grains for reuse. It should be noted that the colour change is not admissible, so you should always use raw material of the same colour, such as blue or green, in which case you can obtain some different colour intensities from the original, by adding chemical dyes.

The manufacturing process involves the production of the yarn, which goes through several methods, including, for example, using a die followed by a proper weaving process to produce the net with the desired characteristics.

To assess the potential of lost fishing gear for the plastics value chain, a wide range of experiments is known, from pretreatment, fibre shredding and washing to high-temperature thermal evaporation of polymers (in Recycling Options for Derelict Fishing Gear, 2018—MARELITT Baltic, the project was co-funded by the Interreg Programme for the Baltic Sea Region 2014–2020).

In summary, the following obstacles to material recycling were encountered in the case of gillnet-dominated DFG samples:

- A high degree of residual contamination with sediments, lead, and organic matter ($> 37\%$ by weight).
- Complex multicomponent material mix containing at least four types of polymers.
- Knots, loops, and twists are retained down to very small grain sizes of less than 1 mm.
- Contamination with wood and rubber fragments prohibits uniformity.
- Material mix is expected to result in inhomogeneous melts.
- Diversity in polymer melting points results in the coking of the least thermally stable polymers.
- Recycles are expected to have a high degree of brittleness and fracture points.
- Lead is spread throughout the samples during processing, introducing toxicity.

For gillnet-dominated, entangled DFG, thermal processing will be more efficient in most cases and is therefore recommended. This situation is different for end-of-life gillnets. As fishermen systematically remove swim and sink lines for reuse, the net material is well-suited for recycling as nets are composed of high-value polymers PA6 and PET. While not part of the MARELITT Baltic project and recycling trials, end-of-life fishing nets are already recycled into yarn, e.g. by Ecoalf in Spain or Aquafil in Slovenia. The Chilean company Bureo regranulates fishing nets to produce skateboards and other beach items. The Danish company Plastix also hosts a small re-granulation unit for PA6 materials. Although not directly addressing

DFG, these efforts support the value chain for gillnets, which might help avoid losses from fishing port collection sites in the marine environment.

It must be noted that recycling of DFG or end-of-life fishing gear does not automatically imply a contribution to the maritime circular economy. This is only the case when the resulting recycling products are designed to have a high recycling potential again after their life span.

Logistics is a high-risk and high-impact issue for the BCpbP, involving various stakeholders. Supply chain management is the main challenge facing this project, as there is a need to ensure that enough quality materials are recycled.

Without quality materials, industry may not be receptive to joining the platform as it may be too costly. The absence of adequate technical means may also be a risk.

Moreover, the project faces three additional challenges:

- The transport and storage of collected materials (fishing nets and marine litter).
- The possible lack of cooperation with entities already established and willing to manage the handling and recycling of similar products, e.g. imported.
- The appearance of low-cost products available on the market comes from countries with cheaper labour costs and low environmental standards [1].

11 Management Structure

Leaders: For each section, a senior expert with a solid scientific background matching the WP topic is designated to ensure the performance and progress of the work package concerning the deliverables and milestones. The leaders are responsible for coordinating the partners' work on the different tasks within the respective section. The leaders also ensure that deliverables are being produced on time with optimal quality and according to the best value-for-money relation. The leaders remain ultimately responsible for the timeliness and quality of all tasks and deliverables within the respective section.

Advisory Board: The consortium will be supported by an advisory board composed of third-party experts in areas relevant to the project. They provide suggestions on open issues, understand the cross-cutting and complex problems, and evaluate the ongoing work and project results. This group of experts will organise their activity through coordination meetings (face-to-face and web conferences), whose frequency will depend on the project requirements.

Ethics Committee: The project will handle sensitive data; therefore, it will take all measures to comply with the national and EU legislation on ethical issues, data protection, and privacy. The project will include an ethical committee within the advisory board constituted by independent auditors of legal experts.

12 Conclusions

According to data extrapolated by BCpbP, based on the fishing activity carried out in the fishing ports of the Cascais Captaincy, it is estimated that around 2400 tons of fishing gear are destroyed or abandoned every year in Portugal alone.

This pilot project aimed to establish a robust supply chain capable of (1) creating value from marine waste, (2) promoting sustainable design alternatives to current consumption patterns, and, at the same time, (3) promoting awareness and literacy for ocean sustainability—with a sustainable impact on the Blue Economy.

To this end, in the fishing ports of Cascais and Ericeira, the fishing community was invited to place their fishing nets and artefacts, as well as the marine waste collected on board their boats, in Maritime Ecopoints created for this purpose, with the motto of another project, also underway in Portugal, called "Fishing for a sea without waste" promoted by Docapesca—Ports and Quotas with the support of APLM—Portuguese Marine Waste Association. It also had the support of the "Ericeira Business Factory" business incubator run by Mafra City Council.

A business model was drawn up following a successful application in January 2020 to the European Commission's BlueInvest Readiness Assistance, with management and operational support from PwC Luxembourg, and was one of the projects selected (out of 200 proposals). This resulted in the initial design of a marine litter management platform—in cooperation with Docapesca, joined by the Cascais and Mafra municipalities, as well as the company Tratolixo—Tratamento de Resíduos Sólidos.

Also in line with the initial purpose of this pilot project, there is a commitment to develop mobilisation work in the social field, involving the fishing community, as well as establishing partnerships between associations/institutions/companies with a similar vision, so that in cooperation, some of the actions set out in SDG 14 can be achieved (14.1; 14.2; 14.3; 14.a; 14.b; 14.c).

In 2023, for this case study for a blue circular economy—BCpbP—it received the SDG 14—Protecting Marine Life Award from APEE—the Portuguese Business Ethics Association.

Thus, the main objective of this study, as part of a service design proposal, was to contribute a set of guiding assumptions for the creation, design, and implementation of a platform for the management of fishing gear and end-of-life plastic materials, as a multidisciplinary approach, with the purpose of SDG 14: protecting marine life, as a response to the disastrous and ongoing proliferation of plastics, nets and fishing gear (DFG) in our seas and oceans.

It was also possible to verify that recycling lost fishing gear recovered from the sea is a challenging and time- and effort-consuming process:

- Requires the adoption of local solutions for a recycling chain.
- Manual sorting is required; a first preprocessing stage, use of small shredders, and/or a precleaning bath near the fishing port, some of the recovered ropes and nets would be valuable raw materials for pelleting.

- For the time being, the only efficient solution to the problems identified with the recycling of DFG is to collect old, discarded, and recovered fishing nets at the fishing ports.
- That it is crucial to mobilise fishermen to generate value from DFG, as well as to pass on to them the added value generated by the processing industry.
- It is also possible to see the importance and relevance of a platform as a monitoring and management interface, as a means of interaction between all the partners involved in the whole process—like a customer relationship.

Finally, as a collaborative and value-creating process—service design, which aimed to involve fishing communities, the scientific community, managers, and decision-makers from public and private entities, as well as NGOs, and which, according to the results obtained, from a blue economy perspective, the involvement and mobilisation of everyone also contributed to raising awareness and joining forces from a perspective of mutual aid between different organisations in favour of the health of marine ecosystems.

References

1. Aaker DA (2017) Strategic market management, 11th edn. Wiley, New York. ISBN: 978-1-119-39220-0
2. Aronowitz S, Henry G (1991) Postmodern education: politics, culture and social criticism. Universsity of Minnesota Press, Minneapolis
3. Ballon P, Pierson J, Delaere S (2005) Test and experimentation platforms for broadband innovation: examining European practice. Available at SSRN: https://ssrn.com/abstract=1331557
4. Barboza L, Frias J, Booth A, Vieira R, Masura J, Baker J, Foster G, Guilhermino L (2019) Chapter 18 - microplastics pollution in the marine environment. In: World seas: an environmental evaluation, 2nd edn., vol. 3. Academic Press, Cambridge, pp. 329–351. https://doi.org/10.1016/B978-0-12-805052-1.00020-6
5. Brink P, Schweitzer JP, Watkins E, Howe M (2016) Plastics marine litter and the circular economy. Institute for European Environmental Policy for the MAVA Foundation, London, pp. 1–17
6. Brink P, Schweitzer JP, Watkins E, Janssens C, De Smet M, Leslie H, Galgani F (2018) Circular economy measures to keep plastics and their value in the Economy, avoid waste and reduce marine litter. Economics Discussion Papers (3). http://www.economics-ejournal.org/economics/discussionpapers/2018-. http://www.economics-ejournal.org/economics/discussionpapers/2018-3
7. Chesbrough H (2003) Open innovation: the new imperative for creating and profiting from technology. Harvard Business School Press, Boston
8. Chesbrough H (2006) How to thrive in the new innovation landscape: open business models. Harvard Business School Press, Boston
9. Docapesca (2021) A Pesca por um mar sem lixo. Sustentabilidade. http://www.marsemlixo.com/o-projeto/
10. World Economic Forum, Ellen MacArthur Foundation and McKinsey & Company (2016) The New Plastics Economy: Rethinking the future of plastics
11. Ellen MacArthur Foundation (2019) Circular economy model's diagram. Ellen Macarthur Foundation, pp. 1–120. https://ellenmacarthurfoundation.org/circular-economy-diagram
12. Fadeeva Z, Van Berkel R (2021) Unlocking circular Economy for prevention of marine plastic pollution: an exploration of G20 policy and initiatives. J Environ Manage 277:111457. https://doi.org/10.1016/j.jenvman.2020.111457

13. Geisendorf S, Pietrulla F (2017) The circular economy and circular economic concepts–a literature analysis and redefinition. Thunderbird Int Bus Rev 60:771–782. https://doi.org/10.1002/tie.21924
14. Geissdoerfer M, Morioka SN, Carvalho MM, Evans E (2018) Business models and supply chains for the circular Economy. J Clean Product 190:712–721. https://doi.org/10.1016/j.jclepro.2018.04.159
15. Global plastic recycling market report 2021 rise in the demand for plastics in diverse industry practices. https://www.globenewswire.com/en/news-release/2021/08/23/2284743/28124/. en/Global-Plastic-Recycling-Market-Report-2021-Rise-in-.the-Demand-for-Plastics-in-Diverse-Industry-Practices.html
16. Jambeck J, Denise Hardesty BD, Brooks AL, Friend T, Teleki K, Fabres J, Beaudoin Y, Bamba B, Francis J, Ribbink AJ, Baleta T, Bouwman H, Knox J, Wilcox C (2018) Challenges and emerging solutions to the land-based plastic waste issue in Africa, Marine Policy, vol 96, pp 256–263. https://doi.org/10.1016/j.marpol.2017.10.041
17. Lin Y, Xiao W (2017) Implementing a smart data platform-how enterprises survive in the era of smart data. O'Reilly, Beijing
18. Mazzucato M (2021) Mission economy: a moonshot guide to changing capitalism. Penguin, Allen Lane-Penguin, London
19. Mitchell W, Piller F, Tseng M (eds) (2007) Extreme customisation: proceedings of the 2007 MCPC conference. MIT Design Lab, Cambridge
20. Morseletto P (2020) Targets for a circular economy. Resour Conserv Recycl 153:104553. https://doi.org/10.1016/j.resconrec.2019.104553
21. Parliamentary Office of Science & Technology (2016) Marine microplastic pollution. POST Note 528:1–5
22. Plastics Europe (2018) Plastics – the facts 2018: an analysis of European plastics production, demand and waste data. https://www.plasticseurope.org/application/files/.6315/4510/9658/Plastics/the/facts/018/AF/web.pdf
23. Rosa CA, Pestana G (2019) The Matrioskas Sequence - a methodological approach for managing knowledge and innovation. In: ECKM'2019 – 20th European conference on knowledge management
24. Shivika S, Subhankar C (2017) Microplastic pollution, a threat to marine ecosystem and human health: a short review. Environ Sci Pollut Res 24(27):21530–21547. https://doi.org/10.1007/s11356-017-9910-8
25. Stolte A, Schneider F (2018) Recycling options for derelict fishing gear
26. Tuomi I (2005) The future of learning in the knowledge society: disruptive changes for Europe by 2020; Background paper prepared for DG JRC/IPTS and DG. https://otrasvoceseneducacion.org/wp-content/uploads/2018/02/ TheFutureOfLearningInTheKnowledgeSociety.pdf. Accessed 7 Jul 2021
27. Tuomi I (2018) The impact of artificial intelligence on learning, teaching, and education: policies for the future. JRC Science for Policy Report- European Commission 2019
28. United Nations Environment Programme (2016) UN environment annual report 2016 – UNEP. Engaging people to protect the planet. https://www.unenvironment.org/annualreport/2016/index.php?page
29. UN Environment (2017) Marine litter socio economic study, united nations environment programme, Nairobi
30. United Nations Environment Programme (2021) Understanding the state of the ocean: a global manual on measuring SDG 14.1.1, SDG 14.2.1 and SDG 14.5.1. UNEP, Nairobi. https://wedocs.unep.org/20.500.11822/35086
31. Von Hippel E (1998) Economics of product development by users: the impact of "sticky" local information. Manage Sci 44(5):629–644

MIX
Papier aus verantwortungsvollen Quellen
Paper from responsible sources
FSC® C105338

If you have any concerns about our products,
you can contact us on
ProductSafety@springernature.com

In case Publisher is established outside the EU,
the EU authorized representative is:
Springer Nature Customer Service Center GmbH
Europaplatz 3, 69115 Heidelberg, Germany

Printed by Libri Plureos GmbH
in Hamburg, Germany